P9-ASN-790

OCEANOGRAPHY

Exploring the Planet Ocean

O c e a n u s
E u r o p a
Euxine
Caspian Sea
L i b y a
A s i a
Arabian Gulf
O c e a n u s

OCEANOGRAPHY

Exploring the Planet Ocean

J. J. BHATT

Rhode Island Junior College

D. VAN NOSTRAND COMPANY

NEW YORK CINCINNATI TORONTO LONDON MELBOURNE

OCEAN
Pearl of
our hope . . .

Tear of
our despair . . .

D. Van Nostrand Company Regional Offices:
New York Cincinnati

D. Van Nostrand Company International Offices:
London Toronto Melbourne

Library of Congress Catalog Card Number: 77-95210

ISBN: 0-442-20698-4

Published by D. Van Nostrand Company
135 West 50th Street, New York, N.Y. 10020

10 9 8 7 6 5 4 3 2 1

Preface

Oceanography is a rapidly developing science. Although less than a decade ago it was considered a rival to space and environmental sciences, it is now merging with these two fields, becoming far more complex in its range and goals than ever before. In recent years, in fact, two new fields, space oceanography, from space sciences, and environmental oceanography, from environmental sciences, have evolved. Presently, the earth's resources technology satellite (ERTS) and the United Nations' Earthwatch Program are applying remote sensing techniques to ascertain economic and environmental resources, an early indication of trends toward twenty-first century oceanography.

This book is intended for a one-semester course in beginning oceanography for students with a limited background in science. I have had two aims: (1) to introduce students to fundamental principles, concepts, and processes of oceanography and (2) to enable them to relate this basic knowledge to such contemporary problems as the tapping of marine resources, environmental pollution, and the management of the world's oceans. Of the over 250 illustrations in the text, a selected number have been reproduced in the accompanying *Instructor's Manual* to serve as transparency masters suitable for overhead projection.

A special feature of the book is a marginal glossary that runs side-by-side with the text, supplying definitions of selected terms as they occur. A more complete glossary of oceanographic terms appears at the end of the book.

I would like to thank many people who extended their help and cooperation: Vickey Briscoe and Nancy Green of Woods Hole Oceanographic Institution; William Ludwig, Lawrence Sullivan, Joseph Morley, and Tsunemasa Saito of Lamont-Doherty Geological Observatory of Columbia University; Nelson Fuller and Thomas Wiley of Scripps Institution of Oceanography, University of California at San Diego; John Skerry, Merton Ingham, Harold "Wes" Pratt, and John Casey of the Narragansett Laboratory and Roger B. Theroux of Woods Hole Laboratory, National Marine Fisheries Service — NOAA; Paul E. Hargraves and Jean Guy Schiling of the Graduate School of Oceanography of the University of

Rhode Island at Narragansett; John Imbrie, Department of Geology at Brown University; Elwyn Wilson of the Oceanographic Data Center — NOAA, Washington, D.C.; Joanne David, Publications Branch — NOAA, Rockwell, Maryland; and Marjorie Dalechek, U.S. Geological Survey, Denver, Colorado.

I am thankful to the following reviewers for their constructive criticism: R. Gordon Pirie of the University of Wisconsin, Robert Stevenson of the University of South Dakota, Fred H. Tarp of Contra Costa College, and Joseph Zawodny of Miami-Dade Community College. I am impressed with the cooperation and efficiency of my publisher D. Van Nostrand Company. In particular, I owe my sincere thanks to Edward Lugenbeel, Bruce Williams, Larry Cone, and Tawnya Kabnick for their excellent cooperation and help.

I am grateful to Deborah Linderman, Patricia, Carrie and Andrew Karn, and Liz Malloux for their help with the preparation of the manuscript.

I am indebted to my wife Meena for giving me both the encouragement and the privacy to reflect and to write. My three-year-old son has made my writings more meaningful. Finally, the inspiring thought of Longfellow "[Ocean] that divides and yet unites mankind" blends elegantly with my professional pursuit, global philosophical outlook, and with my poetic interest.

J. J. Bhatt

Contents

5 Plate Tectonics and the Origin of Ocean Basins

6 The Physical Properties of Seawater

7 Ocean Circulation

8 Ocean Waves and Tides

9 Shoreline Processes and Sediments

10 Chemical Oceanography

11 Life in the Ocean

12 Marine Ecology

13 Biological Resources

14 Physical Resources

15 Environmental Oceanography

16 Ocean Management and Conservation

Chapter Opening Photographs

Opposite page 1: H.M.S. *Challenger (Challenger* Office Report, Great Britain, 1895)

Opposite page 25: Sea landscape (Photo courtesy Dick Huffman from Monkmeyer)

Opposite page 45: Apparent formation of stars within the Nebula M-8 in the constellation Sagittarius (Photo courtesy of Lick Observatory)

Opposite page 57: "Sandfall" in the Cape San Lucas submarine canyon, Baja California (Photo courtesy of Scripps Institution of Oceanography)

Opposite page 75: Looking north over Sinai; Red Sea, Dead Sea, Sea of Galilee, Mediterranean Sea (Photo courtesy of NASA)

Opposite page 95: FLIP (355-foot *FL*oating *I*nstrument *P*latform) in vertical position, allowing scientists a stable platform from which to carry out oceanographic research (Photo courtesy Scripps Institution of Oceanography)

Opposite page 109: A network of buoys for oceanic survey (Photo courtesy Woods Hole Oceanographic Institution)

Opposite page 123: Sand waves (Photo courtesy Hugh Rogers from Monkmeyer)

Opposite page 145: Rocks at the Ovens near Lunenburg, Nova Scotia (Photo courtesy Jim Cron from Monkmeyer)

Opposite page 171: Rosette sampler (Photo courtesy Woods Hole Oceanographic Institution)

Opposite page 187: Ocean life at great depths (Photo courtesy Scripps Institution of Oceanography)

Opposite page 217: The fragility of life at great depths (Photo courtesy of National Marine Fisheries Service — Woods Hole, Massachusetts)

Opposite page 239: Harvesting from the ocean; a SCUBA diver setting up a midwater trawl (Photo courtesy of Oceanic System)

Opposite page 259: Offshore oil rig (Photo courtesy U.S. Department of Interior)

Opposite page 277: The *Argo Merchant* on the verge of sinking (Photo courtesy U.S. Coast Guard, First Coast Guard District, Boston)

Opposite page 293: The submersible ALVIN ready to be launched (Photo courtesy of Woods Hole Oceanographic Institution)

1 The Development of Oceanography As A Science

1-1 OCEANOGRAPHY DEFINED

Oceanography *The scientific study of the oceans and seas.*

THE TERM *oceanography* IS A COMBINATION OF TWO Greek words: *oceanus* and *graphos*. Thus oceanography means the description of oceans. The field of oceanography, however, involves more than just the description of oceans, for it entails various ways of scientifically studying the world's oceans (Figure 1-1).

Oceanography is currently at the center of many contemporary scientific investigations, particularly those related to exploration and exploitation of seafood, minerals, and energy. As a result, fundamental scientific knowledge of oceans extends to include their management and conservation of their wealth. Oceanography has thus become a global responsibility, because it is now realized that the destinies of man and ocean are inextricably bound. Between now and the year 2000, the interaction between man and ocean will be critical to both. The positive interaction could bring many known and still unknown rewards to mankind, but scientists fear a possibly devastating negative interaction. The science of oceanography is, then, more than the scientific study of oceans, because it also aims at solving crucial human problems.

1-2 THE SIGNIFICANCE AND USES OF OCEANS

The earth is unique in the solar system for its enormous quantities of water. Oceans cover approximately 140 million of the total of 200 million square miles of the earth's surface. That is, the world's oceans

Figure 1-1. A space perspective of the earth as viewed from the moon. (Photo courtesy of NASA.)

occupy 70 percent of the earth. In terms of volume, oceans contain 350 million cubic miles of water. Moreover, the oceans contain 3.5 percent dissolved salts, equivalent to 165 million tons of salts per cubic mile. Thus, amazingly, water is an unquestionable georesource; in itself it constitutes the largest single *mobile ore* on the face of the earth.

The importance of oceans is staggering. First, oceans overwhelmingly influence nearly all the surface processes of the earth. They regulate the water cycle* and the flow of carbon dioxide gas in the environment (see Chapter 12). Second, ocean waters not only have supported life, but also have guaranteed its dynamic growth over hundreds of millions of years of evolution, thus enabling life to exist today in countless forms, shapes, and sizes ranging from microorganisms to whales. Third, oceans regulate geologic processes of *weathering* (the breakdown of geologic material such as rocks and minerals) and of *erosion* (which involves transportation of this weathered material). They provide a sink for billions of tons of sediment carried by the world's rivers. In the

Weathering *The distintegrating or decomposing of earth's material through physical, chemical, and biological processes.*

Erosion *The transportation of weathered (broken) material by a moving agent such as wind, water, or ice.*

*The movement of water from the ocean begins, for example, with evaporation followed by condensation and finally as surface (and underground) runoff, returning to the ocean with time.

oceans, sandstone is formed from land-derived sediment and limestone from the bodies of marine organisms. These rocks hold much of the world's petroleum. Marine rocks and sediments also hold extensive metallic deposits of iron, manganese, and nonmetals such as sulfur, phosphates, and salt rocks.

Because of their large volume and rapid fluidity, oceans are remarkably self-purifying and have thus managed to survive uncontaminated for so long. Recently, however, man's actions have threatened this unique self-purifying capacity of the ocean (see Chapter 15). If they are not plundered to destruction, oceans could offer mankind an important energy source, for they contain large quantities of heavy hydrogen (deuterium), which is indispensable in the production of nuclear (fusion) energy. They can also provide energy from volcanoes, especially in the Pacific Ocean where they are systematically distributed along its rim, and from midoceanic ridges; heat flows from the ocean floor; and from waves, notably from tidal waves and currents.

Experts contend that desalinated water from oceans could increase terrestrial production of food. When, therefore, the process of *desalination* of seawater is fully developed for commercial uses, water shortages in many parts of the world, particularly in arid and semi-arid regions still plagued by food shortages, would certainly be alleviated. As it is, each cubic mile of the ocean fringing landmasses is relatively more enriched in organic matter than its counterpart on land. Because this organic matter assures a natural and abundant supply of fertilizers, marine farming has been an attractive enterprise in sea-bound lands such as Japan and New Zealand. Oceans also contain millions of tons of fish and other related protein-rich seafoods which offer a great hope to solve the world's food problems.

Desalination *Process of converting seawater into freshwater.*

Should terrestrial living conditions become intolerable at some future date, because, say, of a nuclear holocaust, uncontrolled environmental pollution, or overcrowded population, the planet Ocean may offer final shelter. Ocean engineers and scientists are at present experimenting with various technological projects, such as man's ability to live in great depths of water for prolonged periods of time, and the construction of undersea parks and shopping plazas for a limited population.

Oceans have provided man a powerful avenue of exploration and a source of inspiration throughout history. They furnish the cheapest means of bulk transportation of the world's goods, since oceangoing vessels can transport a ton of commodity for a few cents, whereas shipping by rail or by air costs much more. Ocean shipping still is the main source of linkage of four billion people of our interdependent world.

From a scientific point of view, the world's oceans have provided crucial testing grounds for many modern discoveries such as seafloor spreading, the systematic distribution of volcanic and earthquake zones along the rim of the Pacific, and the magnetic properties and ages of bottom sediments, particularly those found on either side of the Mid-Atlantic Ridge. These and other related discoveries, which are treated more fully in Chapter 5, have thrown significant light on the riddles of the origin of oceans and continents.

1-3 HISTORY OF OCEANOGRAPHY

The Classical Era

The exact beginnings of oceanography are not known, because ancient man did not keep systematic records, even personal diaries or logbooks. It is known that prehistoric Polynesians and Indians made difficult long-distance sea voyages. Archeologists contend that East Indian traders had a fairly good knowledge of monsoon currents, because sea voyages were quite common in the Indian Ocean as far back as 3000 B.C. Around 1500 B.C., Phoenicians sailed in the West, frequently to the Straits of Gibraltar, and in the East as far as the Persian Gulf, primarily for trading. During this time, Greeks and Phoenicians knew the Mediterranean Sea fairly well. In 600 B.C., the Phoenicians commonly journeyed by sea to North Africa. By 400 B.C., the art of scientific observations of the oceans then included the daily phenomenon of tides, which were explained in terms of phases of the moon. Earlier thinkers had conceived the world as an island surrounded by the ocean (Figure 1-2). Historical study revealed that in 135 B.C. the Sardinia Sea was studied with sounding technique.

Claudius Ptolemy (A.D. 150), the father of the view that held earth to be the center of the universe, was also aware of the extent and size of the Atlantic and Indian oceans. According to him, these oceans were once enclosed seas similar to the Mediterranean (Figure 1-3). About this same time, navigational instruments, notably the *compass* and the *astrolabe*, were invented in China.

Astrolabe *An instrument once used to measure the altitude of heavenly bodies, now replaced by the sextant.*

Between A.D. 800 and 1000, Scandinavian Vikings made frequent voyages on the North Atlantic, discovering Iceland and Greenland. Historians contend that Leif, the son of Eric the Red, reached the North American continent in A.D. 1000.

The Pre-*Challenger* Era

Oceanography can be divided into eras dated with reference to a major scientific expedition aboard the *Challenger* made in the 1870s (Table 1-1). During the Middle Ages in Europe, the progress of ocean-

Figure 1-2. Hecateus' map of the world (circa 500 B.C.). (*Source:* M. R. Cohen and I. F. Drabkin, 1948, *A Source Book in Greek Science,* New York: McGraw-Hill.)

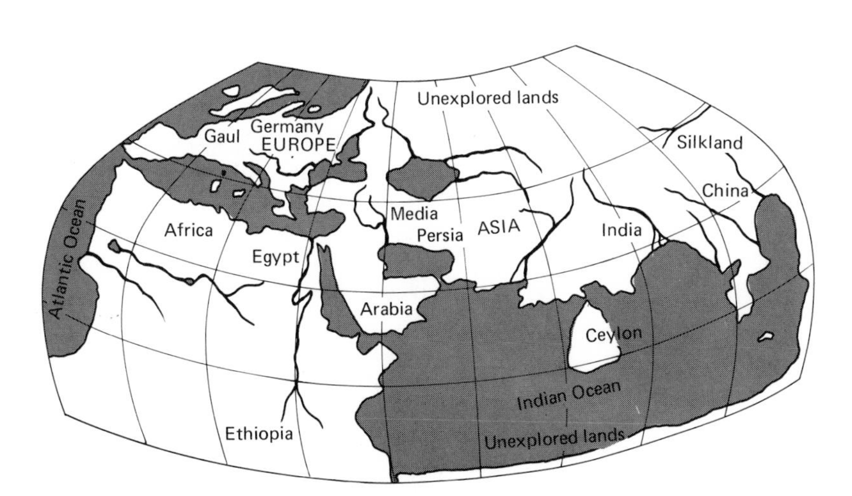

Figure 1-3. Ptolemy's map of the world (circa 150 A.D.). (*Source:* D. E. Smith, 1923, *History of Mathematics,* Volume I, Boston: Ginn.)

TABLE 1-1 Selected Oceanic Expeditions and Related Events

The classical era	c. 3000 B.C.	East Indian sailors and traders know secrets of monsoon currents and carry on oceanic trade and commerce in the Indian Ocean.
	c. 1500 B.C.	Polynesians and Indians know art of long-distance sea voyages.
	c. 500 B.C.	Phoenicians know Mediterranean Sea well.
	c. 300–400 B.C.	Greeks study Mediterranean Sea and publish maps of the oceans.
	1000 B.C.	Vikings sail regularly into the North Atlantic. Leif Ericson voyages across Atlantic Ocean and reaches Canada.
The Pre-*Challenger* era	1492	Christopher Columbus crosses Atlantic Ocean and discovers New World.
	1498	Vasco da Gama crosses Atlantic Ocean and, via Cape of Good Hope, continues his journey into Indian Ocean to discover India.
	1519	Ferdinand Magellan circumnavigates globe.
	1768–1777	Captain James Cook maps New Zealand and the South Seas.
The *Challenger* era	1832–1836	Charles Darwin sails with *HMS Beagle* as naturalist. His observations of marine life in Galápagos islands lead to formulation of theory of evolution and to theory of the origin of coral reefs.
	1872–1876	Wyllie Thomson heads *HMS Challenger*. Collected vast data on physical, chemical, and biological aspects of the world's oceans.
The post-*Challenger* era	1893	Fridtjof Nansen heads voyage of *Fram* to study Arctic Ocean.
	1925	*R/V Meteor* studies ocean floor using echo-sounding method. *R/V Dana* and *R/V Albatross* study Indian and Atlantic Oceans, respectively.
	1960	First submersible, *Trieste,* descends to deepest point in ocean in Marianas Trench.

TABLE 1-1 Selected Oceanic Expeditions *(Cont.)*

The *Glomar Challenger* era	1968	*Glomar Challenger,* a deep-sea drilling vessel, is commissioned by United States for Deep-Sea Drilling Project.
	1970	Many submersibles are used in ocean-related work. International Phase Of Ocean Drilling begins a new era of deep-ocean drilling.

related activity was retarded, at least until the end of the fifteenth century when Christopher Columbus (1492) discovered the New World and when Vasco da Gama landed on the west coast of Malabar (1498). Both these men pioneered long-distance sea voyages. In the following century, Ferdinand Magellan left Spain and circumnavigated the globe. Magellan used a 100 to 200 fathom sounding line to measure the depth of the ocean, but he was never able to reach the bottom.

Sounding *A measured depth of water.*

Fortunately some of the earlier explorers, notably Columbus and Magellan, left personal notes and logbooks containing records of their observations of the oceans. Their writings inspired subsequent generations to explore the ocean world even more vigorously. In particular, William Dampier described in 1700 with great detail maritime meteorological aspects of oceanography in his publication, *A Discourse of the Winds.* Captain James Cook, during his Pacific voyages (1768–1779), constantly logged tides and currents and mapped New Zealand, the South Seas, and the northwest coast of North America. Joseph Banks, a young naturalist on the ship, collected many new animals and plants. And in the United States Benjamin Franklin made the first chart of the Gulf Stream in 1770 and thus improved Atlantic voyages for mail-carrier ships (Figure 1-4). In these ways progress in physical and biological oceanography began.

The *Challenger* Era

At the beginning of the nineteenth century, Sir John Ross and his nephew Sir James Clark Ross successfully measured the depth of Baffin Bay, Canada, and studied the natures and distribution patterns of marine organisms and sediment. Between 1839 and 1843 James Ross similarly studied the Antarctic. The Rosses are also known for attaching a "deep sea clamm" of their own design to the end of the sounding line. With this innovation they were able simultaneously to measure the depth of the sea and to obtain samples from the bottom. They dredged various organisms and mud from depths of 1.75 kilometers in Baffin Bay and from over 6 kilometers in the Antarctic.

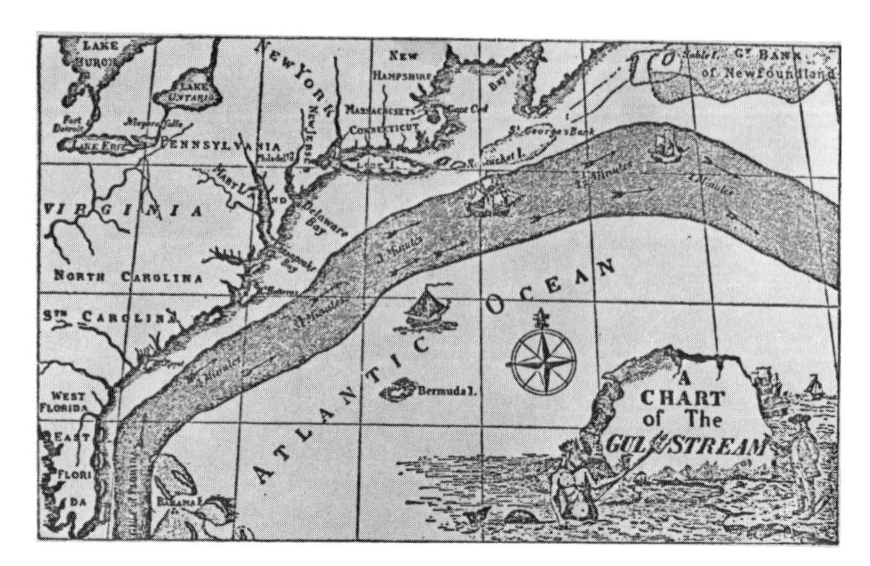

Figure 1-4. Benjamin Franklin's chart of the Gulf Stream. (*Source: Transactions* of the American Philosophical Society, *2*, Philadelphia, 1786.)

Throughout the nineteenth century, progress in oceanography continued. One milestone event was the 1830 voyage of the *Beagle* with young Charles Darwin aboard. Darwin's observations in the Galápagos Islands off the coast of Peru led in subsequent years to the formulation of one of the greatest concepts in natural science, the *doctrine of evolution*. Darwin also explained the origin of reefs in the ocean. In subsequent years Edward Forbes, a naturalist and geologist, pioneered the study of bottom-dwelling plants and animals of the sea. He divided the sea population into eight zones according to a scale of increasing habitation depth. From his studies he concluded that marine plants and animals are largely confined to the surface and decrease in number with increasing depth.

The field of oceanography advanced significantly when Matthew Fontaine Maury, a lieutenant in the U.S. Navy, published his *The Physical Geography of the Sea* in 1855. Maury described in great detail the nature of the Gulf Stream Current, which he astutely called "a river under the sea." His book contained technical information on hydrography and meteorological conditions, as well as maps and charts showing winds and current flows. Maury was one of the principal organizers of the first international conference of oceanography at Brussels in 1853. At that conference, it was decided that standard techniques be followed in making observations of nautical and meteorological conditions at sea.

In 1865, John Forchhammer, a Norwegian geologist, published findings of his more than 20 years of chemical analyses of sea water samples brought from many parts of the ocean. According to him, the total salt content of seawater, although there are relative differences geographically, is virtually constant in the proportions of its major salts. Forchhammer's conclusion, the *law of relative proportions* (now known as the *Forchhammer principle*), was verified by William Dittmar, who was abroad the *Challenger* (see Chapter 10).

Wyville Thomson, a naturalist and geologist who had studied the continental slope along the northwest boundaries of Europe, headed the *Challenger's* four-year (1872–1876) voyage around the world.

The voyage of the research ship H.M.S. *Challenger* (Figure 1-5) epitomized nineteenth-century progress in oceanography. The *Challenger* expedition was launched to learn more about the ocean world, including its topography, biology, geology, chemistry, and physics. It covered over 100,000 kilometers in four years (Figure 1-6). By the end of the voyage, scientists had collected a great amount of data relating to the general nature of the ocean. It took two decades to organize these data, which ultimately resulted in the publication of 50 large volumes. Indeed, the *Challenger's* expedition provided impetus for further development of oceanography.

Although the *Challenger* conducted many depth soundings, the results of these soundings were not satisfactory by modern standards. The soundings were taken at sparse intervals, one sounding for each 6000 square miles of the oceanic area. At best this was a hit-or-miss sounding of the oceanic floor topography and consequently, the *Challenger* report rendered a misleading picture of the ocean bottom. According to the *Challenger* report, the ocean floor was smooth and paved with a fine-grained flourlike material. As we shall see, the case was quite the opposite.

Figure 1-5. The H.M.S. *Challenger* marked the new era of scientific exploration of the world's oceans. (*Source: Challenger* Office Report, Great Britain, 1895.)

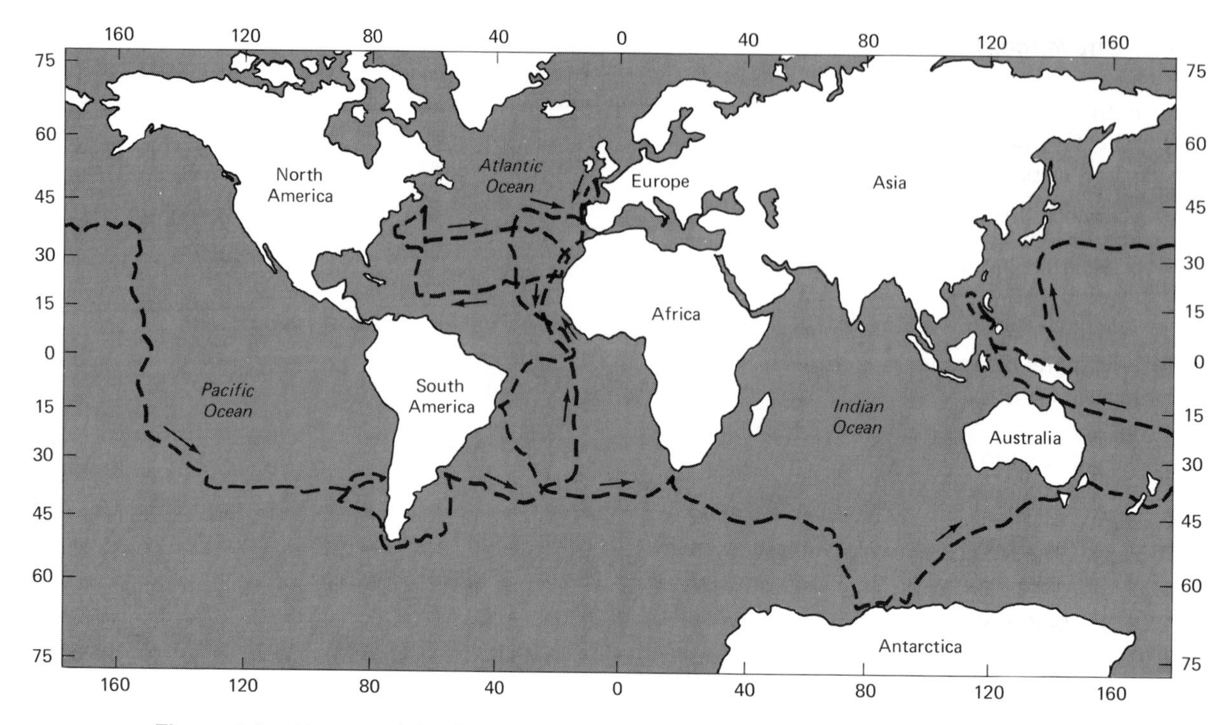

Figure 1-6. Voyage of the *Challenger* (1872–1876). The *Challenger* undertook the first scientific exploration of the world's oceans. It journeyed over 100,000 kilometers and collected a vast amount of data, resulting in the publication of 50 large volumes on oceanography.

In 1877, the United States commissioned the research ship *Blake* under the direction of Alexander Agassiz (son of the famous geologist, Louis Agassiz). One of the important results of the *Blake*'s expedition was the improvement of research techniques. Agassiz and his associates invented various sampling and measuring devices, for example, the "Agassiz trawl," which was a significant improvement over then existing trawls.

Near the turn of the nineteenth century, Fridtjof Nansen, a Norwegian oceanographer, headed the *Fram* expedition of 1893 to 1896 to try to solve some of the mysteries of the Arctic Ocean. He explained circulation patterns by observing the flow of pack ice caused by wind and current; he learned that the currents in the Arctic Ocean are formed as a consequence of a difference in density between the two principal water masses, the Arctic and the Atlantic. In oceanography Nansen is best known for developing a special sample-collecting container, the *Nansen* bottle, which could measure temperature and salinity at varying depths (see Chapter 10). Major oceanographic publications up to the nineteenth century are summarized in Table 1-2.

Nansen bottle *A specially designed bottle to obtain samples of ocean water from beneath the surface.*

The Post-*Challenger* Era

In the first quarter of the twentieth century, three historic voyages, those by the German *Meteor,* the Danish *Dana,* and the *Carnegie,* provided further impetus for the field of oceanography. The *Meteor* took over 70,000 soundings of the ocean floor. As a result, it was later realized that the seafloor was as irregular, rugged, and diversified as the land. Thus one of the great contributions of the *Meteor* voyage was to eliminate the incomplete picture of the ocean floor begun by the *Challenger*. The 1920–1922 voyage of the *Dana* in the Indian Ocean was highlighted by the discovery of a midoceanic ridge, the Carlsberg Ridge. The research ship *Carnegie* studied the Albatross Rise in the Pacific in 1929.

Between 1925 and 1950, the research ships *Discovery I* and *Discovery II* studied the food and feeding habits, migration, breeding, and growth and mortality patterns of whales of the Antarctic Ocean. Following World War II, the voyage of the Swedish *Albatross* helped explain the

TABLE 1-2 Major Oceanographic Publications Up to Nineteenth Century

Hugo Grotius	1608	*Mare Liberum* (translated into English from the original Latin text in 1916 by R. V. D. Maguffin). The author discusses the freedom of the seas.
Benjamin Franklin	1786	Published the first chart of the Gulf Stream in *Transactions of the American Philosophical Society.* Philadelphia, *2,* pp. 314–317.
William Scoresby, Jr.	1820	*Account of the Arctic Region with a history and description of the Northern Whale Fisheries.* Edinburgh, Scotland: Constable.
Charles Darwin	1831–1836	*The Voyage of the Beagle.* New York-Anchor. First to explain origin of reefs and, in subsequent years, formulated the theory of evolution.
Nanthiel Bowditch	1832	*The New American Practical Navigator.* New York: Blunt.
M. F. Maury	1855	*The Physical Geography of the Sea.*
C. W. Thompson	1873	*The Depths of the Sea.* London: Macmillan.
William Dittmar	1884	*Report on the Scientific Results of the Voyage of H.M.S. Challenger.* Vol. 1. One of the early pioneers of developing chemical oceanography.

nature of currents and marine life, improved bottom-coring techniques, and was one of the earlier ships to use the seismic echo technique for the determination of seabed sediment thickness.

In the 1950s, the Swedish ship *Galatha* successfully dredged along the bottom of the Mindanao Trench. About the same time, Russian oceanographers studied the Kurile Trench. In other words, the study of the deeper parts of the ocean began on an international level.

In 1960 the U.S. ship *Trieste,* a manned "submersible," as part of a deep-ocean submarine test, not only touched the deepest part of the ocean in the Marianas Trench, which is 11,022 meters deep, but also discovered the presence of life at such prodigious depth. Man realized for the first time the magnitude of influence exerted by massive and slow-moving downward currents that transport oxygen and nutrients to support life at such an immense depth. One way to assess the development of the field of oceanography is to note the changes in the design of research ships, laboratory equipment, and general working conditions of oceanographer (Figures 1-7, 1-8).

The *Glomar Challenger* Era

In 1968 the *Glomar Challenger,* a highly sophisticated deep-sea drilling vessel, was launched by Scripps Institution of Oceanography in La Jolla, California. As a participant in the Deep-Sea Drilling Project (DSDP), a part of the National Science Foundation Ocean Sediment Coring Program, its major objective was to study the earth's crust by collecting long cores of sediment taken from the deeper regions of the world's oceans (Figure 1-9).

Between 1968 and 1973 the *Glomar Challenger* drilled 450 holes, cored at about 300 sites, and traveled over 275,000 kilometers. Because it has the capacity to undertake scientific exploratory tasks in almost all parts of the sea floor, the *Glomar Challenger* is a powerful research tool. The findings of the *Glomar Challenger* are unquestionably impressive considering its short time of operation. It has so far provided to geological oceanography supporting evidence in favor of such modern concepts as seafloor spreading, continental drift, and global plate tectonics (see Chapter 5).

One of the latest scientific studies of the oceans is the International Phase of Ocean Drilling (IPOD), a joint undertaking of scientists from several countries to drill deeper into the earth's crust than ever before. The IPOD project is aimed at gathering important information on mineral formation and at deciphering basic processes that govern the shaping of the earth's surface.

Oceanography is now a highly sophisticated and multidisciplinary field of science. One criterion of its rapid progress is the successful im-

(a) (b)

Figure 1-7. One hundred years of oceanography. (a) Oceanographers at work aboard ship 100 years ago. (*Source: Challenger* Office Report, Great Britain, 1895.) (b) Oceanographers at work aboard ship today. (Photo courtesy of Woods Hole Oceanographic Institution.)

plementation of projects of worldwide stature such as the *International Geophysical Year* (IGY), DSDP, and now IPOD. Furthermore, rising trends in ocean mining, petroleum, environmental, and engineering fields also once again reveal the accelerated progress of oceanography.

One of the most impressive trends in modern oceanography apart from global oceanic research and the deployment of sophisticated ships such as the *Glomar Challenger* (see Chapter 5) is the network of hundreds and sometimes thousands of sensors of different shapes, sizes, and specifications to measure all types of parameters—physical, chemical, biological, ecological, geochemical (Figures 1-10, 1-11, 1-12). Now scientists can study oceans from outer space by using remote-sensing techniques such as satellites equipped with various types of sensors,

(a)

(b)

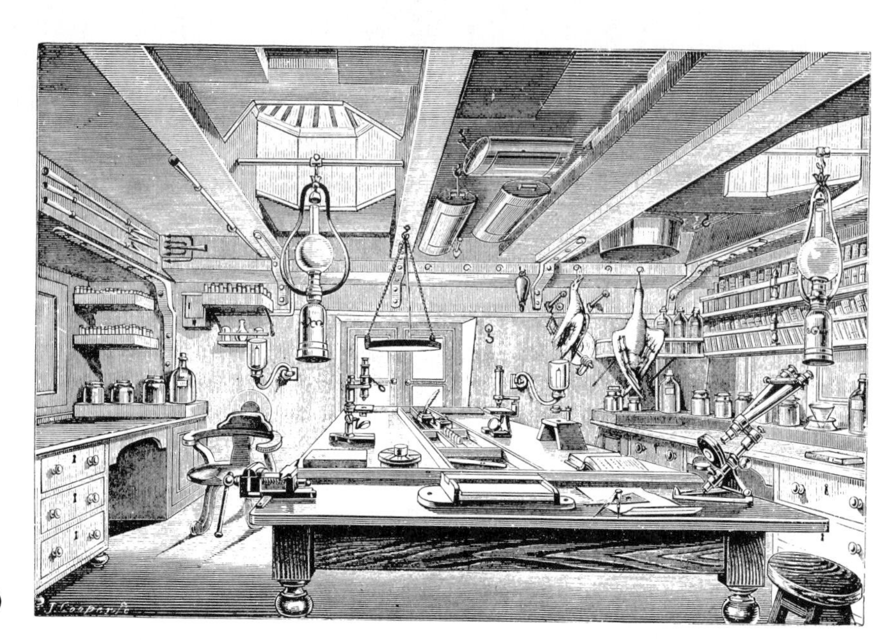

Figure 1-8. One hundred years of oceanography. (a) Oceanographic laboratory aboard a modern research ship. (Photo courtesy of Woods Hole Oceanographic Institution.) (b) Oceanographic laboratory aboard the H.M.S. *Challenger* 100 years ago. (*Source: Challenger* Office Report, Great Britain, 1895.)

Figure 1-9. The *Glomar Challenger*, a research vessel of the Deep-Sea Drilling Project, developed by the Scripps Institution of Oceanography. The 10,500-ton vessel is 130 meters long, has a beam of 21 meters and a loaded draft of 6.5 meters. Its drilling derrick stands 63 meters above the waterline. The *Challenger* conducts drilling operations in the open ocean using dynamic positioning to maintain position over a hole (see Chapter 9). (Photo courtesy of Scripps Institution of Oceanography, University of California.)

including cameras (Figure 1-13). A recent breakthrough is the concept of RUM—remote underwater manipulator (Figure 1-14). The RUM was designed by the petroleum industry in an attempt to resolve the difficulty of repairing and maintaining offshore drilling rigs, particularly in deeper parts of the sea. A RUM may be designed in different ways—in the form of a robot (the petroleum industry's version) or as a submersible (the U.S. Navy's method). In addition to their use of oil-field maintainance and repair, RUMs can be deployed in the study of deep-ocean life, marine mining, and salvaging of sunken ships, and in conducting deep-ocean experiments. Thus the modern technology that has

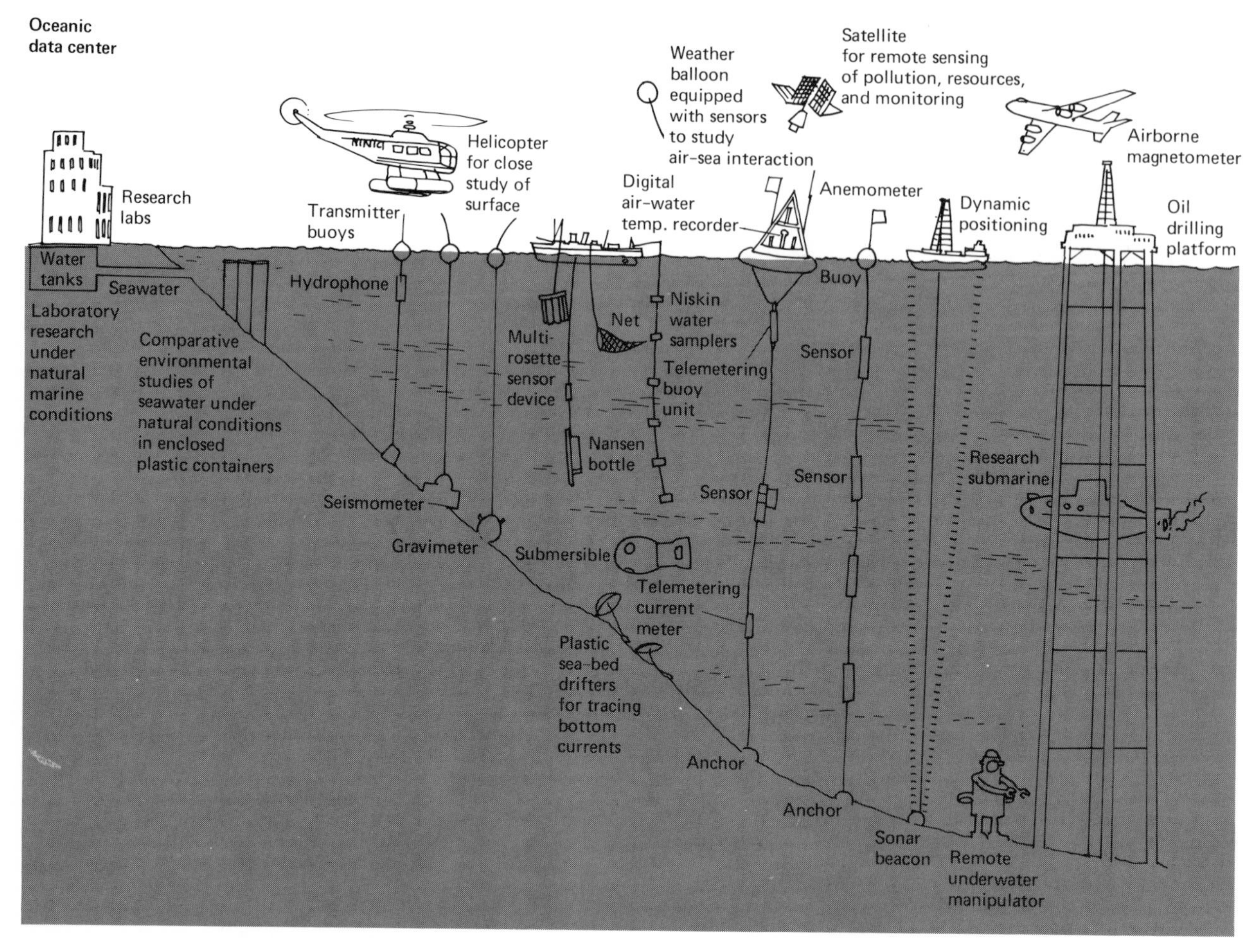

Figure 1-10. Modern oceanographic-research tools and their applications.

sprung largely from the space technology of the 1960s and 1970s is accelerating the growth of oceanography.

But a word of caution must be raised. Although current knowledge of the oceans is increasing at a stupendous rate, we still know little about them. In 1963 and 1966, respectively, the U.S. nuclear-powered submarines *Thresher* and *Scorpion* were lost in waters of less than 3000 meters for reasons not yet fully determined (Figure 1-15). In 1967 the *Torrey Canyon*, a U.S. oil tanker, grounded mysteriously in the English Channel and spilled over 100,000 tons of crude oil. The following year, in California's Santa Barbara Channel, an oil well blew off unexpectedly because of cracking in the seafloor, once again releasing thousands of gallons of crude oil. In 1976 and 1977, a half dozen oil tankers released millions of gallons of oil near or within U.S. waters. In some of these cases, exact causes have not been ascertained. (see Chapter 15).

Figure 1-11. Oceanographers employ such modern underwater sampling devices as this rosette of multibottle array system and Niskin water-sampling bottle. This instrument can be used along with STD probe to measure temperature, conductivity (salinity), and depth simultaneously. Among other advantages, this instrument reduces on-station sampling time. (Photo courtesy of General Oceanics, Inc.)

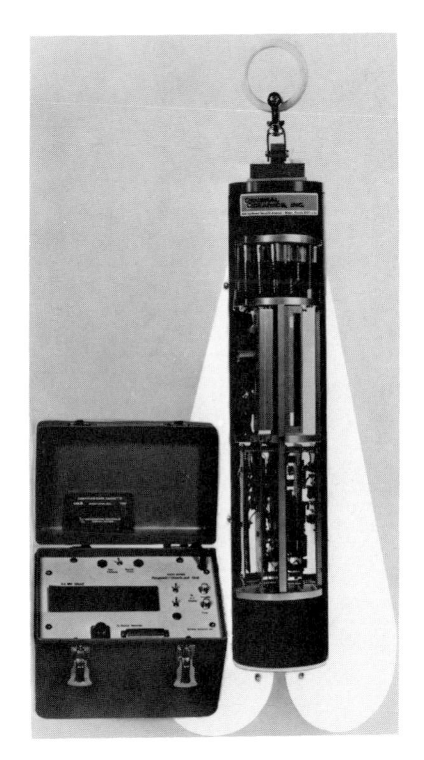

Figure 1-12. Magnetic tape records winged current meter with temperature sensor. This instrument measures current speed and direction versus time. It can measure current speeds from 0 to 200 centimeters per second and can operate to depths of 5000 meters. The tape cassette can store 20,000 readings. (Photo courtesy of General Oceanics, Inc.)

1-4 DIVISIONS OF OCEANOGRAPHY

Oceanography is often organized for convenience into four primary divisions: physical, chemical, geological, and biological.

Physical oceanography deals largely with the study of the physical properties of ocean waters, such as distribution of salinity, temperature, and density; and the nature of oceanic currents, tides, and waves. *Geological oceanography* emphasizes the origin and evolution of oceans through the passage of geologic time, the structure of ocean basins, the

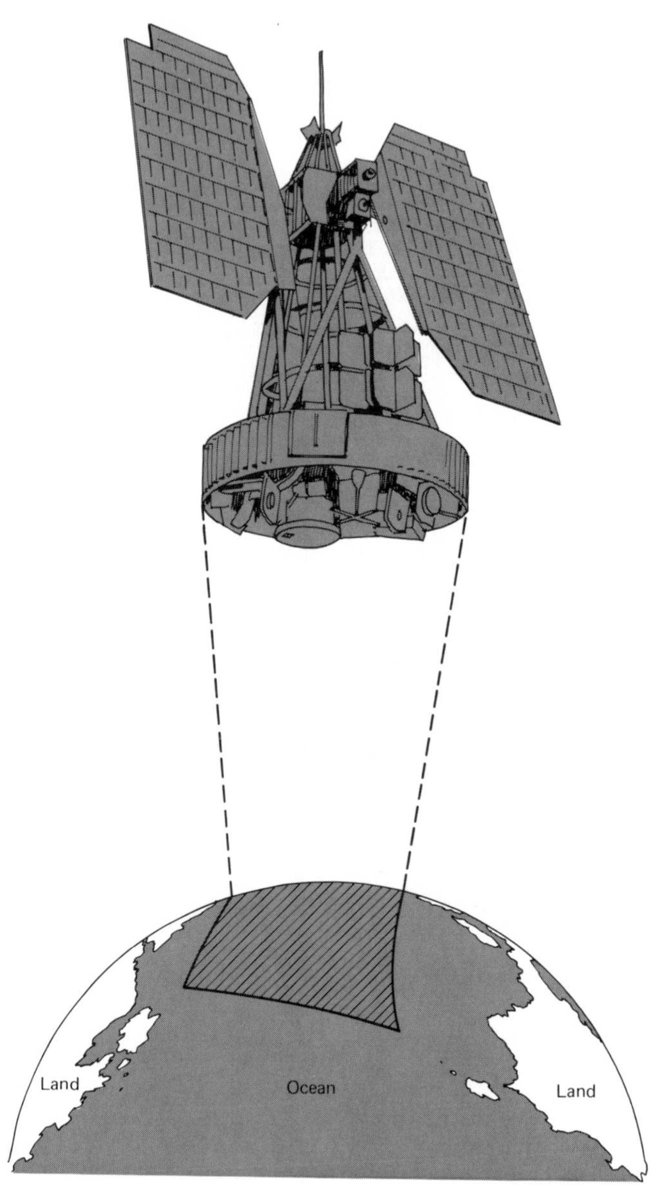

Figure 1-13. Satellites such as this earth's resources technology satellite (ERTS) are the principal tool of space oceanography. The satellite observation provides information concerning the nature, origin, and concentration of minerals, energy, and pollutants, as well as overall monitoring of the ocean.

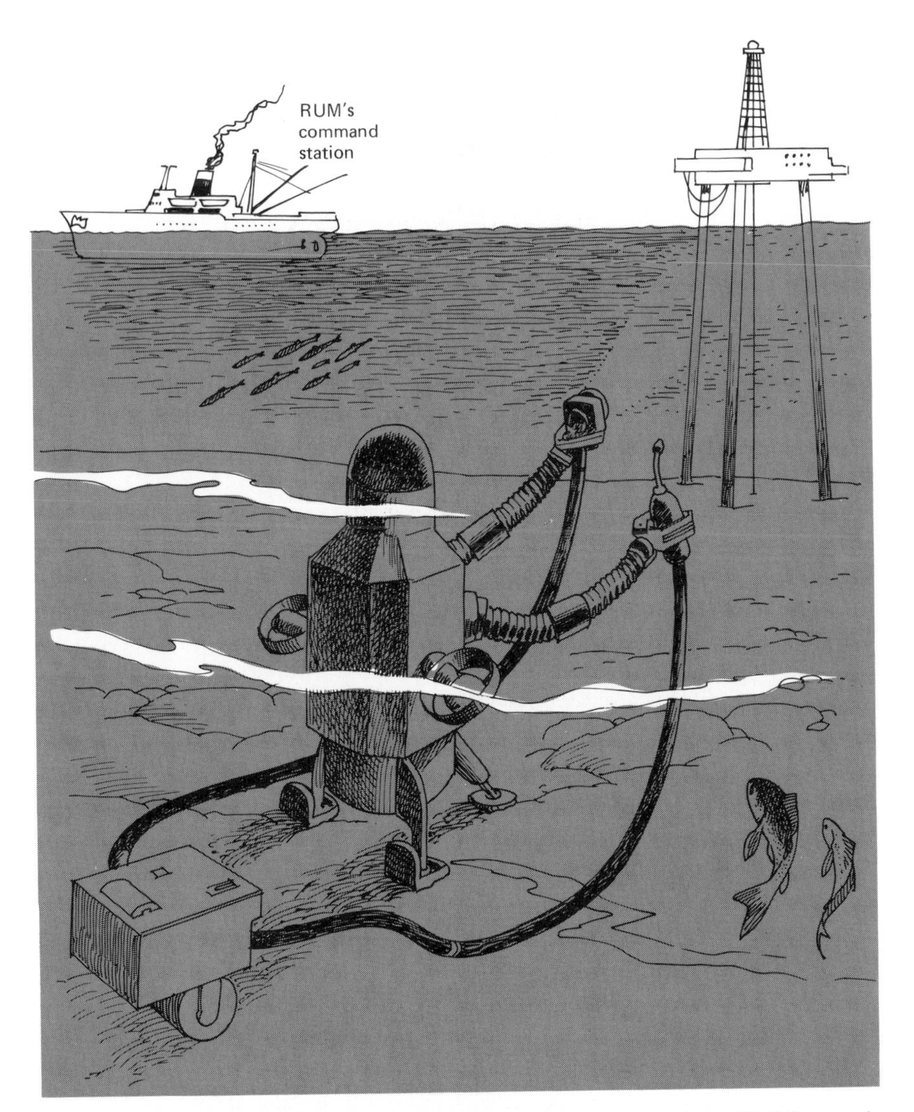

Figure 1-14. Artist's concept of a remote underwater manipulator (RUM) at work in the deep ocean. The RUM would be equipped with television cameras for eyes and a sonar pinger for ears. A hydrophone system would permit the RUM to locate objects on the ocean floor about 500 meters away. Its arms would be packed with a socket wrench, pincher claws, and a valve wrench.

distribution of sediments and fossils on the ocean floor, and the formation of geologic material in deep and shallow oceans. *Chemical oceanography* studies the distribution of chemical elements on the ocean floor and of the general chemistry of seawater. *Biological oceanography* is the study of marine animals and plants, particularly their ordering and distribution at various depths. *Marine pharmacology* is an important

Figure 1-15. Deep-towed vehicle used in *Thresher* search. (Photo courtesy of Woods Hole Oceanographic Institution.)

offshoot of biological oceanography and is a relatively new field of oceanography. It is concerned largely with the identification, extraction, and development of rare vitamins and medicinal drugs from marine plant and animal life. It is thought that some of these vitamins and drugs will some day help to cure cancer and other diseases. Because certain vitamins are found exclusively in sea plants and animals, marine pharmacologists are interested in finding ways to utilize these rare nutrients in order to supplement man's imperfectly balanced diet.

Ocean technology, also referred to as *ocean engineering,* is another rapidly growing new branch of oceanography. Ocean technology deals with, among other things, the construction and design of sophisticated submarines, the construction of giant supertankers, the innovation and development of equipment and instruments for marine petroleum and mining industries, and advanced oceanic research. Other applied and practical fields of oceanography are also becoming important: the international Law of the Sea, marine managment and planning, marine public policy and program, environmental oceanography, and marine fisheries.

1-5 EDUCATION AND CAREER OPPORTUNITIES

As mentioned earlier, oceanography is a rapidly expanding field, and the number of students choosing careers in this field is high. The number of graduates in oceanography has increased in recent years. Several leading institutes of higher learning offer oceanographic degrees, including Scripps Institution of Oceanography at La Jolla, California, Woods Hole Oceanographic Institution at Cape Cod, Massachusetts, and Lamont-Doherty Geological Observatory of Columbia University in New York state.

Most oceanographers have strong backgrounds in other fields of science such as geology and geophysics, physics, chemistry, and biology. Oceanographers find job opportunities for teaching and research with academic institutions; federal agencies such as the Department of the Interior, the Environmental Protection Agency, or the Department of Defense; other governmental and private industrial research laboratories; and with petroleum and mining companies. In recent years job opportunities have gradually expanded in the fields of ocean engineering, marine electronics, marine fisheries, and environmental oceanography. It appears that educational and career opportunities in coming years will increase in oceanography, but by the year 2000 the greatest demands will be for ocean technologists, engineers, and pharmacologists.

SUMMARY

1. Oceanography, the scientific study of the ocean, is a multidisciplinary science involving experts from various scientific fields.
2. The mobile ocean occupies 71 percent of the earth's surface, regulates many natural processes including weather conditions and water cycles, and comprises a storehouse of seafoods and energy that man hopes to obtain for his future survival.
3. Progress in oceanography was slow until the nineteenth century. Until then greater importance was attached to commerce and trade rather than to vigorous scientific study of the oceans. The historic

voyages of the *Beagle,* the *Challenger,* and the *Meteor,* and in modern times by the *Trieste* and the *Glomar Challenger* have significantly helped to advancing the study of oceanography.

4. Today oceanography is a well-established and rapidly growing science. New discoveries are constantly being translated into practical uses. Sources for food and energy are being put to new uses. How well man uses these resources will greatly influence his future.
5. Oceanography is subdivided into geological, chemical, physical, and biological oceanography. Applied and practical fields include development in ocean engineering, marine fisheries, and marine pharmacology.
6. Education and career opportunities in oceanography are promising. A rising interest in oceanography has increased competition for these jobs.

Suggestions for Further Reading

Burgess, R. F. 1975. *Ships Beneath the Sea: A History of Subs and Submersibles.* New York: McGraw-Hill.

Committe on Oceanography. 1972. *Oceanography 1960 to 1970,* No. 11. *A History of Oceanography of the United States.* Washington, D.C.: National Academy of Sciences, National Research Council Committee on Oceanography.

Deacon, Margaret. 1971. *Scientists and The Sea: 1650–1900.* New York: Academic Press.

Eberhart, Jonathan. 1975. "Ever Downward Beneath the Ocean Deep." *Science News, 107,* 9–15.

Hammond, A. L. 1970. "Deep Sea Drilling: A Giant Step in Geological Research." *Science, 170,* 520–521.

Linklater, E. 1972. *The Voyage of the Challenger.* Garden City, N.Y.: Doubleday.

Pirie, R. G. 1973. *Oceanography: Contemporary Readings in Ocean Sciences.* New York: University Press.

Schlee, Susan. 1973. *The Edge of the Unfamiliar World: A History of Oceanography.* New York: E. P. Dutton.

2

The Planet Ocean

2-1 INTRODUCTION

THREE MAJOR COMPONENTS OF THE EARTH INCLUDE THE hydrosphere, the lithosphere, and the atmosphere. Oceans constitute the hydrosphere; the lithosphere is a layer of rocks; the atmosphere is a layer of air. The biosphere, or realm of life, encompasses parts of these three layers. It extends several meters into the soil (lithosphere), several hundred meters (over 6 miles) into the atmosphere, and more than 11,000 meters into the ocean. The world's oceans occupy roughly 71 percent of the earth's surface (Figures 2-1, 2-2).

Biosphere *The total sphere of life in, on, and above the earth.*

The highest point on land, Mount Everest (8,900 meters), has long been awesome. Only in recent years has the awesomeness attached to Everest been equalled by the discovery of the deepest spot on earth, the Marianas Trench (11,022 meters below sea level) of the Pacific Ocean (Figure 2-3). The average height of the land is roughly 0.75 kilometer, but the average depth in the ocean is about 4 kilometers. In fact, if all the irregularities of the land were leveled off, the oceans would flood and submerge all lands to an average depth of 2.5 kilometers. Thus the all-covering ocean on our planet would still be extremely deep.

Oceans are predominant in the Southern Hemisphere, often referred to as the *marine hemisphere*. Landmass, on the other hand, dominates in the Northern Hemisphere, which is known as the *land hemisphere* (Figure 2-4). The land hemisphere includes the continental masses of Eurasia, North America, and Africa above the equator. The marine hemisphere comprises South America, Africa below the equator, Australia, New Zealand, and Antarctica (Figure 2-5). (Important statistical facts about the earth and ocean are presented in Tables 2-1 and 2-2.)

2-2 THE EARTH'S INTERIOR

Seismic studies have shown that earth's interior is divided into three distinct layers: the core, the mantle, and the crust (Figure 2-6).

Seismic *Relating to earthquakes.*

Figure 2-1. The face of the planet Ocean. (Photo courtesy of National Marine Fisheries Service.)

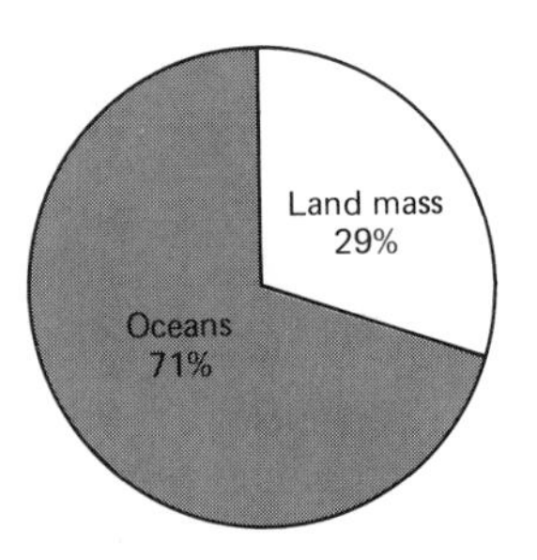

Figure 2-2. Distribution of oceans and landmasses.

These divisions may be simply compared to a soft-boiled egg: the core corresponds to the yolk, the mantle to the white, and the crust to the hard, thin shell.

The Core

The core of the earth begins 2,900 kilometers below its surface and extends up 3,480 kilometers to the center of the earth. Seismic studies have in fact shown the existence of two cores: the *outer liquid core* and the *inner solid core*. This deduction is based on known physical properties of certain seismic waves: *compression waves* (*P* waves) act to compress and

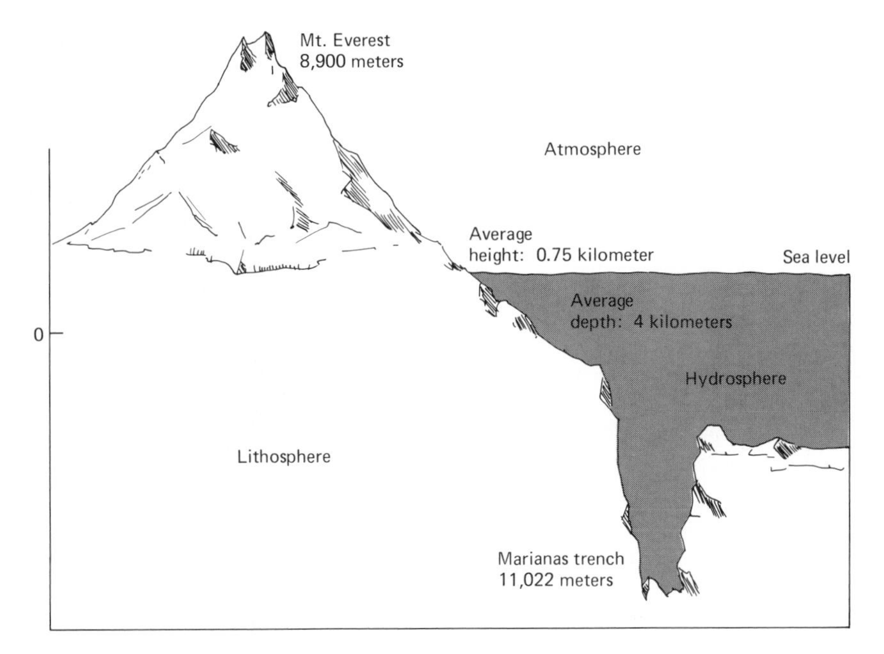

Figure 2-3. Basic data on planetary elevation and depth.

dilate rock as they propagate forward; *shear waves* (*S* waves) can shake rock laterally as they propagate forward. P waves move through solid and liquid media, whereas S waves move only through solid media. Seismologists have noted that compression waves can pass through solid inner and outer liquid cores but that shear waves stop at the outer core. This selective penetration of the two kinds of waves led to the inference that there are two cores at the center of the earth (Figure 2-7).

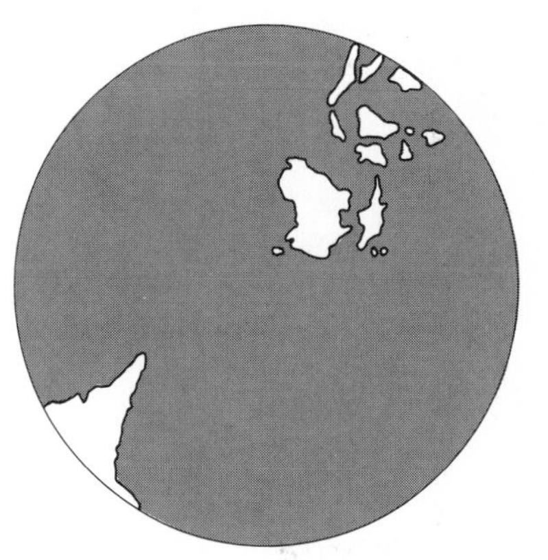

Figure 2-4. The earth's hemispheres.

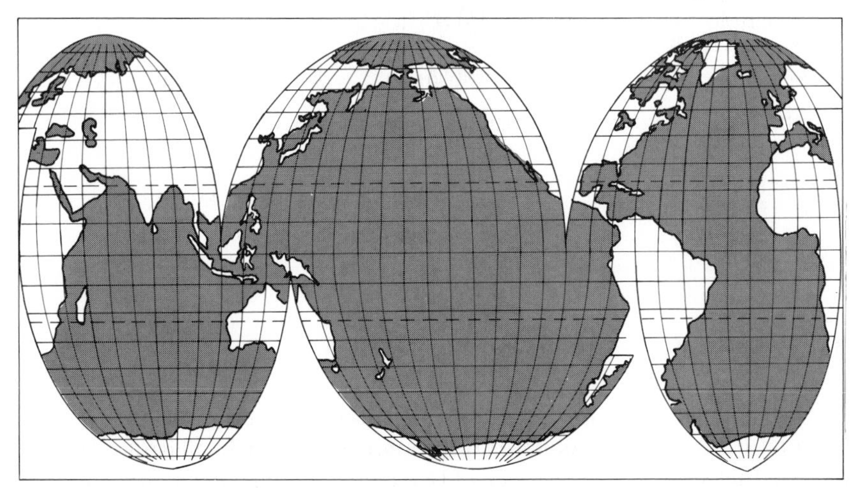

Figure 2-5. Goode's homalographic projection of land and oceans.

TABLE 2-1 Some Statistics Concerning the Earth and Oceans

Earth	
Mass	5.976×10^{27} grams
Volume	1.083×10^{12} cubic kilometers
Area	510.100×10^{6} square kilometers
Equatorial radius	6378.163 kilometers
Polar radius	6356.177 kilometers
Oceans	
Volume	1.350×10^{9} cubic kilometers
Area	362.033×10^{6} square kilometers
Mean depth	3729 kilometers

Core *Innermost portion of the earth. Also a cylindrical sample of sediments.*

The inner core is probably comprised of an alloy of iron and nickel. The iron alloy constitutes about 80 to 85 percent of the inner core; nickel and other unknown elements comprise the remainder. Although a few scientists contend that the inner core may be composed of compressed hydrogen, it is more accepted that the core is primarily composed of iron and nickel.

The Mantle

Mantle *The intermediate layer between the earth's core and crust.*

The intermediate layer of the earth's interior is the *mantle.* It accounts for 84 percent of the earth's volume and 67 percent of its weight. The mantle is complex in structure; it is about 2,900 kilometers thick and is subdivided into 10 or more layers. The mantle is composed of minerals rich in silica, magnesium, and iron. The amount of magnesium increases toward the surface, and the amount of iron increases toward the core. The upper mantle is hotter than the lower mantle, because it has an extensive distribution of radioactive minerals such as uranium, thorium, and potassium. Three subdivisions are recognized in the upper mantle: the *mesophere,* the *asthenosphere,* and the *lithosphere.* The asthenosphere, a hot, plastic, semiliquid layer, is sandwiched between a hard layer of mesosphere (below) and a rigid layer of lithosphere (above). The asthenosphere is 100 to 400 kilometers thick. The upper region of the asthenosphere is the *low-velocity layer;* and the boundary between the mantle and the crust is the *Mohorovičić discontinuity (Moho).*

Mohorovičić discontinuity (MOHO) *Seismic discontinuity indicating compositional change between the crust and mantle of the earth. It is found beneath the crust at depths ranging from 5 to 50 kilometers.*

TABLE 2-2 Land and Water Hemispheres

Name	*Area (in square kilometers)*	*Percent*
Oceans	361,059,000	70.8
Continents	148,892,000	29.2
Total	509,951,000	100.0

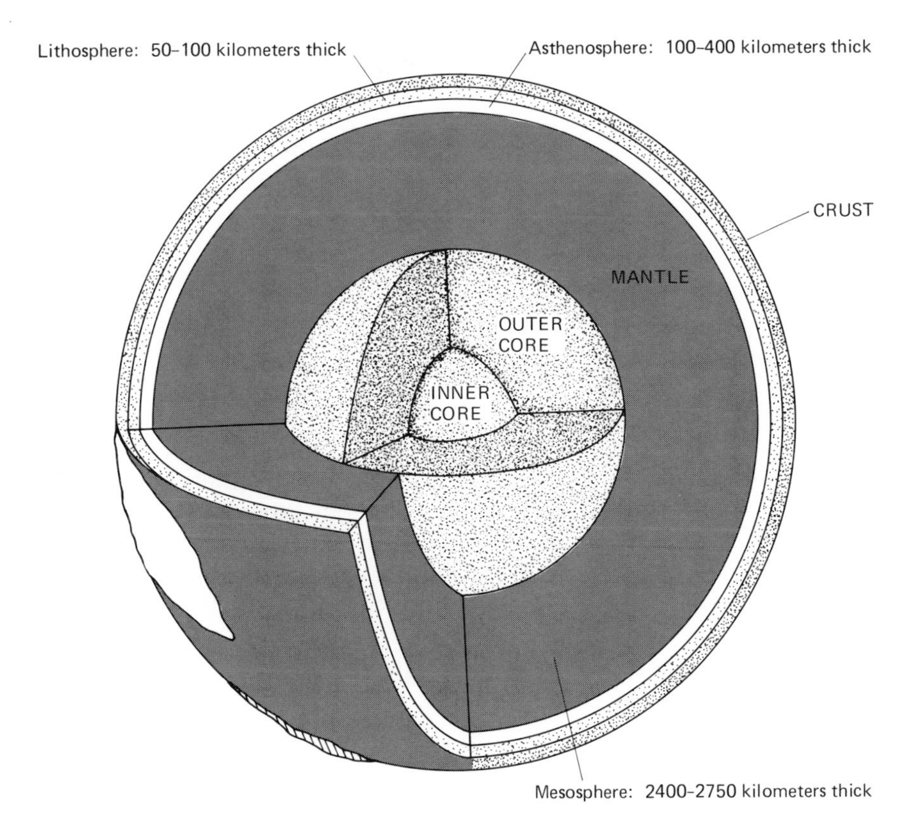

Figure 2-6. The structure of the earth's interior with its principal layers and sublayers. Inner core: 1320 kilometers. Outer core: 2160 kilometers. Mantle: 2900 kilometers. Crust: 5 to 40 kilometers.

This boundary is marked by a systematic change from the lighter material of the crust to the heavier material of the mantle (Figure 2-8).

The Crust

The *crust* of the earth rests on the Moho and is 6 to 50 kilometers thick. The crust is thicker under continents and diminishes progressively under ocean basins. It may be 50 kilometers under mountainous regions of continents and may be less than 5 kilometers under oceans. In certain parts of the Pacific Ocean the crust is virtually absent.

The crust is divided into two principal layers: the lower layer, *sima,* carries heavy concentrations of silica and magnesium, whereas the upper layer, *sial,* has heavy concentration of silica and aluminum. Sima is largely basaltic, with a density of 2.9, and occurs under oceans and continents. Sial is chiefly granitic, with a density of 2.6, and occurs mainly under continents (Figure 2-8).

Crust *Uppermost and thinnest segment of the earth. Thickness varies from 5 to 50 kilometers. Enriched in basaltic content in the oceanic crust and granitic in the continental crust.*

Sima *A layer of the crust beneath continents and oceans rich in silica and magnesium.*

Sial *A layer of the crust beneath all continents rich in silica and aluminum.*

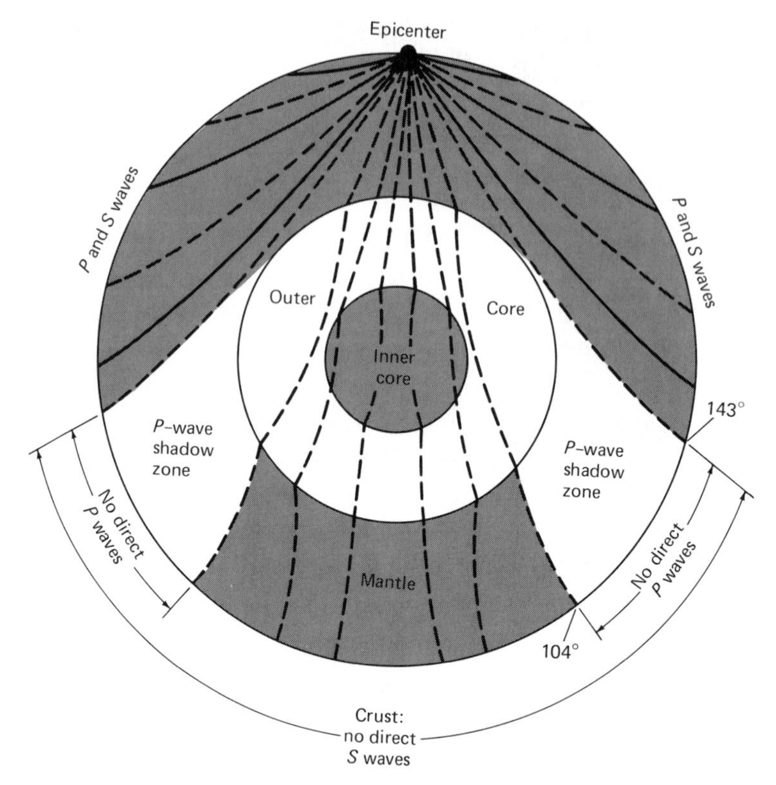

Figure 2-7. Seismic studies of the earth's interior show two cores: the inner solid core and the outer liquid core. Dashed lines indicate *P* waves. Solid lines indicate *S* waves.

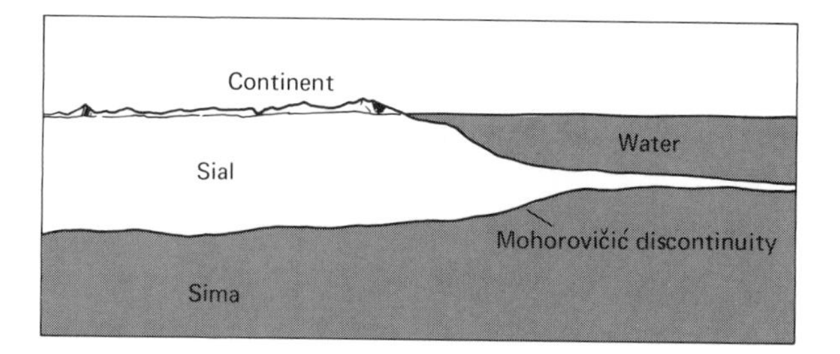

Figure 2-8. Distribution of crust under the continents and oceans. *Sial* is granitic crust rich in silica and aluminum. *Sima* is basaltic crust rich in silica and magnesium. Note the relatively thinner portion of the crust under the ocean basin.

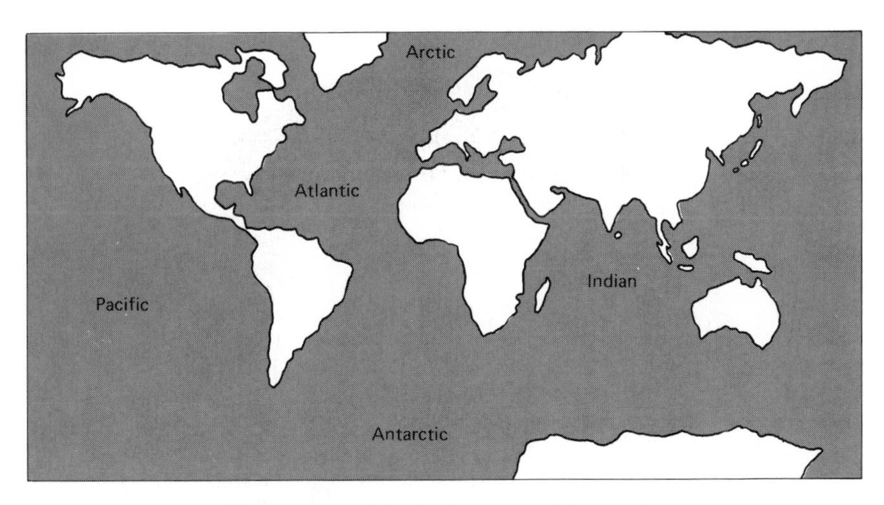

Figure 2-9. Principal oceans of the world.

2-3 THE WORLD'S OCEANS

The five major oceans in the world are the Pacific, the Atlantic, the Indian, the Arctic, and the Antarctic. The Pacific, South Atlantic, and Antarctic occupy approximately 90 percent of the Southern Hemisphere. The Indian, North Atlantic, and Arctic cover the watermass in the Northern Hemisphere (Figure 2-9, Table 2-3).

TABLE 2-3 Basic Facts about Major Oceans

Ocean	Area *(in millions of square kilometers)*	Volume *(in millions of cubic kilometers)*	*Depth (in meters)*	*Temp. (°C)*	*Salinity (%)*	*Remarks*
Pacific	166,241	696,184	4,188	3.36	34.62	Coldest, deepest, and largest ocean, occupying more than half of the volume of ocean basins.
Atlantic	94,314	337,210	3,736	3.72	34.76	Receives large amount of sediment from many rivers such as the Amazon, the Congo, the Mississippi.
Indian	77,118	284,608	3,872	3.73	34.90	Receives large amount of sediment from the Indus, the Ganges, and the Brahmaputra.
World	362,033	1,349,929	3,729	3.52	34.72	Oceans occupy 70 percent of the earth's surface.

SOURCE: From H. W. Menard, and S. M. Smith, 1966, "Hypsometry of Ocean Basin Provinces," *Journal of Geophysics Research, 71,* 4305; and R. B. Montgomery, 1958, "Water Characteristics of Atlantic Ocean and of World Ocean," *Deep-Sea Research, 5,* 146.

The Pacific Ocean

The Spanish explorer Vasco Nuñez de Balboa discovered and named the Pacific Ocean; "pacific" means peaceful. The Pacific is the largest, coldest, and deepest of all oceans. It occupies about one-third of the total surface area of the earth and contains 696 million cubic kilometers volume of water. Most of the Pacific flow is very deep where average depth exceeds 5000 meters. The Pacific Ocean forms a circular-shaped basin; along its periphery are numerous volcanoes, faults, and trenches that form the *ring of fire*. The western border of the Pacific is irregular and encompasses several marginal seas:* the Sea of Okhotsk, the Sea of Japan, the East China Sea, the Yellow Sea, the South China Sea, the Sea of the East Indian Archipelago, the Coral Sea, and the Tasman Sea. The Bering Sea and the waters of the Antarctic Ocean mark the north and south boundaries of the Pacific. Hundreds of scattered volcanic islands, including Hawaii, in the Pacific Ocean together form an area comparable to that of the United States.

The Atlantic Ocean

The ancient Romans named the Atlantic after the Atlas Mountains of North Africa. These mountains marked the limit of the then-known world, the Mediterranean Sea. "Atlantic" probably designated their belief that the ocean lay beyond the Atlas Range. The Atlantic Ocean is divided from the Indian Ocean by 20°E, which crosses Cape Agulhas and from the Pacific Ocean by 67°W, which runs through Cape Horn to the South Shetland Islands. The Atlantic Ocean is oblong from north to south and is irregularly shaped. It is approximately one-half the size of the Pacific and narrows considerably between the landmasses of western Africa and eastern Brazil. These protruding continental bulges on both its sides separate the ocean into the North and South Atlantic.

The North Atlantic is surrounded by several marginal seas and bays, including the Mediterranean Sea, the Baltic Sea, the North Sea, the Black Sea, the Bay of Biscay, Baffin Bay, Hudson Bay, the Gulf of Mexico, and the Caribbean Sea. The South Atlantic is bounded by west Africa, eastern South America, and the north Antarctic Ocean. The Mississippi, the Amazon, and the Congo rivers empty millions of kilograms of sediment into the Atlantic Ocean annually.

*The seas are relatively much smaller and shallower bodies of salt water that are confined to submerged portions of continents.

The Indian Ocean

Historically the Indian Ocean was a well-known commercial route that linked the trade centers of Ujjan, India, Baghdad, and Alexandria. One contributing factor to the proliferation of sea trade among these cities was the monsoon currents that helped the sailors to undertake long voyages in the Indian Ocean.

Monsoon *Arabic for season; referred to the winds of the Arabian Sea in the Indian Ocean that flow from the ocean to the land (i.e., from southwest) during summer and from land to the ocean (i.e., northeast) in winter time.*

The Indian Ocean is roughly triangular in shape and is enclosed by the Indian subcontinent, the east coast of Africa, and the northern and western coasts of Australia. India's three largest rivers, in terms of volume of water and length, the Ganges, the Indus, and the Brahmaputra, discharge the world's largest amount of terrestrial sediment into the Indian Ocean. The periphery of the Indian Ocean includes the Red Sea, the Persian Gulf, the Arabian Sea, and the Bay of Bengal. Madagascar and Sri Lanka are the largest-sized islands in the Indian Ocean.

The Arctic Ocean

The Arctic Ocean, the smallest and shallowest of all oceans, exists literally "on top of the world." Its basin is almost circular shape and is bordered by Russia, Scandinavia, Greenland, Canada, and Alaska (Figure 2-10). The North Pole is located about 500 miles from its center.

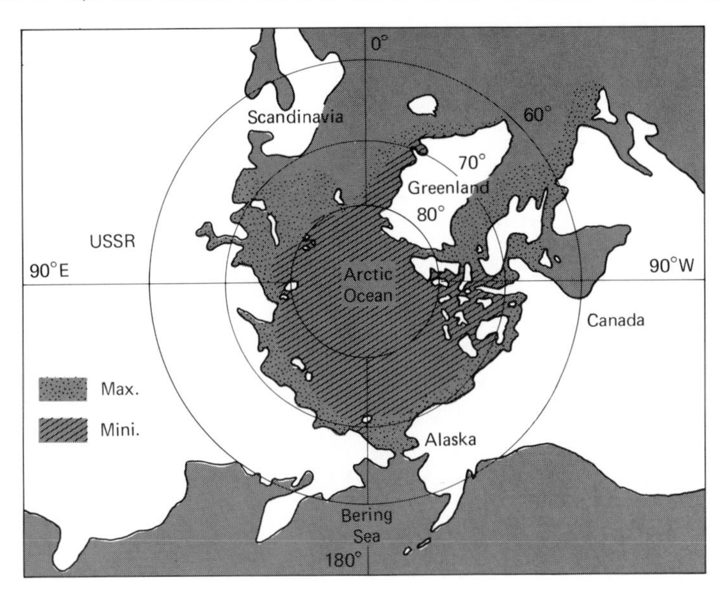

Figure 2-10. Arctic Ocean, with average boundaries of sea ice in autumn and spring.

Floating masses of ice, the *polar ice pack,* cover most of the ocean during most of the year. The topography and structure of the Arctic Ocean are dominated by three submarine ridges, which result in a number of basins and deeps. The Lomonosov Ridge is a prominent structure in the Arctic Ocean.

The Arctic Ocean is separated from the Pacific by the Bering Straits. The ocean connects with the Atlantic via small seas on either side of Greenland; the Greenland and Norwegian seas and the Denmark Strait provide the connection to the east and Baffin Bay, Davis Strait, and the Labrador Sea to the west.

The Antarctic Ocean

The Antarctic Ocean, at times referred to as the "southern ocean," dominates the entire Southern Hemisphere with no landmass interruptions. The Antarctic Ocean extends to 40° S, touching the southern tips of Africa and Australia. It thus encompasses approximately 75 million square kilometers and accounts for over 20 percent of the total area of all oceans.

The Antarctic Ocean was intensively studied during the International Geophysical Year in 1957. Since then the United States, Russia, Australia, New Zealand, and other countries have maintained research stations for continuing study of the Antarctic Ocean. This ocean is in the path of the west winds, which trigger the powerful *westwind drift,* a west–east current of the upper layers of water.

2-4 THE WORLD'S MAJOR SEAS

The Mediterranean Sea

The Romans originally named the Mediterranean Sea *Mare internum,* which in Latin means inland sea. They later gave it its present name Mediterranean, to imply "middle of the earth." The Mediterranean Sea is bounded by Europe, Asia, and Africa. The Strait of Gibraltar not only marks its western boundary but also provides it with an outlet to the Atlantic Ocean. In the southeast the Suez Canal connects the Mediterranean with the Red and Arabian seas and with the Indian Ocean. To the northeast the Dardanelles and the Bosporus link it with the Black Sea (Figure 2-11). The Mediterranean Sea has several arms. The Adriatic, the Ionian, and the Tyrrhenian seas surround much of Italy. The Aegean Sea, another arm, is located east of Greece. And another arm, the Black Sea, extends northward between Turkey and Russia. The Mediterranean Sea contains many large islands, among them Crete, Cyprus, Sardinia, Sicily, and the Balearic Islands.

The depth of the Mediterranean Sea varies from approximately 300 meters at the Strait of Gibraltar to over 5000 meters in the Ionian Sea. Mediterranean and Atlantic waters flow in opposite directions at the Strait of Gibraltar.

The North Sea

The North Sea is a northwestern extension of the Atlantic Ocean that occupies an area between Great Britain and continental Europe. It surrounds Britain, including the Orkney and Shetland Islands, Denmark and part of Norway, the Strait of Dover, part of France, Belgium, Holland, and Germany, and the Arctic Ocean. It extends 11,000 kilometers from the Strait of Dover to the Shetland Islands and 650 kilometers from Scotland to Denmark. The North Sea covers an area of 475,000 square kilometers. The shores of all countries touching the sea are indented with bays, fjords, and estuaries.

The Caribbean Sea

The Caribbean Sea is a westward extension of the North Atlantic Ocean. It is partially enclosed by the Antilles, by Central America, and

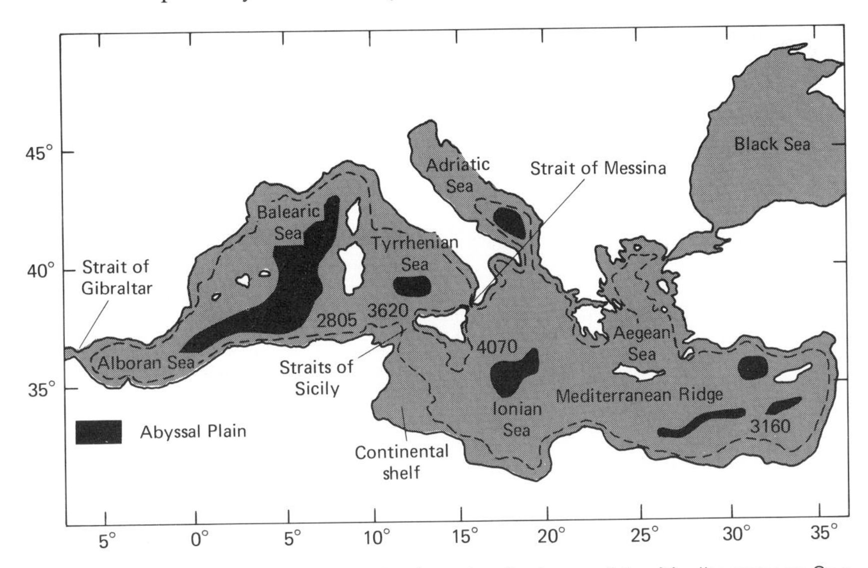

Figure 2-11. The bathymetry and submarine features of the Mediterranean Sea. Depths in meters. (*Source:* R. Fairbridge, ed., *Encyclopedia of Oceanography,* © 1966 by Litton Educational Publishing, Van Nostrand Reinhold Company.)

by South America. It is a typical tropical sea with warm, shallow waters where temperatures are usually over 25°C. It extends for 2800 kilometers from Guadeloupe to Honduras and for 1400 kilometers from Cuba to Panama. It thus occupies over 1,200,000 square kilometers. A strong surface current of warm water from the Atlantic Ocean flows through the Caribbean Sea to the Gulf of Mexico, aided by the prevailing easterly trade winds. The warm waters nourish widespread coral reef growth here.

The Baltic Sea

The Baltic Sea, between Sweden and Russia, provides a passage to the North Sea and subsequently into the Atlantic Ocean. It is approximately 1500 kilometers long and 640 kilometers wide and marks 8000 kilometers of coastline. The Gulf of Bothnia, which marks the northern extension of the Baltic Sea, separates Sweden from Finland.

The Red Sea

The Red Sea is an extension of the Indian Ocean and divides the Arabian Peninsula from northeastern Africa. The Red Sea is about 2,100 kilometers long and about 320 kilometers wide. Its average depth is about 600 meters. It covers 68,000 square kilometers and is comparable in size to California. The waters of the Red Sea are extremely salty because of a combination of prevailing high winds and extreme heat (temperatures often rise above 38°C). The Suez Canal to the west connects it with the Mediterranean Sea, and to the east the Gulf of Aden joins the sea first with the Arabian Sea and ultimately with the Indian Ocean.

2-5 COASTAL WATERS

Estuaries

Estuary *The mouth of a river valley where marine influence is manifested as tidal effects and increased salinity of the river water.*

Estuaries are a vital natural resource of man. They are used for commercial, industrial, and recreational activities, and play an important role in the natural cycles of fish, animal, and plant life. One of the most important features of estuaries is the free mixing of fresh- and seawater, consequently permitting a variety of organisms, from microscopic species to fish, birds, and mammals, to flourish.

Estuaries provide excellent grounds for clam and oyster farming (see Chapter 13), and often shrimp move from the sea to estuarine nursery areas. Salmon and striped bass journey upstream through estuaries to spawn. Estuaries are highly populated bodies of water supported by luxuriant growths of a variety of plants from algae and eelgrass to mangroves.

An estuary is a partially enclosed body of water connected with the ocean and characterized by the mixing of fresh- and seawater because of runoff (Figure 2-12). San Francisco Bay, Chesapeake Bay, and Puget Sound are well-studied estuaries.

Four major types of estuaries include drowned river valleys, fjords, bar-built estuaries, and tectonically formed estuaries (Table 2-4).

Fjord *A long, narrow, deep, U-shaped inlet that usually represents the seaward end of a glacial valley that has become partially submerged after the melting of the glacier.*

Drowned river valleys are estuaries formed during the Ice Age, particularly when glaciers were thawing, subsequently causing the sea level to rise worldwide. The flooding of lowlands resulted in drowning many drainage areas, which developed as estuaries. Drowned river valleys are common in many lowlands, especially in Atlantic and Gulf Coast areas of North America. Specific examples include Chesapeake Bay and Long Island Sound (Figure 2-13).

A *fjord* is a glacially eroded trough that is now occupied by an arm of sea (Figure 2-14). Fjords are characterized by their long, narrow, steep-walled (U-shaped) profiles. Large numbers of fjords are found at latitudes above 38° in both hemispheres, especially along the coasts of Norway, Greenland, Labrador, British Columbia and Alaska, and New Zealand. The deepest known fjord, Vanderford in Antarctica, is over 2000 meters.

When a drowned river valley is closed from open sea because of a bar or a barrier, it forms a *bar-built estuary*. Pamlico Sound in North

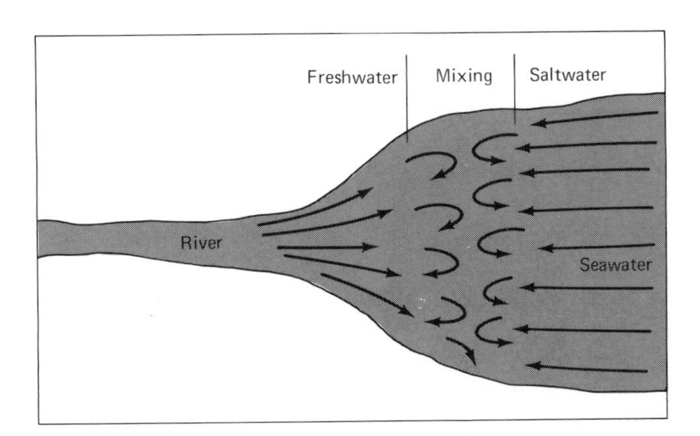

Figure 2-12. Schematic representation of estuarian circulation and mixing of fresh (river) and salt (sea) water. Note that nutrients are retained and are permitted to recirculate within the estuary.

Figure 2-13. Chesapeake Bay estuarine system, Maryland and Virginia. Salinity varies from 5‰ near the mouth of a river to 30‰ near Cape Charles.

Figure 2-14. Quarjak Fjord, Norway. (Photo courtesy of R. B. Theroux, National Marine Fisheries Service.)

TABLE 2-4 Some Major U.S. Estuaries

Location	*Example*	*Type*	*Remarks*
East Coast	Pamlico Sound	Bar-built	Formed by barrier islands separating drowned valleys from ocean; covers about 6,500 kilometers; freshwater supplied by Pamlico and Neuse Rivers.
	Chesapeake Bay	Drowned valley	Formed form the drowning of valleys of Susquehanna River and its tributaries, such as the Potomac and the James; largest in estuarian area coverage (11,000 kilometers).
West Coast	San Francisco Bay	Tectonic	Formed by faulting (or folding) resulting in low area and permitting river flow; covers 1200 kilometers; freshwater supply provided by Sacramento and San Joaquin rivers.
	Puget Sound	Fjord	Freshwater supplied by Skagit and Snohomish rivers.
Gulf Coast	Laguna Madre	Bar-built	Similar to Pamlico Sound; fed by Rio Grande.

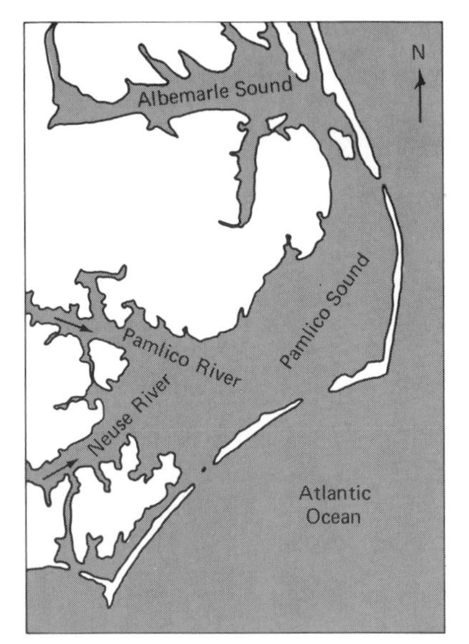

Figure 2-15. Pamlico Sound, North Carolina.

Carolina and Galveston Bay and Laguna Madre in Texas are notable examples (Figures 2-15, 2-16).

Some estuaries are formed as a result of nearby coastal *tectonic* activity such as faulting or a local subsidence to form a basin that becomes a receptacle for river discharge. San Francisco Bay and Drakes Estero in California are classical examples of estuaries of tectonic origin (Figure 2-17).

Salt Marshes

Often an estuary extends into low-lying flat areas known as wetlands or salt marshes. This extension is largely due to the successful sediment trapping and binding ability of certain estuarine plants such as algae (e.g., *Enteromorph*) and marine grasses (e.g., *Zostrea*) and various salt-tolerant plants. A typical salt marsh is abundantly supported by salt marsh grass (e.g., *Spartina*). The periodic submergence and emergence of marshes at high and low tides provide excellent conditions for the growth of marsh grasses, particularly under areas protected from wind and waves. The marsh grass spreads across tidal flats of silt and clay, accelerating further build-up of the salt marsh.

Tidal flats (or tidal marshes) are muddy areas that are alternately exposed and drowned due to low and high tides. Most tidal flats are dissected by many channels following a meandering path (zig-zag pattern-like rivers). At high tide seawater invades the tidal flat through these channels increasing the supply of sediments from the ocean. At low tide much of this sediment is deposited at the bottom of the salt marsh. Near the mouth of the salt marsh, coarser sandy material dominates, but, due to diminishing current action, finer silt and mud material are deposited at the head. High silt content and the current flow in the salt marsh can cause a smothering effect on organisms. The substrate (or the bottom) of the salt-marsh exerts considerable influence on organisms. The muds of the bottom of marshes have a tendency to trap more saline waters as the tide ebbs; thus high-salinity tolerant organisms grow in great numbers, often colonizing this part of the marshland. This in turn helps in trapping and binding more sediments. Consequently these plants modify the salt marsh in time. Mangroves in tropical zones bind sediment efficiently, thus accelerating the modification of marshlands by filling them. Incidently, this sediment-binding

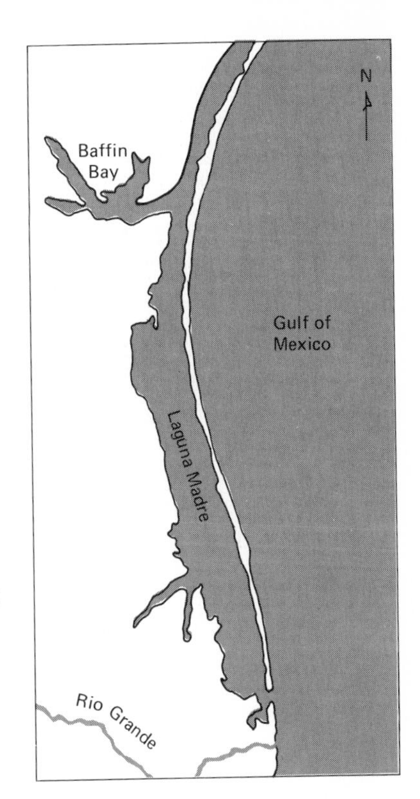

Figure 2-16. Laguna Madre, Texas.

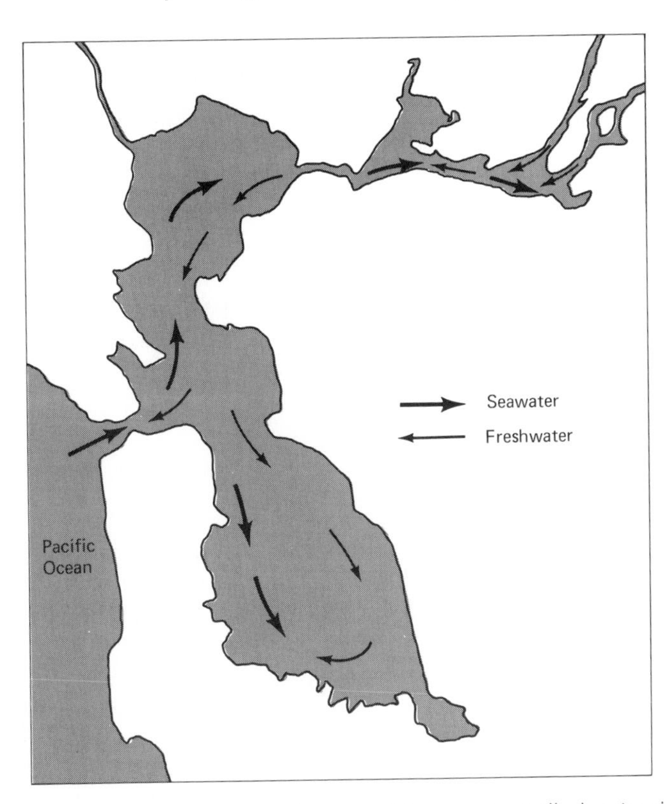

Figure 2-17. San Francisco Bay is one of the most studied estuaries in the world.

ability of plants is one of the most effective natural methods of stabilizing and reclaiming tidal flats in estuaries in many parts of the world, for example, in Great Britain, the Netherlands, Denmark, and New Zealand.

Coastal Lagoons

Lagoon *Body of shallow water characterized by a restricted connection with the sea.*

Coastal lagoons are formed when inlets or embayments are enclosed by depositional barriers of sand (Figure 2-18). Lagoons are generally divided into three zones: (1) the fresh water zone near the mouths of rivers; (2) a salt-water tidal zone at the entrance; and (3) the intermediate transitional brackish (or moderately saline) and tideless water. The general extent of these zones depends greatly upon the supply of either fresh or salt water flowing into the lagoon. At high flood the fresh water zone would proportionately expand, and at high tides (especially coupled with storms) the salt water zone would predominate. The salt content of a lagoon gradually decreases from the entrance to the mouth of a river. When a lagoon is completely cut off from the sea, it may change into a fresh water lake. The protective waters of a lagoon, particularly near the mouth of a river, sometimes receive a greater supply of sediment than that being removed by wave and current action. As a result, a delta is formed.

Figure 2-18. Barrier beach and lagoon. (Photo courtesy of E. Stebinger, U.S. Geological Survey.)

Deltas

Herodotus in the fifth century used the name *delta* to describe the triangular-shaped river deposit at the mouth of the Nile. Besides triangular in shape, deltas may commonly be arcuate or fan-shaped (such as the Rhine delta), digital (such as the Mississippi delta), or estuarine (such as the Susquehanna delta).

Delta *A triangular-shaped deposit at the mouth of a river.*

Deltas, regardless of their shape, form when the supply of land sediments, usually from rivers, is significantly greater than the erosive forces of sea waves and currents. If the sediments so deposited are removed quickly by waves, no delta will form. The Mississippi River deposits over two million tons of sediment daily onto its extensive delta.

SUMMARY

1. The three components of the earth include the hydrosphere (oceans), the lithosphere, and the atmosphere.
2. The average depth of oceans is 4 kilometers; the average height of land is approximately 0.75 kilometer.
3. The highest point on earth is Mt. Everest (8,900 meters above sea level); the deepest point in the ocean is at the Marianas Trench (over 11,022 meters below sea level).
4. The five major oceans are the Pacific, the Atlantic, the Indian, the Antarctic, and the Artic.
5. The Pacific Ocean is the largest, deepest, and coldest ocean.
6. The Atlantic Ocean is roughly half the size of the Pacific and is bordered by many seas.
7. The Indian Ocean is about the size of the Atlantic. Monsoon currents predominate. Three large rivers, the Indus, the Brahmaputra, and the Ganges, discharge significant amounts of sediment into the Indian Ocean.
8. The Antarctic Ocean dominates the Southern Hemisphere.
9. The Arctic Ocean is the smallest ocean. It is covered by the polar ice pack most of the time.
10. Among major seas of the world, the Mediterranean, the North, the Caribbean, and the Red are have been well studied and scientifically and economically exploited.
11. Principal bodies of the coastal waters include estuaries, lagoons, and deltas.

12. An estuary, an embayment of the ocean, is a semi-enclosed coastal body of water linked to the ocean at one end and to a river on the opposite. Mixing of salt- and freshwater is a characteristic feature of an estuary.
13. Estuaries fall into four categories: (1) drowned valleys, (2) fjords, (3) bar-builts estuaries, and (4) tectonic estuaries.
14. Salt marshes are low lying extensions of estuaries. They are characterized by abundant growth of marsh grass. The binding ability of marsh plants facilitates modifications of salt marshes.
15. Lagoons are partially or wholly enclosed by depositional barriers occurring along the coasts.
16. Deltas are land-derived deposits built by rivers near the mouth of a sea. They may be triangular, arcuate, digital, or estuarine.

Suggestions for Further Reading

Duxbury, A. C. 1971. *The Earth and Its Oceans.* Reading, Massachusetts: Addison-Wesley.

Schubel, J. R., and D. W. Pritchard. 1972. "The Estuarine Environment." *Journal of Geologic Education, 20.* 60–68, 179–188.

Shepard, F. P. 1974. *Submarine Geology,* third edition. New York: Harper and Row.

Siever, Raymond. 1975. "The Earth." *Scientific American, 233,* 83–90.

Weyl, P. R. 1975. "The Earth's Mantle." *Scientific American, 232,* 50–63.

3

The Origins of the Earth, Seawater, and Life

3-1 INTRODUCTION

BEFORE WE EXPLORE THE ORIGINS OF LIFE, WE will review fundamentals of geologic time. Evidence from modern radioactive dating methods suggests that the earth is 4.5 billion years old. This time is systematically organized into major and minor time units: eons, eras, periods, and epochs (Table 3-1). The two major *eons* are Cryptozoic (Greek: *cryptos*, invisible; *zoan*, life), and Phanerozoic (Greek-*phaneros*, visible; *zoan*, life). The Cryptozoic eon is also known as the Precambrian Era, which occupies 88 percent (3.9 billion years) of geologic time. The Phanerozoic eon contains three eras: Paleozoic (ancient life), Mesozoic (middle life), and Cenozoic (recent life).

During most of the Cryptozoic eon, life was rare and primitive; in the Phanerozoic eon it became abundant and more advanced. The oldest evidence of life comes from scattered fossils found in Africa, which are about 3.4 billion years old.

3-2 THE ORIGIN OF THE EARTH AND THE SOLAR SYSTEM

At present, the *dust-cloud hypothesis* is widely accepted as an explanation for the origin of the solar system. According to the dust-cloud view, the planets and the sun formed simultaneously from an original vast cloud of gas and dust, which occupied a previously empty region in the Milky Way galaxy. This interstellar mass continued to shrink in response to the pressure of starlight; in shrinking, it formed a cloud,

TABLE 3-1. Geologic Time Scale

Time Units (dates in millions of years before present)						Biological Developments	Physical Developments	Timetable for Continental Drift
Eon	*Era*	*Period*		*Epoch*				
Phanerozoic	Cenozoic			Recent				
		Quaternary		Pleistocene	1	Man	Ice Age	Present day configuration of continents
				Pliocene	13			
				Miocene	26		Himalayan Mountains	
		Tertiary		Oligocene	40	Birds and Mammals		Second fragmentation of Gondwana and Laurasia separated by Tethys Sea
				Eocene				
			70	Paleocene	70			
	Mesozoic	Cretaceous	135					
		Jurassic	180				Rocky Mountains	
		Triassic	220			Reptiles		
	Paleozoic	Permian	270				Ice Age	
		Pennsylvanian (Upper Carboniferous	310			Amphibians	Forest (coal formed)	First fragmentation of Pangaea into Gondwana and Laurasia
		Mississippian (Lower Carboniferous	350			Land plants	Appalachian Mountains	
		Devonian	405			Fishes		
		Silurian	430					
		Ordovician	490					
		Cambrian	570			Marine invertebrates		
Cryptozoic		Precambrian	3500 (?)			Primitive life	Ice Age	
			4500			Origin of life	Primordial air and water Origin of earth	

which subsequently began to collapse under the influence of its own gravity. The continuing increase in pressure caused the temperature of the contracting cloud to rise, which eventually led to the formation of a central radiant mass—the Sun. The planets and satellites formed from dust that was scattered on the periphery of the original cosmic cloud. The exact manner in which all planets were formed is not yet fully known. However, scientists contend that the gas and the dust that was swirling along the rim of the sun broke into turbulent eddies, forming first into protoplanets and eventually into planets (Figure 3-1).

One of the most difficult aspects of any attempt to explain the origin of the solar system is the problem of *angular momentum*. This is the force (or spin) with which a body moves, defined mathematically as the product of its mass and velocity. Although the sun accounts for 99.8 percent of the total mass of the solar system, it accounts for only about 2 percent of the total angular momentum of the solar system because of its slow rotation. Jupiter, on the other hand, although possessing 0.1 percent of the system's total mass, has the largest fraction (about 59 percent) of the total angular momentum. This discrepancy in the relation between angular momentum and mass was not explained by earlier hypotheses concerning the origin of the solar system.

The dust-cloud hypothesis, by contrast, has attempted to provide an explanation for the discrepancy in the angular momentum of the solar system. Three explanations are considered.

First, the discrepancy in the angular momentum of the sun and the planets can be explained in terms of the existence of a magnetic field produced by electrical current. The magnetic field would give rise to ionized gas. Ionized gas behaves differently than ordinary gas, because it remains unchanged by gravity, rotation, and pressure. The magnetic interaction among the sun, the ionized gas, and the electrically charged

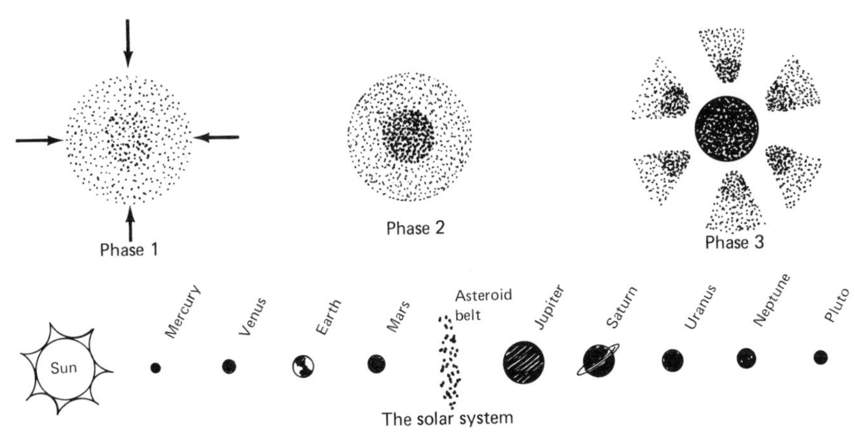

Figure 3-1. Major formative stages of the solar system, as explained by the dust-cloud view.

particles surrounding it would, over a period of time, produce a necessary *breaking effect,* causing the sun to lose most of its angular momentum.

Second, according to some scientists, the sun was originally 100 times brighter than it now is. Consequently it managed to expel gas in far greater amounts than its present *solar winds.* The greater outrush of gas possibly diminished the sun's spin or angular momentum.

Third, the sun at one point in its formative stages contracted to the present orbit of Mercury. At that stage it was in a gaseous state. Subsequently, boulder-like chemical aggregates formed within this gas, expanded, and stretched in all directions from the fast spinning sun. This material (gas filled with chemical aggregates), under the influence of the sun's magnetism, lost much of its spin.

The dust-cloud hypothesis is supported by three salient astronomic observations: (1) Extensive fields of gas and dust, as postulated by the hypothesis to have existed "originally," are still observable in the spiral arms of our galaxy and elsewhere; (2) during a recent seven-year period, astronomers observed the stars being formed in the heart of the Sagittarius nebula in the manner described by the dust-cloud hypothesis

Figure 3-2. Modern astronomic observations have shown that stars are apparently forming within the nebula M-8, in the constellation Sagittarius. Dark specks are considered to be dust clouds condensing into stars, and perhaps planets. (Photo courtesy of Lick Observatory.)

(Figure 3-2); and (3) the dust-cloud hypothesis is supported by firsthand astronomic observations of the youthful star T-Tauri, which is driving excessive gas and dust outward in the fashion postulated to explain the evolution of the solar system as a whole.

The question of the origin of the solar system is still one of the unresolved enigmas of modern science. Advancements in space sciences and technology, including the landing of remote-controlled spacecraft (such as the Viking mission to Mars) on planets will provide additional evidence.

3-3 THE ORIGIN OF SEAWATER

Although the origin of the solar system is still a mystery, the origin of the oceans is now fairly well understood in terms of modern theories of seafloor spreading, global plate tectonics, and continental drift (see Chapter 5).

Three hypotheses attempt to explain the origin of seawater: (1) condensation, (2) weathering of volcanic rocks, and (3) degassing of the mantle.

The *condensation hypothesis* was acceptable in the past when the idea of a molten earth was popular. It was then thought that a primitive atmosphere laden with water in the form of dense clouds of steam and vapor initially existed and that on condensation (with cooling of the earth), torrential rains fell, forming boiling pools, which, as cooling continued, ultimately became oceans and seas (Figure 3-3).

Opponents of this hypothesis maintain that if this postulation is correct, the original constituents, including high quantities of rare gases such as neon, xenon, argon, and krypton, should now be present in the atmosphere but are in fact nearly absent. Neon, for example, has an atomic weight of 20 whereas the atomic weight of water is 18. Why, then, did the atmosphere retain the relatively light compound, water, and not the heavier element, neon? That is, if the atmosphere were unable to retain neon, why could it retain the lighter vapors? At present, the earth holds more than 13,000 cubic kilometers of water in the atmosphere and over 1 billion cubic kilometers in the ocean, but neon is only negligibly present. Proponents of the condensation hypothesis argue that noble gases, such as argon, neon, krypton, and xenon, are chemically inert and are thus quickly lost into space because they cannot form compounds that would bind them. Further, the condensation hypothesis provides neither consideration of the recycling of water nor specification of the rates and volumes of the first torrential rains.

According to the second hypothesis, of the *weathering of volcanic rocks,* as the earth consolidated into volcanic and other types of rocks

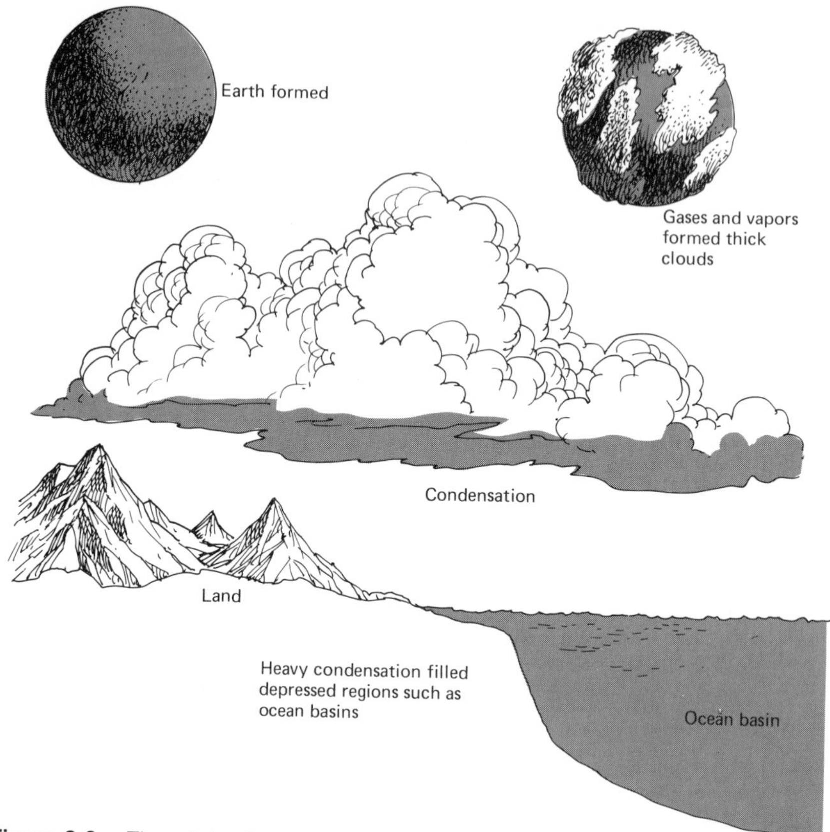

Figure 3-3. The origin of seawater according to the condensation view.

from its previously molten condition, much of the original water was trapped in these rocks. Because of subsequent breakdown of these rocks by wind, water, and related surface processes, the trapped water was released and eventually formed oceans and seas (Figure 3-4).

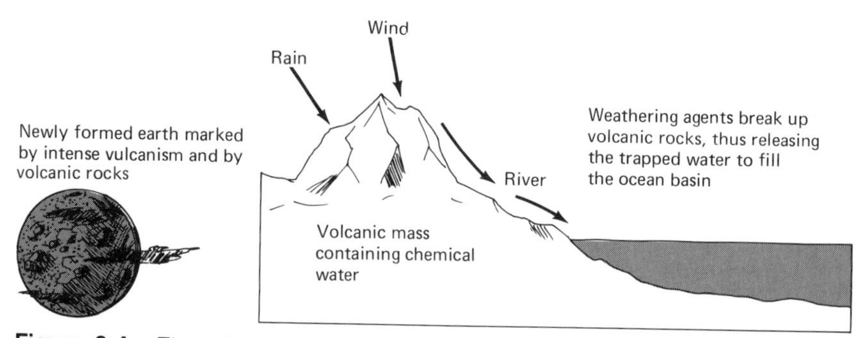

Figure 3-4. The origin of seawater according to the weathering of volcanic rocks.

Experimental and field observations, however, suggest that volcanic rocks contain an inadequate quantity of water to account for the present volume of water in the world's oceans. If all the water so trapped in the volcanic rocks were released, it would provide only less than 50 percent of present ocean water.

Although this hypothesis does not satisfactorily account for the volume of the oceans, it well explains the presence of most cations (ions with positive charge) such as sodium, potassium, magnesium, and calcium. It is now widely accepted that most cations in the oceans came from the decomposition of volcanic rocks.

According to the *degassing* theory, oceans were formed as a direct consequence of the "sweating out" of the water from crustal rocks that were subjected to radioactive heat and vulcanism (Figure 3-5). Soon after its formation, the earth cooled and solidified; gaseous and volatile constituents escaped to the surface in a process known as *degassing*. The release of water into the ocean was gradual. In addition, oceans gradually and continuously received water from volcanoes and geysers. Assuming that excess volatiles are supplied at the rate of less than 1 percent per year, it would explain the present volume of water in the ocean. *Excess volatiles* are elements and compounds that exist in the ecosphere in excess of what is otherwise furnished by the weathering of crustal rocks. Excess volatiles include water, carbon (in carbon dioxide), chlorine, nitrogen, and sulfur (Table 3-2).

Excess volatiles *Elements like water, carbon dioxide, nitrogen, and sulfur that are more abundant in the ocean, atmosphere, and sediments than can be accounted for by breakdown of rocks.*

According to the degassing hypothesis, no more than 6 percent of total excess volatiles could have been present in the early atmosphere or in the ocean. The chief proponent of the view, W. W. Rubey (1955), quantitatively estimated that discharge of hot springs from continents and from the ocean floor is approximately 66×10^{15} grams of water per year. If this annual rate were combined over 4×10^{9} years of the earth's age, the yield would be 2.6×10^{26} grams of water. Although the bulk of this water is recycled and only a fraction of it (0.6 percent) represents newly emergent water, the total present volume of the ocean can be ac-

TABLE 3-2. Excess Volatiles in Atmosphere, Hydrosphere, and Buried Sedimentary Rocks

Substance	*Units of 10^{20} grams*
Water	16,600
Carbon as carbon dioxide	910
Chlorine	300
Nitrogen	42
Sulfur	22

SOURCE: W. W. Rubey, 1955, *Development of the Hydrosphere and Atmosphere*, Geological Society of America, special paper 62.

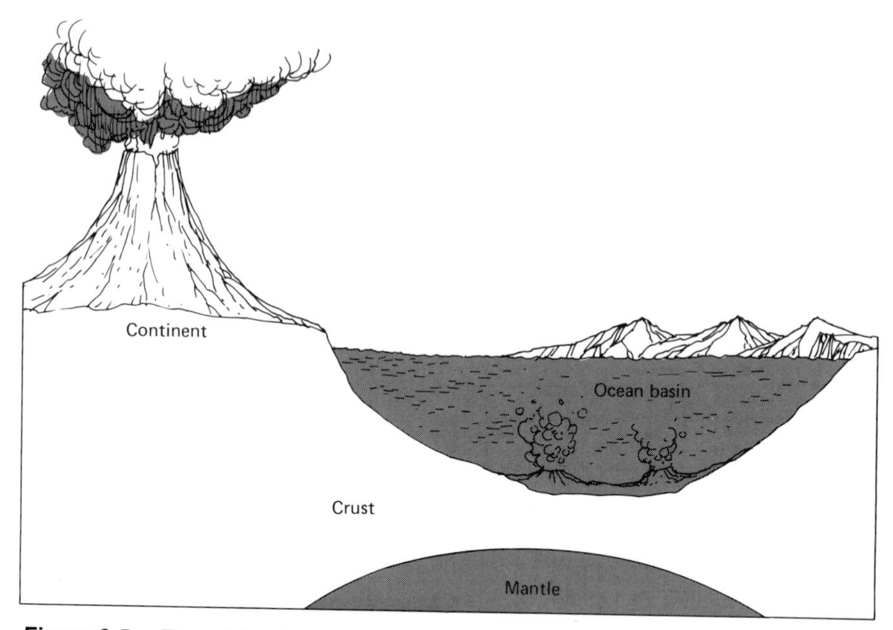

Figure 3-5. The origin of seawater according to the degassing view.

counted for from this source. Another important aspect of the degassing view is the fact that much of free water in the ocean came relatively late in geological time, thereby providing a satisfactory explanation of why oceans are not saltier than they are.

3-4 THE ORIGIN OF LIFE

Life may be defined as any organized aggregate of matter that is able to grow and reproduce. Three views explaining the origin of life on the earth: (1) Life began by a supernatural act of creation; (2) life developed spontaneously from inorganic matter; and (3) life was transported from outer space—the Panspermia view.

That life resulted from a supernatural act of creation by a divine force was a long-standing and consciously held maxim. According to this view, life began abruptly by an act of a divine creator and with it all animals and plants appeared in their respective existing forms. This view lacks empirical proof and is not accepted by the scientific community.

That life originated spontaneously was a widely accepted explanation during the Middle Ages. It was then thought that worms originated from mud, maggots from decaying flesh, and mice from rags. In 1860,

Louis Pasteur proved that maggots developed from eggs laid by flies and not from decaying flesh; but despite evidence of his later experiments and especially in the light of later evidence, Pasteur did not believe in the spontaneous creation of life from inorganic matter.

This later evidence comes in the present century: J. B. S. Haldane and A. I. Oparin independently proposed that life generated spontaneously from inorganic matter. According to these scientists, the primeval atmosphere possessed abundant ammonia, methane, and water vapor and had no free oxygen. When these gases reacted chemically with sunlight, organic molecules, particularly proteins essential for life, formed and were later transported to planetary seas and lakes until some of the shallow waters turned into soups of organic compounds. In this type of environment life supposedly began.

In 1952, Stanley Miller experimentally tested Haldane's and Oparin's hypotheses by spraying a mixture of methane, ammonia, water vapor, and hydrogen over an electric discharge (to simulate solar energy) (Figure 3-6). After a time, Miller found that water in the bottom of the apparatus representing the primeval sea contained certain amino acids that are bases for life. Soon Miller recognized a primordial organic

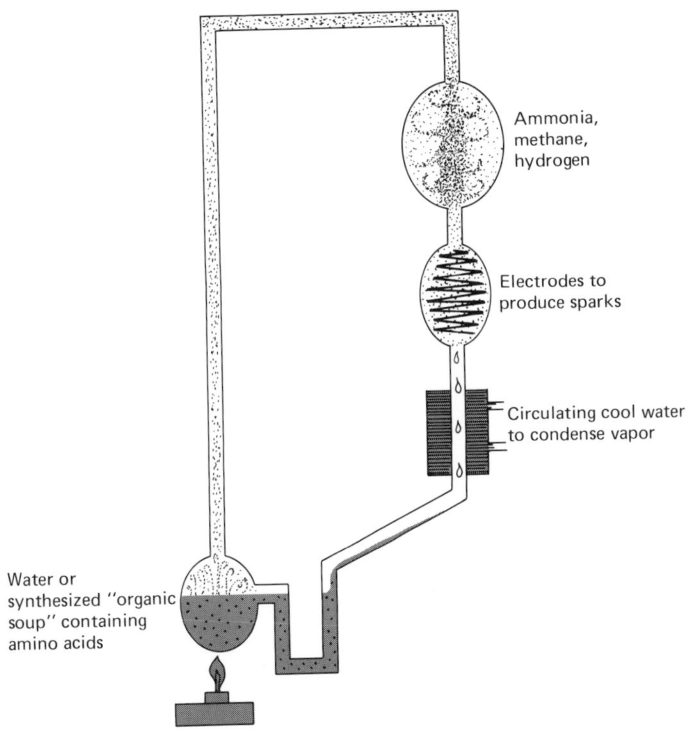

Figure 3-6. Miller's experiment proved that life can be synthesized spontaneously from inorganic matter.

soup–*weltschlamm*–from which planetary life began. At present, the spontaneous creation of life from inorganic matter is a widely accepted hypothesis.

According to the Panspermia view, life is widespread, perhaps universal. Life therefore must have spread from one or a few places of origin throughout the universe. What is implied, of course, is that life on earth was transported from outer space.

Astronomical observations have shown a common and relatively abundant occurrence of chemical compounds such as hydroxyl radicals, ammonia vapor, water vapor, carbon monoxide, hydrogen cyanide, and formaldehyde in the universe. These compounds form dense cloudy concentrations of huge mass, often attaining the size of a star. These abundantly distributed compounds are significant in their close associations with the basic components of life.

Meteoritic analyses, particularly of carbonaceous chondrites, show the presence of a variety of carbon-rich compounds. Murchison meteorite, a chondritic meteorite that fell in Australia, showed the presence of amino acids, thus strongly suggesting the existence of life elsewhere in the universe, a subject that is still controversial.

The Viking mission to Mars has sent back several photographs of the Martian terrain showing the presence of drainage patterns similar to those on earth. Water exists on Mars in the form of ice. However, no conclusive evidence of life was found by the Viking mission.

SUMMARY

1. The earth is 4.5 billion years old. Geologic time is divided into the Cryptozoic and the Phanerozoic eons. The Cryptozoic eon, often referred to as Precambrian, occupies 88 percent of the total geologic column (approximately 3.9 billion years). Life was uncommon during the Precambrian. In the Phanerozoic eon life was very abundant in the shallow regions of the ocean. The Phanerozoic eon is subdivided into eras, periods, and epochs.
2. According to the dust cloud hypothesis, the solar system formed from a vast cloud of gas and dust. Gravitational forces aided in condensing huge quantities of gaseous and dust material. The central region of the cloud became the sun. The gas and dust material around the sun broke into turbulent eddies and formed protoplanets, which eventually formed planets. One of the most difficult features of the solar system to explain is the discrepancy of angular momentum between the sun and the planets.

3. According to the degassing hypothesis, air and water formed because of the escape of water vapor, nitrogen, methane, and other gases from the earth's interior.
4. According to the hypothesis of spontaneous creation, life formed from inorganic matter. Primodial atmospheric gases such as ammonia, methane, and water vapor interacted in the presence of sunlight to form amino acids. The hypothesis of spontaneous creation has been supported by experimental work.

Suggestions for Further Reading

Calvin, M. 1975. "Chemical Evolution." *American Scientist, 63,* 169–177.
Cameron, A. G. W. 1975. "The Origin and Evolution of the Solar System." *Scientific American, 233,* 33–41.
Eicher, D. L. 1976. *Geologic Time,* second edition. Englewood Cliffs, New Jersey: Prentice-Hall.
McAlester, A. L. 1977. *The History of Life,* second edition. Englewood Cliffs,New Jersey: Prentice-Hall.
Miller, S. L., and L. E. Orgel. 1974. *The Origins of Life.* Englewood Cliffs, New Jersey: Prentice-Hall.
Press, Frank, and R. Siever. 1974. *Earth.* San Francisco: W. H. Freeman.
Rubey, W. W. 1955. "Development of the Hydrosphere and Atmosphere, with Special Reference to Probable Composition of the Early Atmosphere." *Geological Society of America,* Special Paper 62.
Stokes, William L. 1974. *Essentials of Earth History,* second edition. Englewood Cliffs, New Jersey: Prentice-Hall.

4

The Profile of the Ocean Floor

4-1 INTRODUCTION

THE FLOORS OF THE OCEANS ARE RUGGED AND complex and include the world's longest mountain ranges, deepest trenches, and largest plains (Figure 4-1). In general, the ocean floor has four divisions: (1) the continental shelf, (2) the continental slope, (3) the continental rise, and (4) abyssal plains and other associated features including oceanic rises, ridges, trenches, and abyssal hills (Figures 4-2, 4-3). Basic information concerning the extent of marine physiographic provinces is summarized in Table 4-1. Major submarine features off the east coast of the United States are shown in Figure 4-4.

4-2 THE CONTINENTAL SHELF, SLOPE, AND RISE

The *continental shelf* borders continents and is characterized by a gently sloping area running from the depth of the low spring tide to a

TABLE 4-1 Percent of Physiographic Provinces in Oceans and Adjacent Seas

Oceans and Adjacent Seas	*Continental Shelf and Slope*	*Continental Rise*	*Ocean Basin*	*Ocean Ridge and Rise*	*Trenches*	*Volcanic Ridges Cones etc.*	*Percent of World Ocean in Each* **Ocean Group**
Pacific	13.1	2.7	43.0	35.9	2.9	2.5	50.1
Atlantic	17.1	8.0	39.3	32.3	0.7	2.0	26.0
Indian	9.1	5.7	49.2	30.2	0.3	5.4	20.5
Arctic	68.2	20.8	0	4.2	0	6.8	3.4
Percent of World Ocean in Each Province	3.1	5.3	41.8	32.7	1.7	3.1	—

SOURCE: H. W. Menard, and S. M. Smith, 1966, "Hypsometry of Ocean Basin Provinces," *Journal of Geophysical Research,* Volume 71.

Figure 4-1. (*At right*) Relief map of the Atlantic Ocean. (Courtesy of Alcoa.)

Figure 4-2. (*Below*) Principal submarine topographic features.

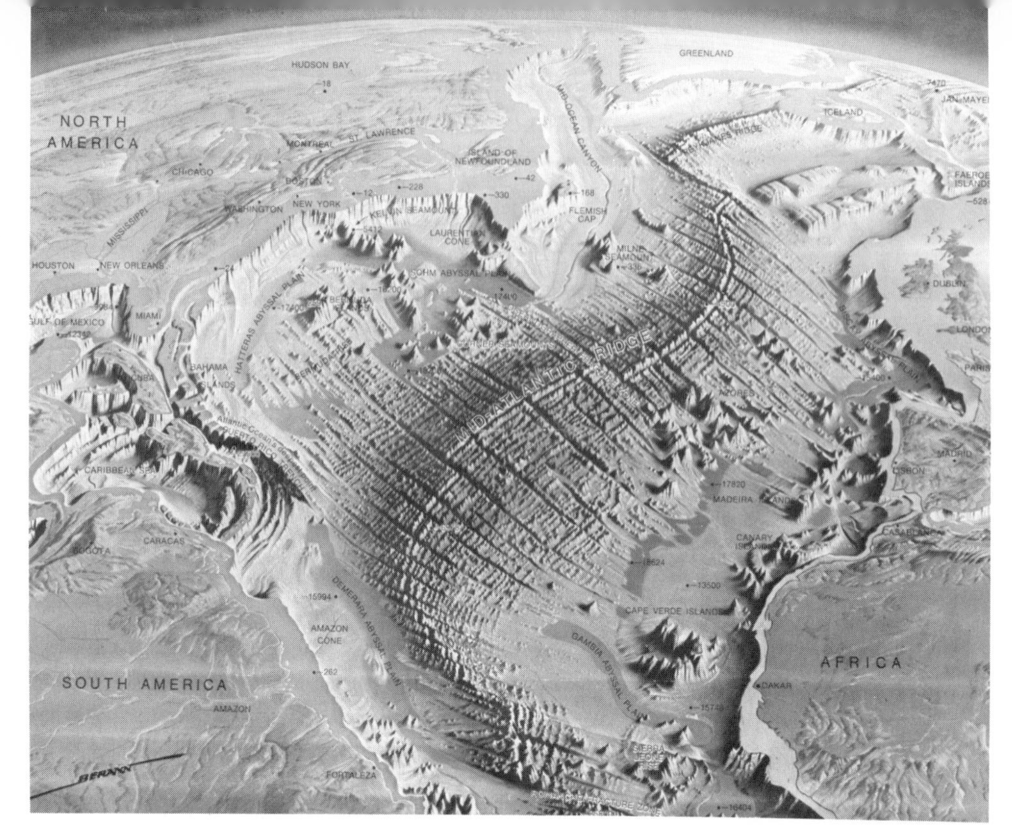

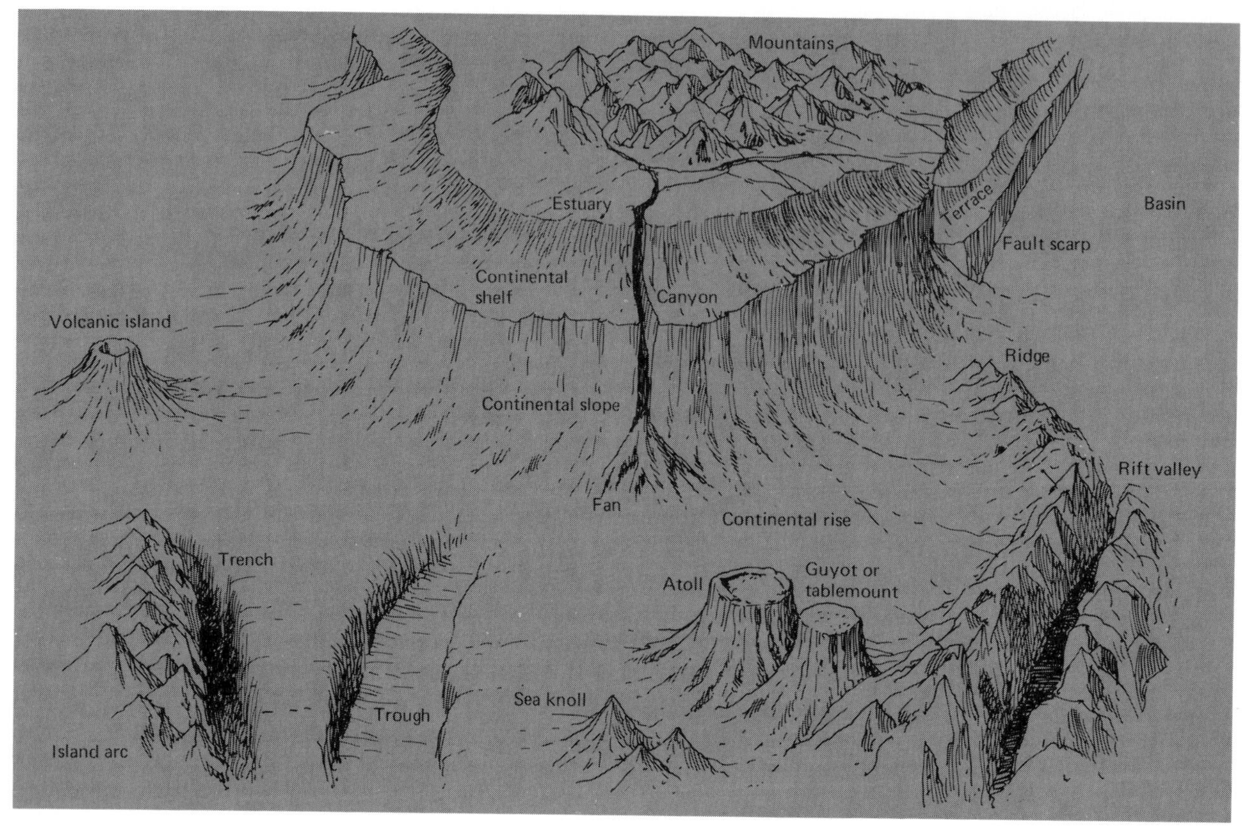

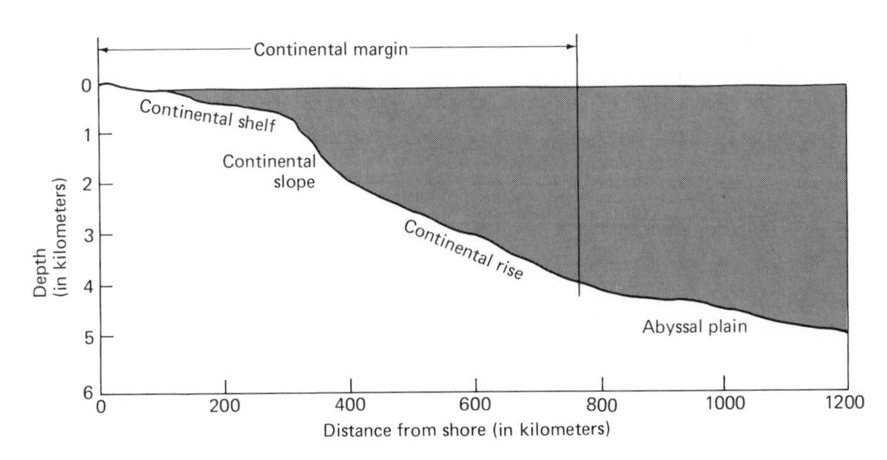

Figure 4-3. Major submarine topographic elements.

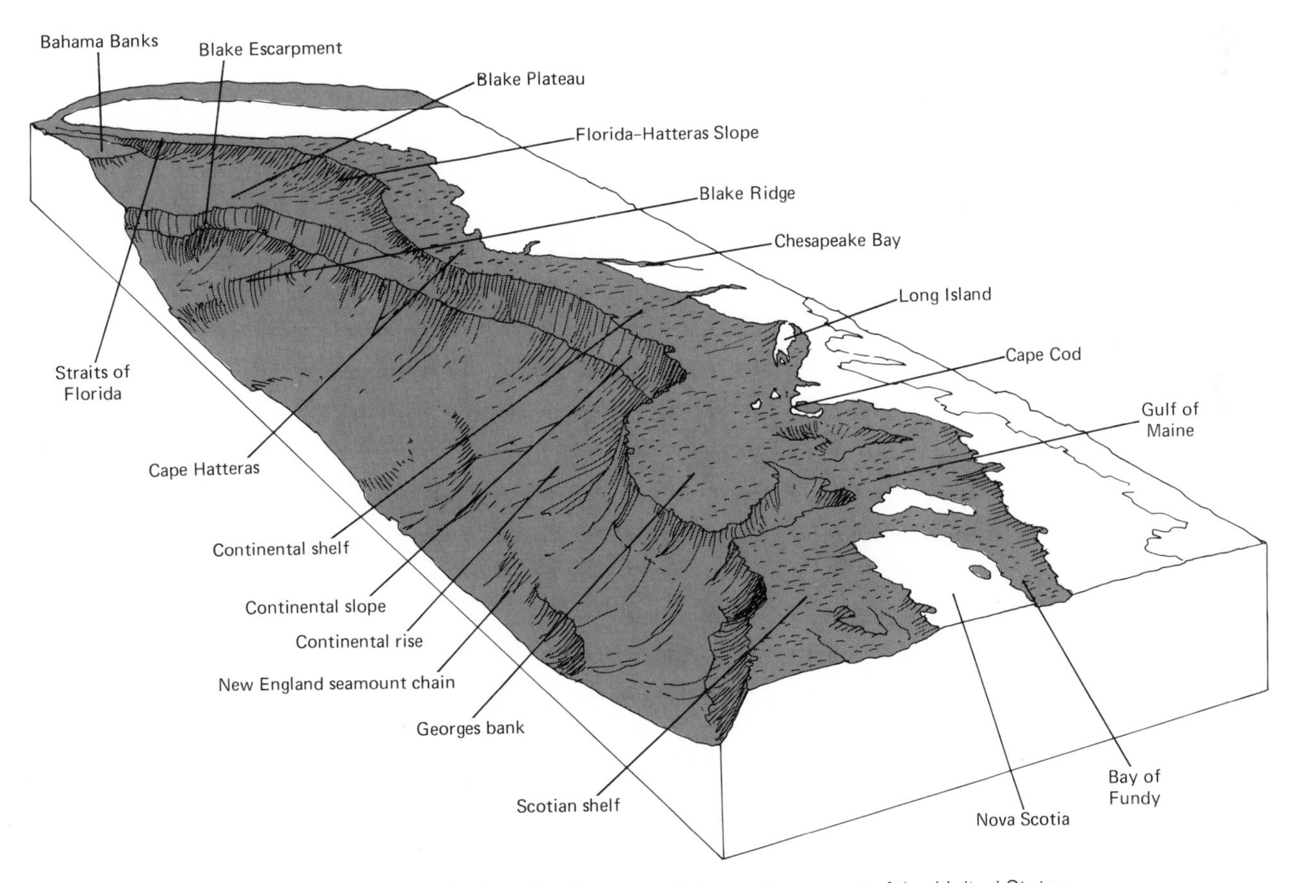

Figure 4-4. Principal submarine features off the eastern coast of the United States.

Continental shelf *A gently seaward sloping surface extending between the shoreline and the continental slope.*

depth of approximately 150 to 200 meters. The shelf extends seaward with an average gradient of 1 in 500 and an average width of 65 kilometers. The continental shelf slopes toward the ocean depths, sometimes gradually, sometimes steeply. As a result, the edge of the shelf varies in width and depth from 20 to 530 meters. Along some landmasses, the shelf is very narrow, but elsewhere it is nearly absent, as in the eastern Pacific. And the shelf may be hundreds of kilometers wide, as it is off the coast of Arctic Siberia.

There are two types of shelf: (1) glaciated shelves, and (2) nonglaciated shelves. Glaciated shelves are characterized by the presence of glacial deposits of gravels and pebbles, deep glacial valleys and fjords, and rising banks along the outer portions. The continental shelf off New England is a notáble example of a glaciated shelf. Glaciated shelves occupy significant portions of offshore areas such as the Georges Bank off New England. Nonglaciated shelves occur off large river flows, and they tend to be considerably wide and shallow (about 100 meters deep). The continental shelf off the Yellow Sea and off the Alaskan part of the Bering Sea are well-known examples. The continental margins that include shelves, slopes, and rises are characterized by the presence of submarine canyons (see Section 4-7). In general the shelf is a broad region covered by land-derived sediments. Most of the sediments in the shelf are derived from the breakdown of the land mass by rivers, wind, or glaciers and the subsequent spreading of sediment by waves and currents in the sea. The shelves of the world are of great importance because they provide the richest fishing grounds to the fisherman and potential mining sites for economic minerals. Much of offshore petroleum is confined in shelves (see Chapters 13 and 14). Many nations are presently concerned about territorial claims to their portion of the continental shelf (see Chapter 16).

Continental slope *A relatively steeply sloping surface running seaward of the continental shelf.*

The region that marks the break in slope from the outer edge of the continental shelf into deeper water is the *continental slope.* The average depth of this marked change in inclination is about 150 meters. The change in the gradient may be essentially gentle from 2 degrees to as high as 5 degrees; in exceptional places, it may be 15 degrees. The continental slope along many coasts of the world is furrowed by deep, canyonlike trenches terminating as fan-shaped deposits at the base.

Continental rise *A gently sloping surface at the base of the continental slope.*

Where the continental slope ends, the gently sloping continental *rise* begins, as shown by the much gentler slope of the latter. With increasing depth the continental rise becomes virtually flat and is then known as the abyssal plain. The continental rise has an average slope of between 0.5 to 1°, and in most cases even less. The general relief of the continental rise is low.

4-3 ABYSSAL PLAINS AND ABYSSAL HILLS

Abyssal plains *Flat, nearly level areas occupying the deepest portions of many ocean basins.*

Abyssal plains are areas of deep-ocean floor found at depths of 3000 to 6000 meters. They occupy about 40 percent of the ocean floor and are present in all major oceans and several seas of the world. They have been found in the Gulf of Mexico, the Bay of Bengal, and the Wedell Sea, and more commonly in the Atlantic and the Indian Oceans. In the Atlantic Ocean the Sohm abyssal plain is well known. It occurs in the northern Atlantic (south of Newfoundland). In general, abyssal plains are more common where land-derived sediments are in great supply. Owing to the large supply of sediment the irregular topography is buried and formed into relatively flat areas, known as abyssal plains. Since more rivers flow into the Atlantic and the Indian oceans than into the Pacific, the supply of sediments in these oceans is much higher, favoring the formation of abyssal plains. Often the ocean floor is characterized by the presence of low relief (300 to 600 meters) areas with diameters of 7 to 10 kilometers. These hills are known as *abyssal hills.* Abyssal hills are formed by materials from the crust beneath the ocean. They are more common in the Pacific Ocean because of the relatively poor supply of sediments.

Abyssal hills *Relatively small topographic features of the deep ocean floor ranging from 600 to 1000 meters high and a few kilometers wide.*

4-4 MIDOCEANIC RIDGES

Midoceanic ridges are submarine mountain ranges a few hundred kilometers wide and hundreds (often thousands) of kilometers in length. They form the longest mountain system on earth, with a global network of narrow mountain ranges of high relief in three major oceans. The total length of the midoceanic ridge system is about 75,000 kilometers. The famous Mid-Atlantic Ridge runs midway from Iceland through the entire Atlantic Ocean and occupies slightly less than one third of the ocean's total area (Figure 4-5). Recent studies have shown the presence of a rift valley extending along the crest of the Mid-Atlantic Ridge. The floor of the rift is between 1300 to 3400 meters below sea level. There are mountains on either side of the rift that have an elevation of about 2000 meters above its floor. These are known as *rift mountains.* The rift valley is a site of earthquake and volcanic activities. Incidently, Iceland is the northernmost area of the Mid-Atlantic Ridge that is now exposed. It was formed by volcanic lava flows similar to those active in the midoceanic ridges.

The *Mid-Indian Ocean Ridge,* in the Indian Ocean basin, forms an inverted Y-shaped mountain range, which runs from the Red Sea and the

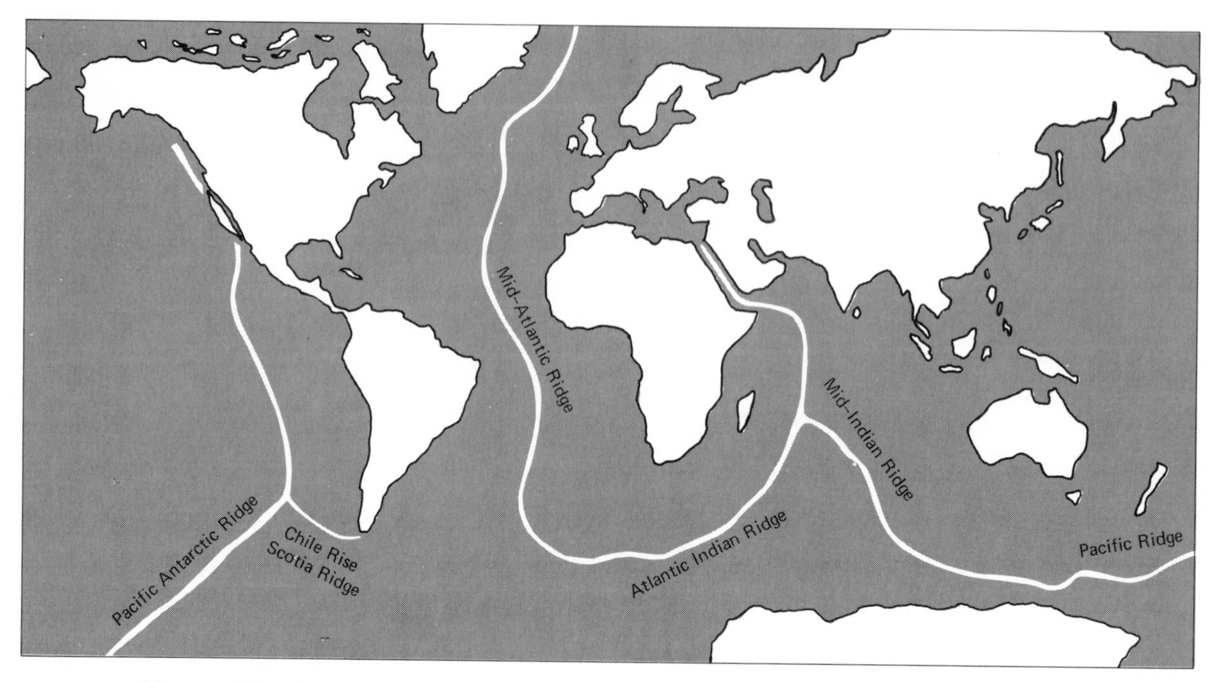

Figure 4-5. Oceanic ridge system forms the world's longest mountain chain (approximately 75,000 kilometers long).

Gulf of Aden to the middle of the Arabian Sea, dividing into two southerly arms. One of the arms extends southwest and links with the Mid-Atlantic Ridge; the other extends southeast between Australia and Antarctica and joins with the Pacific Ridge.

The *Lomonosov Ridge* in the Arctic Ocean runs from the continental shelf of Asia to that of North America. It governs the circulation and exchange of water in the Arctic Ocean.

4-5 SEAMOUNTS AND GUYOTS

Seamount *A submarine mountain rising more than 1,000 meters above the ocean floor.*

Guyot *A flat-topped sea mountain.*

Seamounts are submarine mountains or peaks of volcanic orgin that may reach almost to the ocean surface (Figure 4-6). In general, seamounts have more than 1000 meters local relief. Flat-topped seamounts are known as *guyots* (gē-yo). Oceanographers believe that there are 10,000 seamounts and guyots in the Pacific Ocean alone, mostly concentrated in the North and Central Pacific and the Marianas basin. An explanation of the difference between the sharp crested seamounts and the flat-topped guyots has been postulated. The guyot must have been above sea level at some point in its formative stage. There, it experienced the erosional action of surface waves and currents blowing off its top (or peak). Subsequently, the guyot sank below sea level. The tops of many guyots are usually between 100 to 500 meters below sea level. The depth of seawater above the guyots is attributed to the post-glacial rise in sea level. Another explanation might be that the guyots sink as a consequence of the resettling of volcanic material at their base.

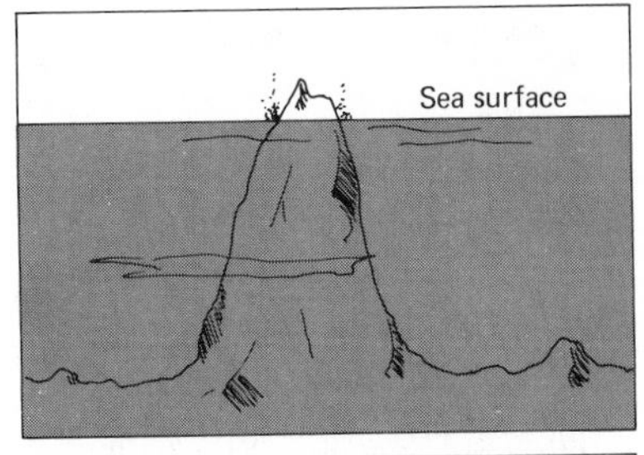

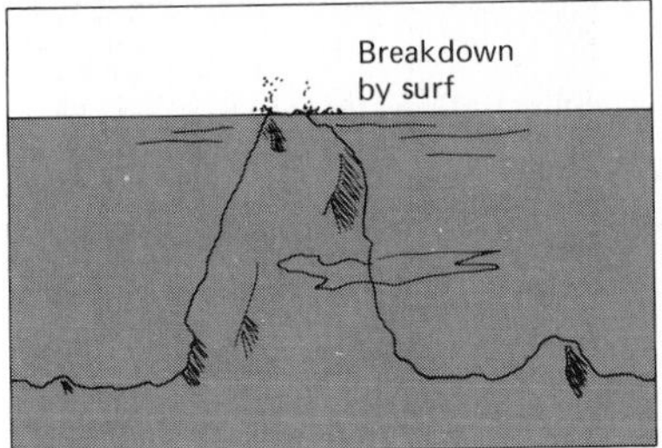

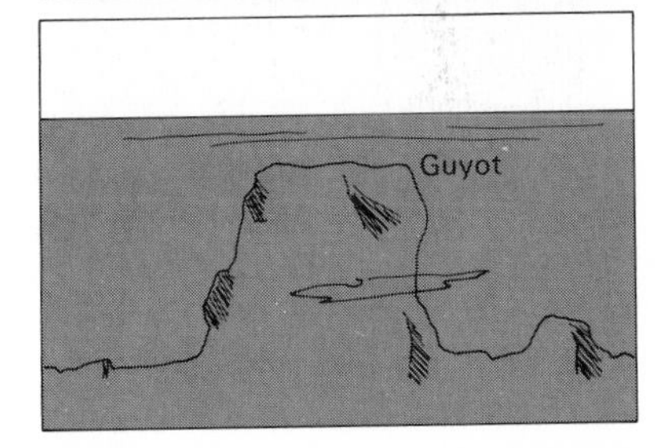

Figure 4-6. Continual attack by surf causes erosion of the top of the seamount. As a result, the seamount becomes a flat-topped guyot.

4-6 OTHER FEATURES

Oceanic Rises

Oceanic rises are large-scale, isolated, nonseismic areas that stand out at great heights above the ocean floor. The Bermuda Rise in the North Atlantic, about 500 kilometers wide and 1000 kilometers long, is a well-known example. One of the noticeable features of the Bermuda Rise is the relatively narrow pedestal (80 kilometers wide and about 125 kilometers long at the base) emerging upward from a depth of 4.5 kilometers at the peak of the rise. The crest of the pedestal is crowned by the luxuriant growth of coral reefs that has formed the island of Bermuda.

Trenches

Trench *A long, narrow and steep-sided depression on the ocean bottom.*

Ocean trenches can be more terrifying than the steep walls of the Grand Canyon. Trenches are much deeper and steeper than the Grand Canyon. The depth of the Grand Canyon is less than a half a kilometer, while trenches can be deeper than 10 kilometers. Trenches are narrow, linear areas on the ocean floor, as shown by a seismic profile of Kurile Trench off Japan (Figure 4-7). These are the deepest areas on the ocean floor, and they extend thousands of meters. Although trenches are found in all major oceans, they are most common in the Pacific Ocean (Figure 4-8, Table 4-2). Among the trenches of the Pacific Ocean, the Marianas Trench is the deepest (11,022 meters). In 1960, the bathyscape *Trieste* descended at one of the points in the Marianas Trench, a site now known as the *Challenger Deep.* Other trenches of the Pacific Ocean include the Tonga–Kermadec Trench (extending from New Zealand to Samoa), the Philippine Trench (also known as the Mindanao Trench), and the Japan Trench. The Peru-Chile Trench runs parallel to the western coast of South America.

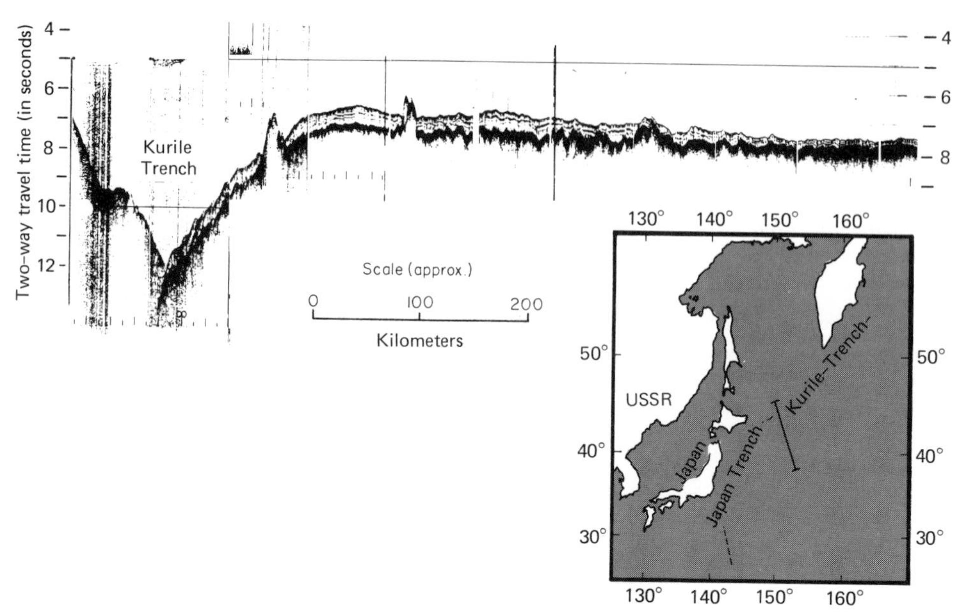

Figure 4-7. Seismic profile of Kurile Trench off Japan. (Photo courtesy of William Ludwig, Lamont-Doherty Geological Observatory of Columbia University.)

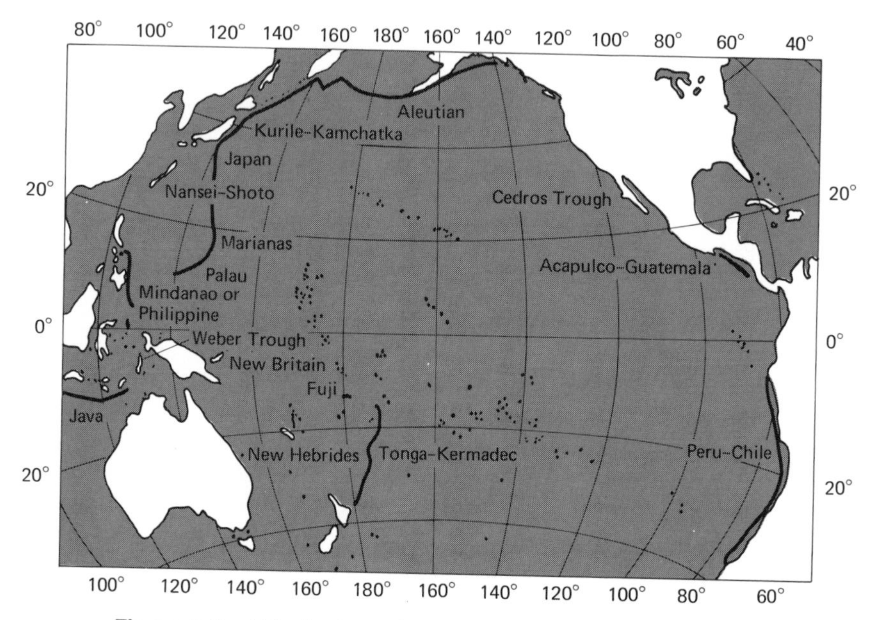

Figure 4-8. Principal trenches of the Pacific Ocean.

TABLE 4-2 Major Oceanic Trenches

Location	*Trench*	*Depth (in meters)*	*Length (in kilometers)*	*Mean Width (in kilometers)*	*Areal Extent (in square kilometers)*	*Volume (in cubic kilometers)*
Pacific Ocean	Kurile–Kamchatka Trench	10,550	2200	120	264,000	1,320,000
	Japan Trench	8,412	800	100	80,000	336,000
	Marianas Trench (including Challenger Deep)	11,022	2550	70	17,850	98,200
	Philippine Trench	10,550	1400	60	84,000	420,000
	Tonga Trench	10,882	1400	55	77,000	415,800
	Kermadec Trench	10,047	1500	60	90,000	450,000
	Aleutian Trench	7,679	3700	50	185,000	673,000
	Middle America Trench (including Acapulco Deep)	6,662	2800	40	96,000	316,800
	Peru–Chile Trench	8,055	5900	100	590,000	2,360,000
Indian Ocean	Java Trench	7,450	4500	80	180,000	666,000
	Vema Trench	6,402	700	25	17,500	56,000
	Mauritius Trench	5,564	1080	30	32,400	95,400
Atlantic Ocean	Puerto Rico Trench	8,385	1550	120	186,000	779,000
	Cayman Trench	7,093	1450	70	101,500	360,000
	Romanche Trench	7,856	300	20	6,000	21,900

Island Arcs

Island arcs are a series of arcuate ridges having a seaward convex-curve-like tendency. They are generally bound by deep-sea trenches. Island arcs stand directly above the ocean basin floor at the base and reach above the ocean surface. They are common along the rims of the Pacific Ocean. Examples of island arcs include the Aleutian islands off Alaska and the Kurile, Japanese, Ryukyu, and Philippine Archipelagoes on the Asian part of the Pacific Ocean. Island arcs often separate relatively shallow seas; for example, the Aleutian island arc separates the shallow Bering Sea from the deeper Pacific basin.

Fracture Zones

The ocean basin floor is characterized by the presence of fracture zones. These zones are formed by faulting, that is, displacement along fractures in the oceanic crust. They are a series of linear volcanic features generally at right angles to the midoceanic ridge system (Figure 4-9). The fracture zones are most prominent in the eastern Pacific basin. The Mendocino and Murray fracture zones are notable examples. The Murray fracture zone is known to have cut into the continent (off Southern California), and the Mendocino fracture zone has displaced the continental shelf off northern California.

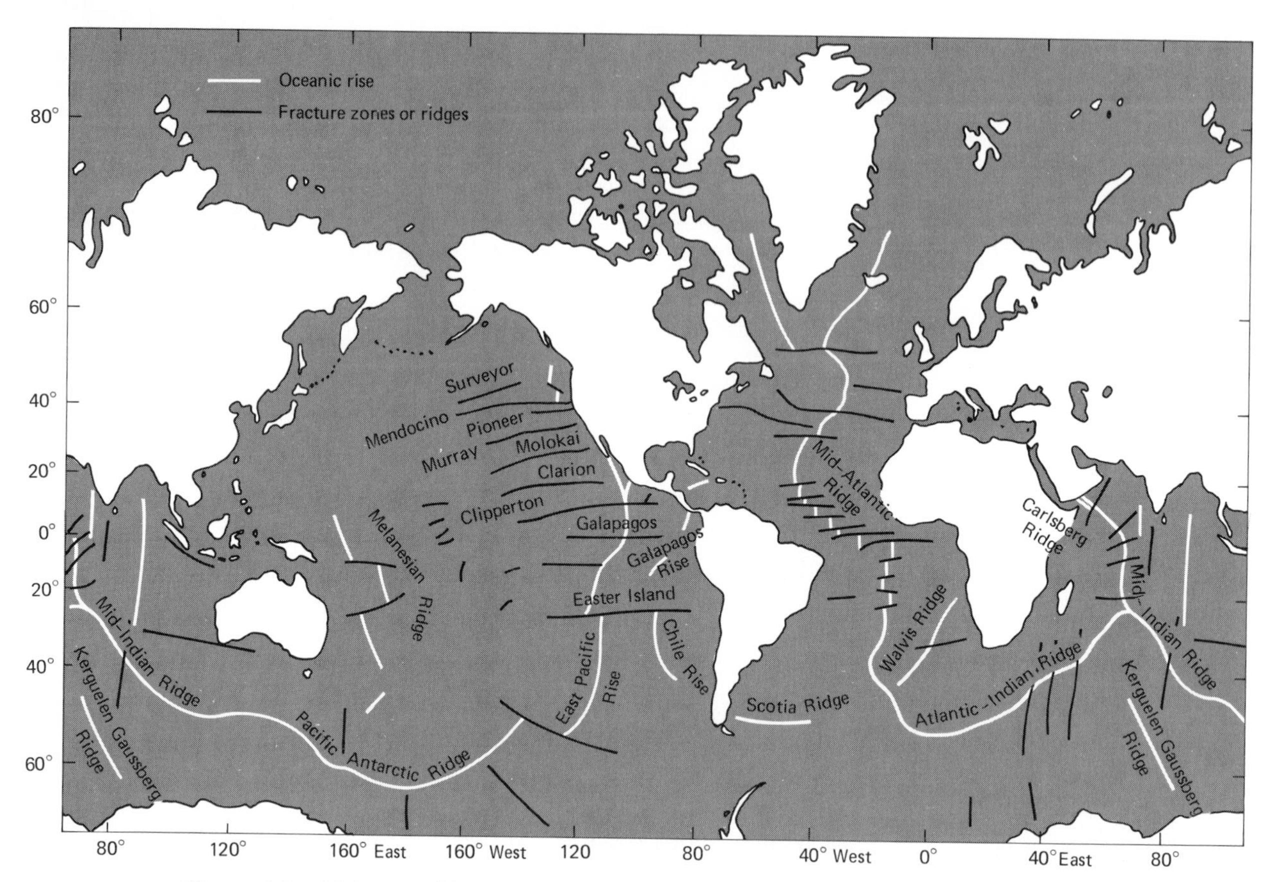

Figure 4-9. Midocean ridge and rise system showing island arcs and fracture zones in the ocean floor. (From H. W. Menard, 1965.)

4-7 SUBMARINE CANYONS AND TURBIDITY CURRENTS

Submarine canyon *A steep, narrow canyon cut into the continental shelf or slope.*

Submarine canyons are deep gorges on the ocean floor. They are characterized by short, steep slopes that form long, concave profiles; by few tributaries; and by heads at depths of a few hundred meters. Some canyons have their heads near the mouths of rivers. Many canyons terminate in fan-shaped cones or in deltas on the seafloor (Figure 4-1). Most canyons are restricted to the continental shelves, slopes, and rises.

Hudson Canyon, one of the best-known canyons in the world, begins near the mouth of the Hudson River and extends into the Atlantic (Figure 4-10). Other canyons are Monterey Canyon, which extends into the Atlantic from the mouth of the Congo River, and La Jolla Canyon of southern California near San Diego, which has been studied by the oceanographers at Scripps Institution of Oceanography (Figure 4-11).

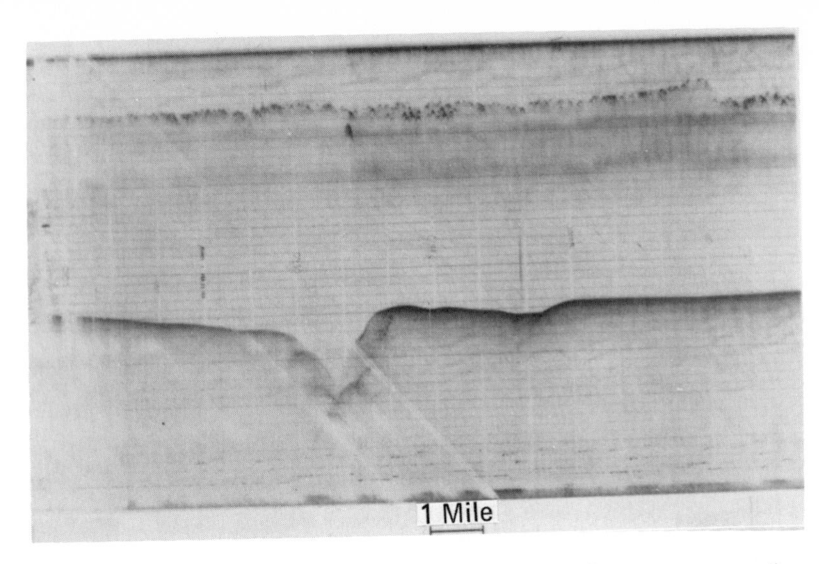

Figure 4-10. Oblique (sounding) profile of Hudson Canyon across the continental rise. The canyon is about 200 fathoms deep, is sharp sided, and is less than 3 kilometers wide. The true bottom is probably 1,820 fathoms, and the real side slopes are masked by side echoes. (Photo courtesy of R. M. Pratt, U.S. Geological Survey.)

Figure 4-11. This 10-meter "sandfall" is in the Cape San Lucas submarine canyon, Baja California. It is fed by currents of sand pouring in from nearby beaches into the canyon. (Photo courtesy of Scripps Institution of Oceanography.)

The origin of submarine canyons is debated among experts; two explanations have been advanced: the subaerial hypothesis and the submarine hypothesis. According to the *subaerial hypothesis*, submarine canyons formed during glacial epochs when sea levels were lowered. Lowering made it possible for rivers to continue their erosional activity in a manner similar to Colorado River erosion in the Grand Canyon. The hypothesis of subaerial origin has been supported by intensive study of La Jolla Canyon where evidence of subaerial activity to a depth of about 1000 meters has been observed. However, the subaerial hypothesis does not satisfactorily explain the origin of deep-water canyons. The major argument against this hypothesis is that it assumes drastic lowering of sea levels to thousands of meters.

Turbidity current *A downslope movement of dense, sediment-bearing water produced when sand and mud on the continental shelf are first dislodged and then let loose in suspension.*

The *submarine hypothesis* emphasizes the action of turbidity currents in the formation of canyons. *Turbidity currents,* particularly their ability to form deep-ocean canyons, were recognized when undersea cables broke during the 1930 earthquake on the Grand Banks off Newfoundland. Gravity is the principal driving force for turbidity currents. Given a steep slope and relatively unconsolidated sediments along the slope, huge quantities of material may be transported (Figure 4-12). The speed of turbidity currents is often far greater than in river floods. Essentially, subaqueous turbidity currents are powerful erosive agents that cut deep submarine canyons and deposit sediments as deltas or fans. Such a submarine fan can be observed at the mouth of Hudson Canyon. Turbidity currents leave other evidence of their activity, especially sedimentary *turbidites,* characterized by their *graded bedding*.

Graded bedding *A special type of stratification in which a gradation in grain size is displayed, with a concentration of coarser grains at the bottom and finer grains at the top.*

The major argument against the submarine hypothesis remains in the doubtful capacity of turbidity currents to erode the material of the shelf and slope. This counter argument stems from the fact that,

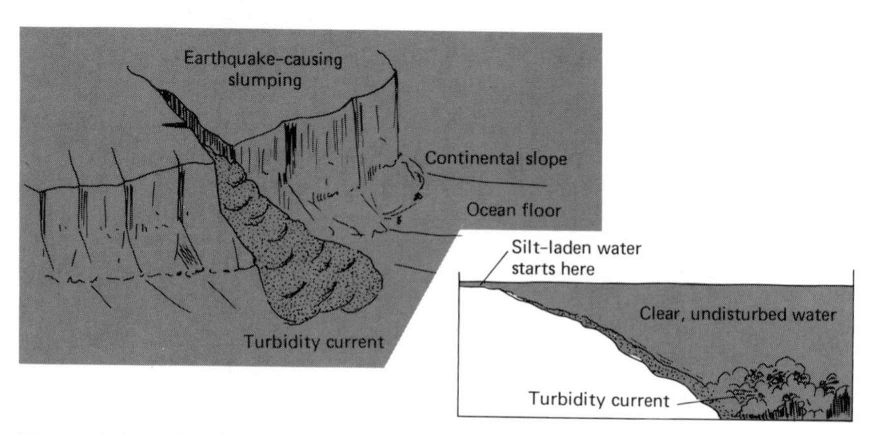

Figure 4-12. (*Left*) Turbidity currents are produced by earthquakes causing slumping, as in the Grand Banks episode of the 1930s. (*Right*) Turbidity current is produced under experimental conditions in a water tank. (*Source:* From a photo by H. S. Bell, Sedimentation Laboratory, California Institute of Technology.)

generally, the canyons are much smaller in size than the fan-shaped deposits at their mouths. Despite this argument, the submarine hypothesis is widely accepted.

4-8 DEPTH-MEASUREMENT TECHNIQUES

Until the nineteenth century, crude methods were used for depth measurement. For example, a lead weight was attached to a rope that was lowered into the sea to determine depth. Later the rope was replaced by piano wire, which was lighter and caused less friction. But the wire did not plunge vertically, because of surface currents and vessel movements. This method did not provide a reliable depth profile of the seafloor. During voyages of the *Challenger* (1872–1876), the wire method was the principal technique for depth determinations.

At the beginning of the twentieth century, an electronic sounding device, the *echo sounder* (or sonic depth finder), was invented. The echo-sounding technique, first applied during the voyage of the *Meteor* in 1925, helped to circumvent the problems of earlier techniques. It provided a continuous and reliable profile of the ocean bottom and showed, in contrast with the profile suggested by the *Challenger,* that submarine topography was quite complex and rugged. In the echo-sounding method, sound from the ship is projected to the seafloor and is bounced back to the ship (Figure 4-13). The time that it takes for the sound to make the round trip can be correlated with distance factors to ascertain water depth. The echo-sounding device is equipped with a precision graphic recorder, so that it not only records the traveling time of sound, but also simultaneously provides a visual, graphic record of the returning sound itself (Figure 4-14). Water depth (D) equals one-half the time of travel times the speed of sound in water (1,460 meters per second). This calculation, however, involves certain corrections because the speed of sound in seawater increases with increasing temperature, salinity, and depth.

Echo sounder *Electronic device employed to determine and record water depth in the ocean.*

Echo sounding *A technique for determining the depth of water by measuring the time required for a sound signal to travel to the bottom and back to the ship that emitted the signal.*

The echo-sounding technique is now routinely employed by oceanographers to prepare depth profiles of the ocean floor, to locate, for example, a seamount, a guyot, or a canyon, and to construct depth (or bathymetric) charts. Echo-sounding techniques can also be used to obtain information about the nature of sedimentary rock layers beneath the ocean floor (Figure 4-15).

A modified version of the echo-sounding device is the side-scanning SONAR (SOund NAvigation and Ranging). SONAR technique continously scans the ocean floor beneath and to the sides of the observation vessel. Sidescanning SONAR is used only in very shallow water or in the deepest parts of the ocean.

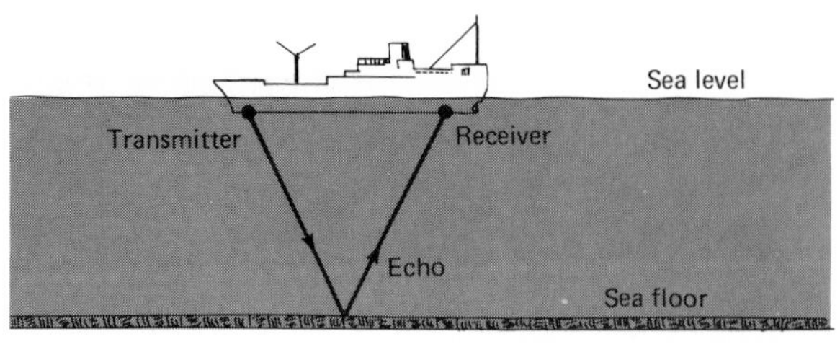

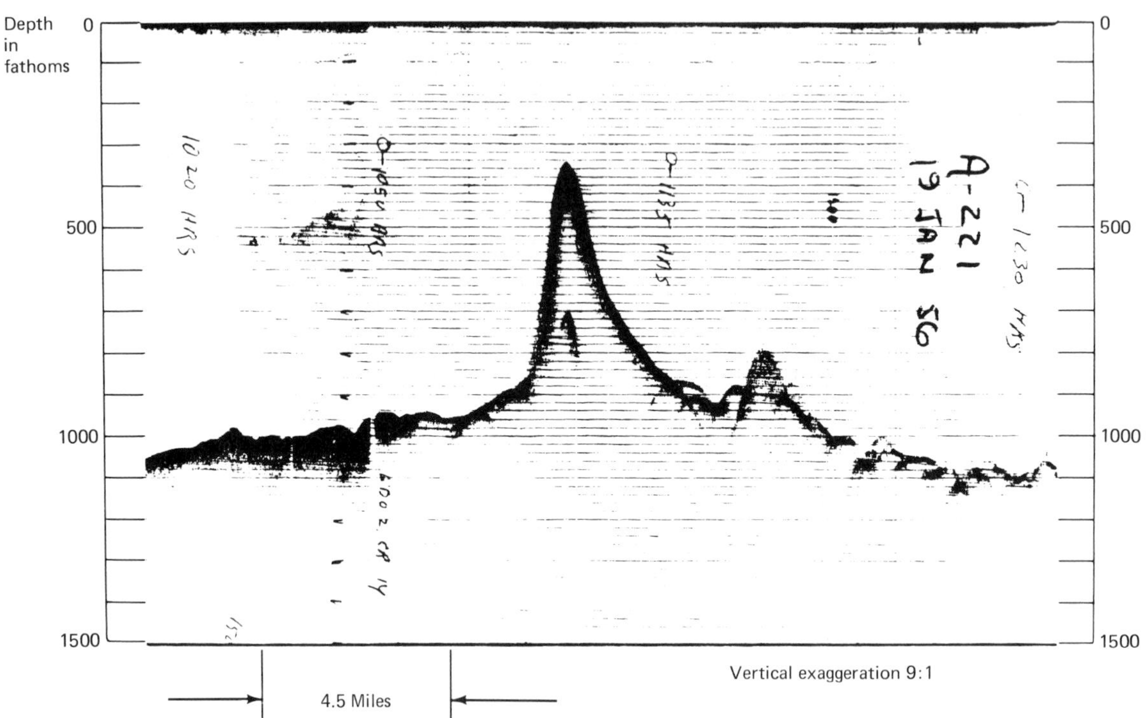

Figure 4-13. Echo-sounding method is routinely used *(top)* to ascertain submarine profiles such as a seamount in the Caribbean Sea *(bottom)*. (Photo courtesy of Woods Hole Oceanographic Institution.)

Figure 4-14. Precision graphic recorder displays a visual graphic profile of the returning sound. (Photo courtesy of R. B. Theroux, National Marine Fisheries Service.)

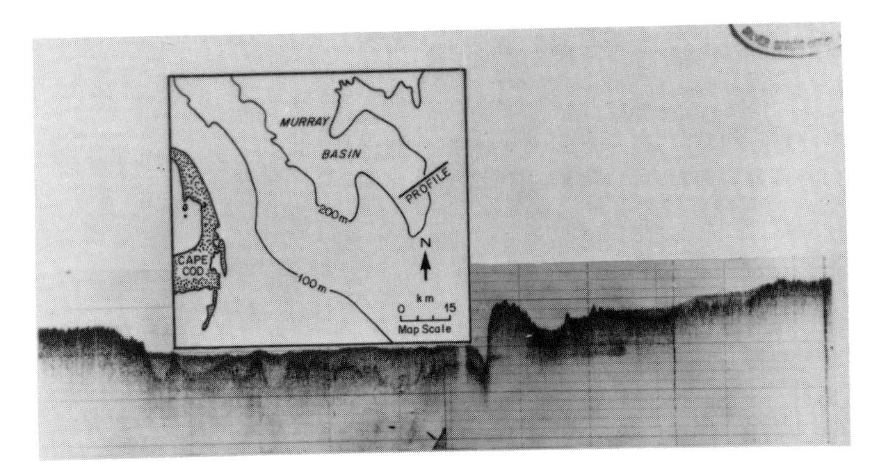

Figure 4-15. Echo-sounding profile across the southern end of Murray Basin. Flat basin floor is 205 meters below sea level. Light gray discontinuous layer below the flat is probably fine-grained sediment, whereas the dark irregular layer to black horizon below it is possibly glacial deposits. (Photo courtesy of J. Schlee, U.S. Geological Survey.)

SUMMARY

1. The ocean floor is rugged and complex. It is subdivided primarily into the continental shelf, the continental slope, continental rise, and abyssal plain.
2. The continental shelf is a seaward extension of the land. It varies in depth from 20 to 530 meters and slopes from the spring tide to a depth of 150 to 200 meters.
3. The continental slope varies in average inclination from 2 to 5 degrees, although its gradient can be as high as 15 degrees. It does not hold sediment well.
4. Abyssal plains cover 40 percent of the ocean floor. Its main topographic features include oceanic rises, seamounts, guyots, and island arcs.
5. Submarine canyons are deep gorges in the sea floor. Relatively shallow canyons may have originated by subaerial weathering that took place during glacial times when sea level was lower. But most scientists adhere to the hypothesis that these canyons were formed by the action of turbidity currents and are confined to the shelves.
6. Well-known submarine topographic features from the world's ocean include the East Pacific Rise and the rim of trenches in the Pacific Ocean; the Mid-Atlantic Ridge in the Atlantic Ocean; and the Mid-Indian Ridge in the Indian Ocean.
7. The echo-sounding method is routinely employed to study submarine topographic features.

Suggestions for Further Reading

Anderson, Alan, Jr. 1975. "Mid-Atlantic Ridge: Diving to the Birthplace of the Ocean." *Science Digest, 77*, 68–74.

Burke, C. A., and C. L. Drake (eds.). 1974. *The Geology of Continental Margins*. New York: Springer.

Emery, K. O. 1969. "The Continental Shelves." In *The Ocean*. San Francisco: Freeman. Pages 41–52.

Emery, K. O., and Elazar Uchupi. 1972. "Western North Atlantic Ocean: Topography, Rocks, Structure, Water, Life, and Sediments." *American Association of Petroleum Geologists*, Memoir 17.

Menard, H. W. 1969. "The Deep Ocean Floor." In *The Ocean*. San Francisco: Freeman. Pages 53–63.

Shephard, Francis P. 1974. *Submarine Geology*, third edition. New York: Harper & Row.

Turekian, K. K. 1976. *Oceans*, second edition. Englewood Cliffs, New Jersey: Prentice-Hall.

Wertenbaker, W. 1974. *The Floor of the Sea and the Search to Understand the Earth*. Boston: Little, Brown.

5 Plate Tectonics and the Origin of Ocean Basins

5-1 DEEP-SEA DRILLING

OCEANOGRAPHERS HAVE RECENTLY MADE A CONCERTED WORLDWIDE ATTEMPT to learn more about the basic geological and geophysical aspects of oceans. This effort is well documented by undertakings of such international ocean-related research projects as the International Geophysical Year (IGY) in 1957; the International Indian Ocean Expedition; the short-term project Mohole, and the International Phase of Ocean Drilling (IPOD), a deep-sea drilling program.

Project Mohole *The United States Marine project to drill a hole into and sample the earth's mantle. It was abandoned in the mid-60s.*

During the early 1960s, American scientists undertook Project Mohole to obtain core samples from the mantle beneath the sea. The project was supported by the National Science Foundation. The barge *Cuss I,* designed for deep-sea drilling, was equipped with a large drilling rig. It drilled 10 experimental holes in waters off San Diego and Guadalupe Island off the west coast of Mexico. Its drills penetrated the seafloor 150 meters. Project Mohole yielded abundant information and fomented useful ideas in furthering the deep-sea technology. In 1968 the Deep-Sea Drilling Project (DSDP), funded by the NSF, was undertaken by Scripps Institution of Oceanography.

The *Glomar Challenger* (Figure 5-1) was the primary research tool of the DSDP. The *Glomar Challenger* is a corvette 123 meters long and 20 meters wide and displaces 10,500 tons of water. Its 44-meter drilling derrick can drill in waters of over 6000 meters and can obtain samples from more than 750 meters below the seafloor. This highly sophisticated ship houses 70 research personnel and has modern facilities, including a satellite navigational system equipped with an onboard digital computer that pinpoints its exact location within one-tenth of a nautical mile. It is also equipped with a weather satellite that transmits

Figure 5-1. *D/V Glomar Challenger.* Since 1968, the *Glomar Challenger* has collected vast amounts of information on the thermal and chemical properties of the ocean floor. The *Challenger's* discoveries contributed a great deal to the development of modern geological oceanography in the 1960s. (Photo courtesy of Scripps Institution of Oceanography.)

photographs directly to the ship for weather prediction and general sea states. The ship is stabilized by a gyroscopically controlled system that enables it to move less than 6 degrees under adverse sea conditions (Figure 5-2, Table 5-1).

During the Deep-Sea Drilling Project, oceanographers collected vast amounts of information on oceanic floors, gravitational and magnetic fields, heat flow, and composition of oceans (Figure 5-3). The most important outcome of the project was the confirmation of the continental drift, the conceptualization of seafloor spreading, and formulation of the theory of plate tectonics. All these have a great bearing on the origin and evolution of the ocean basin.

5-2 CONTINENTAL DRIFT

The first scientific study of the process of continental drift was made in 1885, when Edward Suess, a Swiss geologist, suggested that the

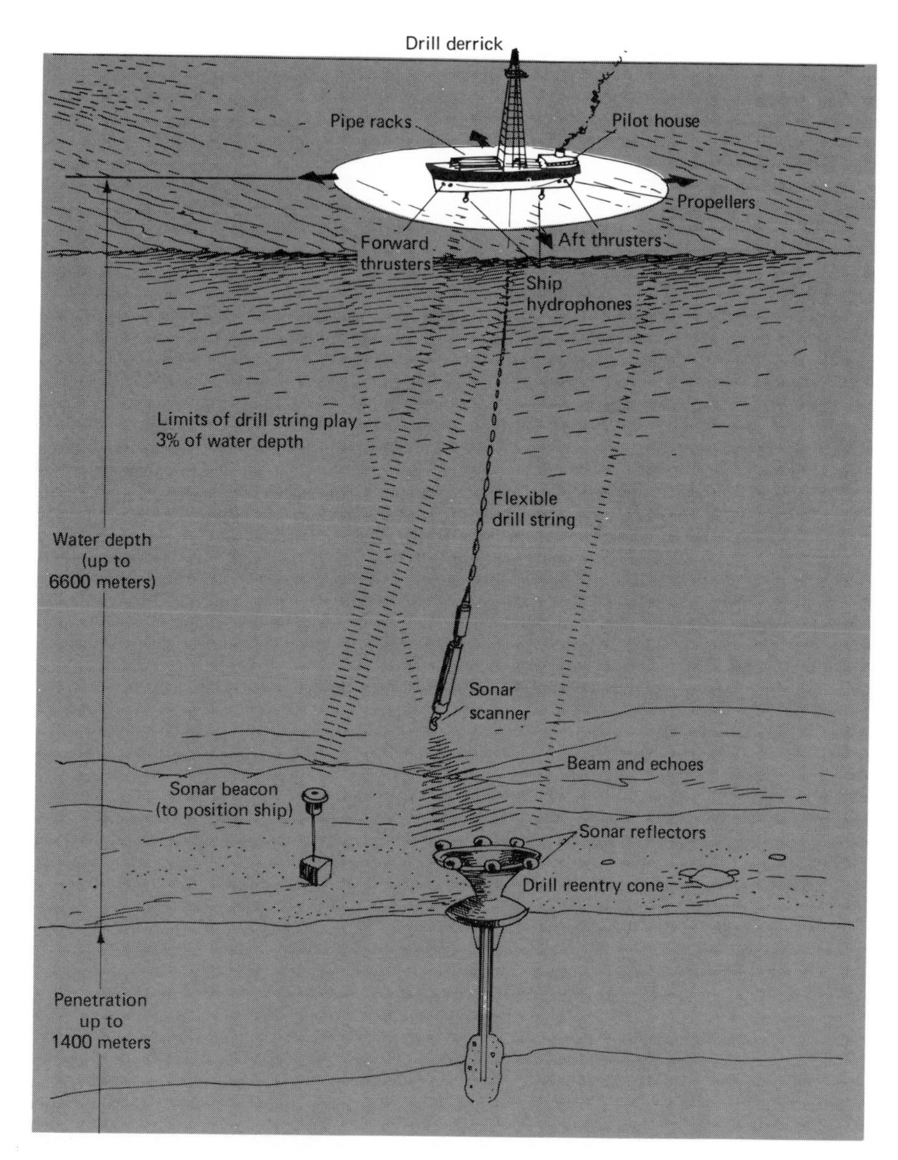

Figure 5-2. Reentry and dynamic positioning of *D/V Glomar Challenger.* The vessel employs highly developed acoustical position-sensing equipment with automatic propulsion-unit control to provide dynamic positioning over a drilling hole. Dynamic positioning refers to holding a station above a sonar sound source placed on the ocean bottom while drilling. Two tunnel thrusters forward and two thrusters aft, along with the ship's two main propellers, are computer controlled to hold the ship's position without anchors in depths up to 6500 meters.

TABLE 5-1 The Glomar Challenger

	Remarks
Operation capabilities	Excellent through dynamic positioning system. Can be steadily positioned in deep water without anchors, and remain at a drilling site for a long time. Deepest water drilling, 6244 meters; longest drill string, 6766 meters.
Performance	Between 1968 and 1973, achievements included: About 450 holes drilled and cored at about 300 sites. Over 120,000 meters of aggregate penetration. Over 50,000 meters of sea-floor cored. About 30,000 meters of core recovered. Over 275,000 kilometers voyaged.
Examples of scientific results	Ocean basins young (less than 250 million years old) Seafloor spreading verified. Continental drift theory verified. Mediterranean Sea dried up 5 to 10 million years ago. Glaciation in Antarctica in effect for past 20 million years. Discovery of metals such like iron and manganese on deep-ocean floor.

SOURCE: National Science Foundation, *Deep-Sea Drilling Project: Ocean Sediment Coring Program,* U.S. Government Printing Office 19740-528-489.

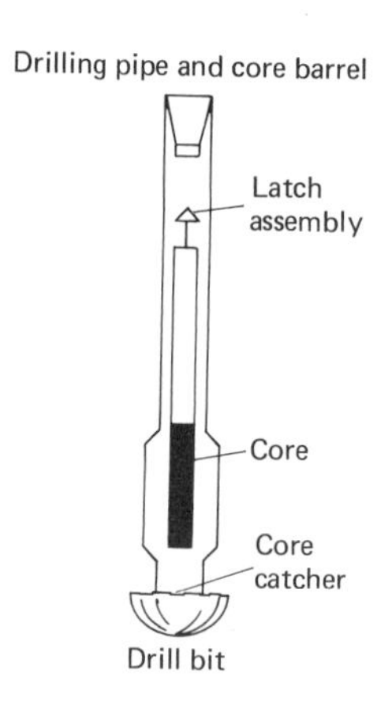

Figure 5-3. *Left:* A deep-sea core in a laboratory aboard the research ship *R/V Knorr. Right:* Deep-sea drilling equipment. (Photo courtesy of Woods Hole Oceanographic Institution.)

continents of South America, Africa, Antarctica, Australia, and India were once joined in a single landmass called *Gondwana.* In 1912, Alfred Wegener published his *The Origin of Continents and Ocean Basins.* In this classic book, Wegener asserted that all continents were once parts of a single supercontinent, *Pangaea.* In subsequent geologic time Pangaea supposedly fragmented into two giant continents that were separated by the ancient *Tethys Sea* (Figure 5-4). The northern continent of *Laurasia* encompassed North America and Eurasia, and the southern continent of Gondwana comprised the continents earlier suggested by Suess. Further fragmentation occurred, thus forming modern continents. According to contemporary accounts of continental drift, the first breakup of the Pangaea began 200 million years ago and the second about 135 million years ago (Figure 5-5).

Gondwana *Theoretical ancient southern continent encompassing India, Australia, Antarctica, Africa, and South America.*

Pangaea *The hypothesized ancient single supercontinent that comprised all continents.*

Tethys Sea *An ancient sea that divided Laurasia from Gondwana along the present Alps–Himalayan mountain belt.*

Laurasia *Theoretical ancient northern continent encompassing North America and Eurasia.*

Proponents of the theory of continental drift adhere to at least two convincing geologic evidences, including the matching of coastlines, particularly those between the southern Atlantic coasts of South America and Africa, and the similarity of sedimentary deposits of coal (350 million years), of glacial deposits called tillites (250 million years), and the occurrence of similar fossil plants and animals in Africa, South America, Antarctica, and India (Figure 5-6).

Critics of the theory of continental drift maintained that this was merely circumstantial evidence and insisted on gathering more conclu-

Figure 5-4. The supercontinent of Pangaea broke into two continents that were later separated by the Tethys Sea.

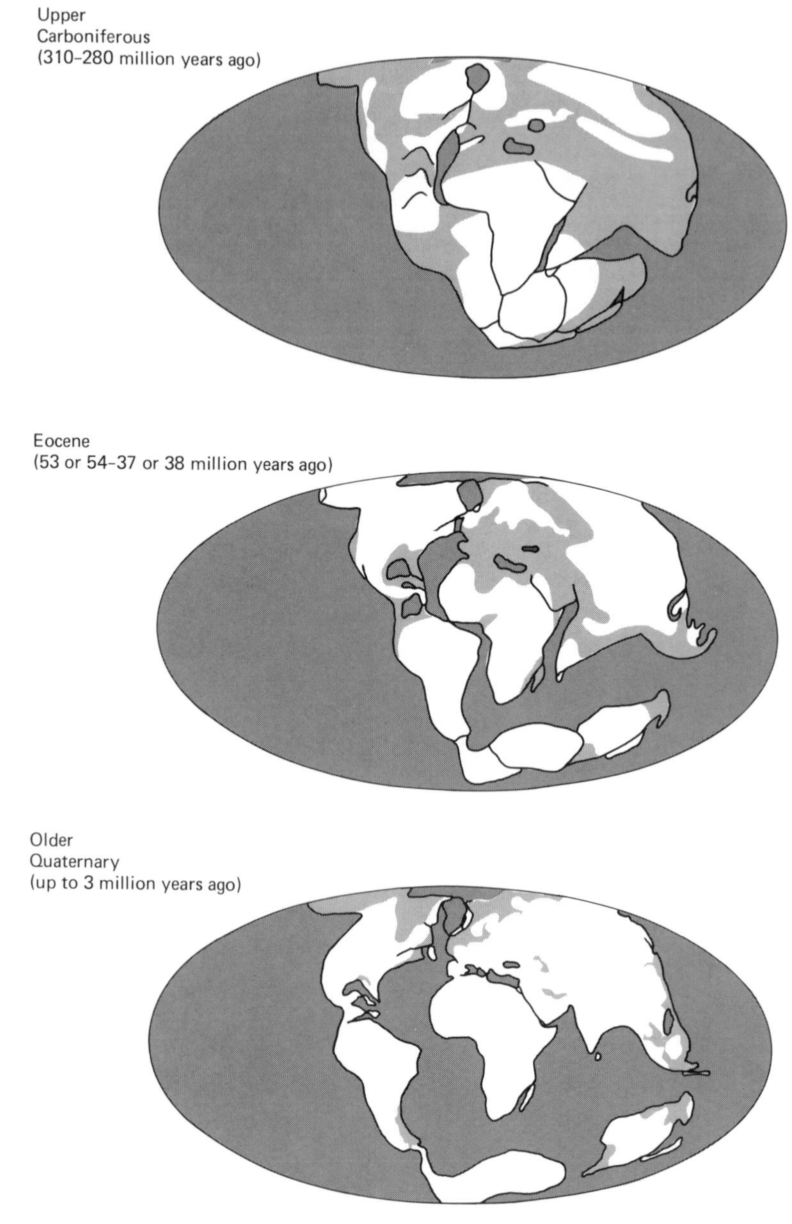

Figure 5-5. The pattern of continental drift, as envisioned by Alfred Wegener.

Continental drift *The concept that, due to the weakness of the suboceanic crust, continents can drift on earth's surface much as ice floats through water.*

sive evidence to explain the "true" mechanism of the drifting of the continents. As a result, the whole problem of continental drift became quite controversial. Recent interest in continental drift was rejuvenated when discoveries of paleomagnetism and seafloor spreading were made in the 1950s and 1960s. The computerized continental-drift reconstruc-

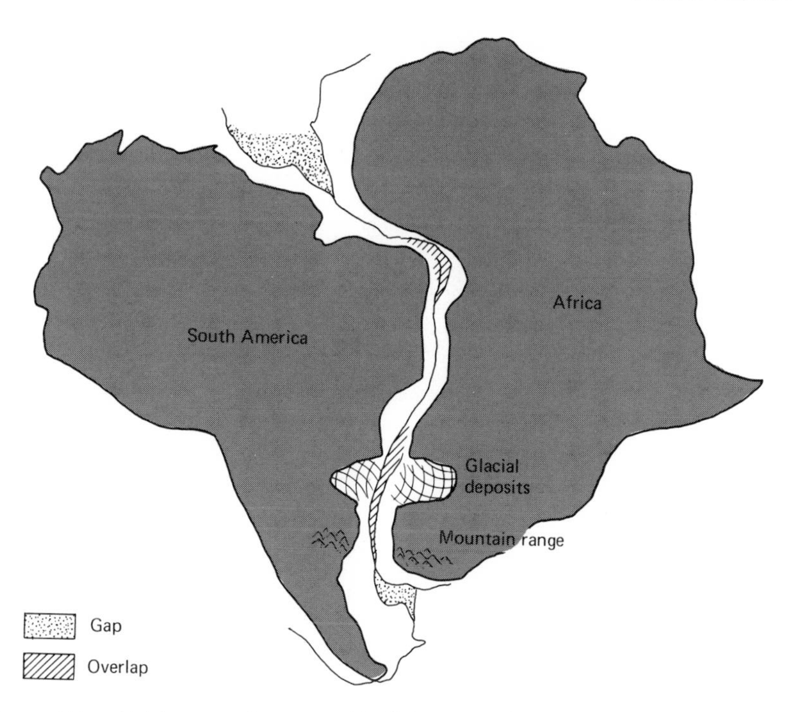

Figure 5-6. Evidences of geographic fit between South America and Africa. The continuity of glacial deposits and mountain ranges in these two continents was the major supporting evidence added by proponents of the theory of continental drift. (*Source:* After Bullard et al., 1955.)

tion fit between Africa and North and South America is among the most modern evidences supporting the theory of continental drift (Figure 5-7).

5-3 PALEOMAGNETISM

Paleomagnetism is the study of variations in the earth's magnetic fields as recorded in ancient rocks (Figure 5-8). Soon after an iron-rich liquid such as basaltic lava, for example, emerges at the ocean bottom, its iron particles become magnetized; that is, iron atoms are oriented in the direction of the earth's magnetic field. The orientation of such atoms also occurs when iron-rich particles settle in water. In recent years, intensive research in paleomagnetism has shown that the earth's magnetic field has reversed several times in the geologic past (Figure 5-9) and that the positions of the poles appear to have shifted in relation to the continents (Figure 5-10). If the earth's poles seem not to have been stationary, the implication is that the continents themselves must have

Paleomagnetism *The residual magnetism in fossilized rocks.*

Basalt *A dark-colored igneous rock commonly found on the ocean floor, mineral enriched or relatively high in iron and magnesium content.*

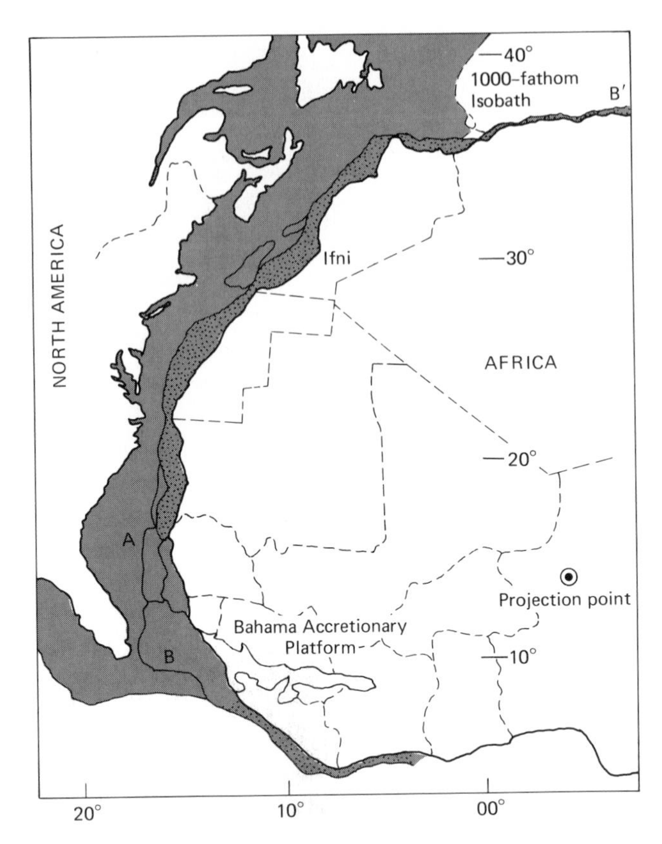

Figure 5-7. A computerized continental-drift reconstruction fit between Africa and North and South America. (*Source:* Redrawn from R. S. Dietz and J. C. Holden, 1975, *Journal of Geophysical Research, 75.*)

drifted and hence now stand in different positions with respect to the poles. Paleomagnetic studies of samples collected from North America and Europe indicate that these two continents once were joined together. Paleomagnetic and radioactive dating studies in Brazil and Gabon, West Africa, have shown that rock samples from these locations are almost identical in composition, geologic structure, and age, suggesting that Africa and South America too were once joined together.

5-4 POLAR WANDERING

Paleomagnetic studies of rocks of different ages have shown that the position of the earth's magnetic poles shifted a number of times in the geologic past and resulted in the concept of *polar wandering*. Polar

Figure 5-8. Oceanographers use a magnetometer in the study of magnetic properties of iron-bearing oceanic sediments and rocks. (Photo courtesy of Woods Hole Oceanographic Institution.)

wandering is a misnomer, because it does not mean the shifting of the poles over the earth. The poles have remained fixed, but the continents and seas beneath them have moved. The mobility of the crust on which land and oceans rest results from a slow shifting of solid lithosphere or its slippage on the plastic layer of asthenosphere beneath it; consequently, land and seas change their position in relation to the poles.

Polar wandering *The apparent shifting of the magnetic poles throughout geologic time.*

5-5 SEAFLOOR SPREADING

In the early 1960s the concept of seafloor spreading was proposed to offer a satisfactory explanation for the movements of continents. According to this hypothesis, the entire seafloor itself is mobile and is

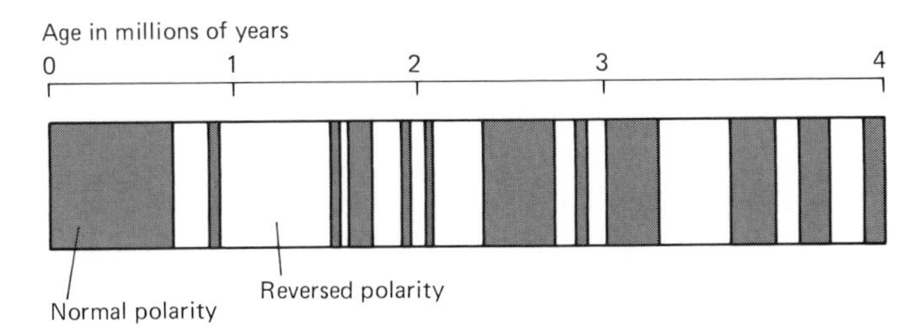

Figure 5-9. Reversal of polarity during past 4 million years. Based on combined paleomagnetic and radiometric dating techniques. (*Source:* After A. Cox, 1973, *Plate Tectonics and Geomagnetic Reversals*, San Francisco, Freeman.)

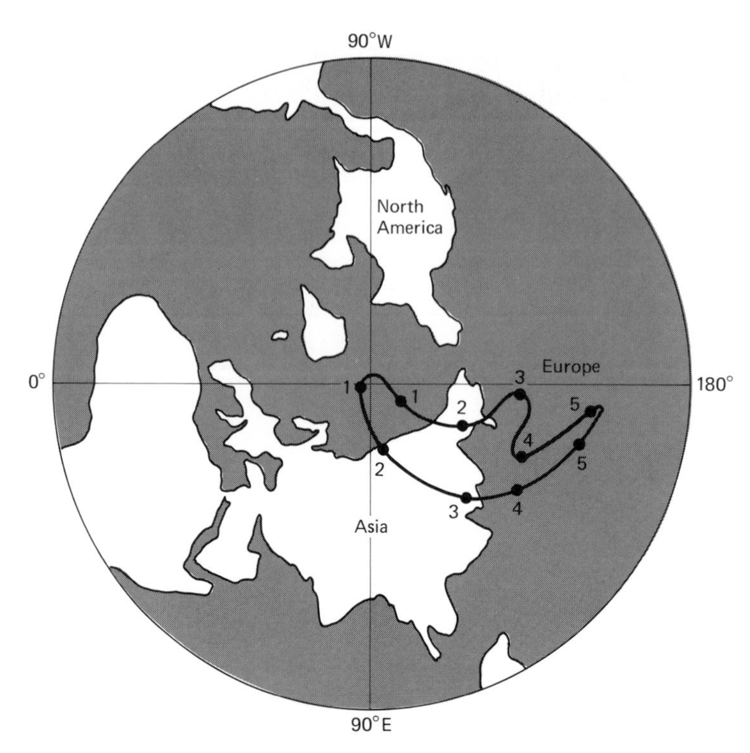

Figure 5-10. Paleomagnetic studies of rocks from North America and Europe reveal that these two continents were once joined. Arrows show the paths journeyed by the magnetic poles throughout geologic time. (Adapted from A. Cox and R. R. Doell, 1960, "Review of Paleomagnetism," *Bulletin of the Geological Society of America, 71.*)

moved by convection currents in the mantle. The seafloor moves along midoceanic ridges, in opposite directions from either side of them. The entire process is analogous to the movement of a conveyor belt. Fresh crustal material is formed at ridges and moves away gradually, until it is transported to trench areas where it moves downward and is melted once again into the earth's interior (Figure 5-11).

Magnetic anomaly *Distortion of the regular pattern of the earth's magnetic field resulting from the various magnetic properties of local concentrations of ferromagnetic minerals in the earth's crust.*

Studies of long, linear, magnetic anomalies—that is, bands of rocks of alternating magnetic polarity on the seafloor—have showed that such linear magnetic patterns paralleled those of midoceanic ridges and were symmetrical on either side (Figure 5-12). It was therefore deduced that volcanic material had been forced up through the center of the ridge and that it flowed steadily away on either side as fresh rising lava from beneath pushed the older lava along the rift. On cooling, lava magnetized and thus maintained the polarity of the earth's magnetism. These magnetic studies were supported by radioactive determination of the ages of samples obtained from either side of the Mid-Atlantic Ridge. Radioactive dating showed that seafloor spreading was taking place at

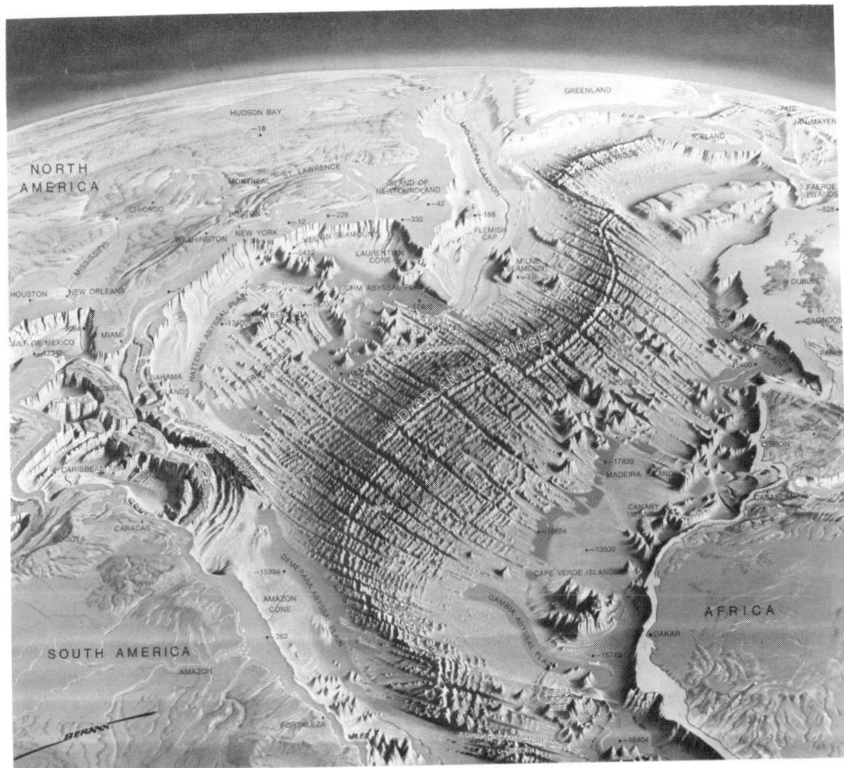

The Atlantic Ocean Floor

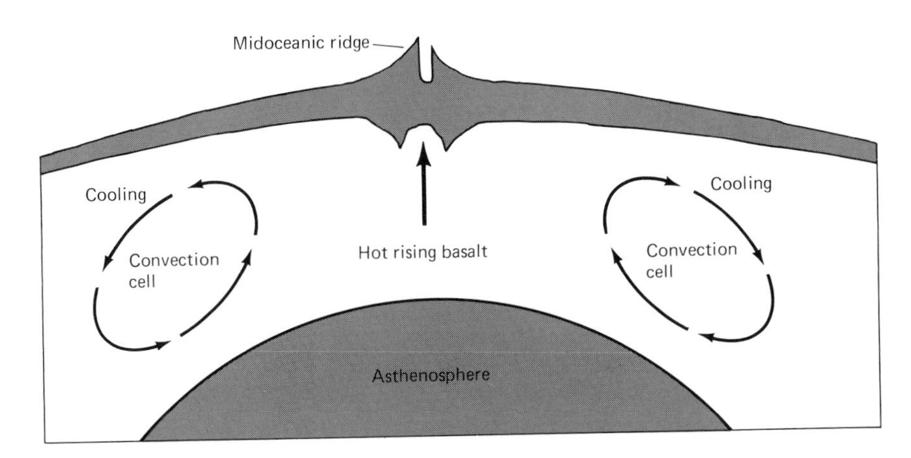

Figure 5-11. The cause and effect of sea-floor spreading. (Refer to Figure 4-1).

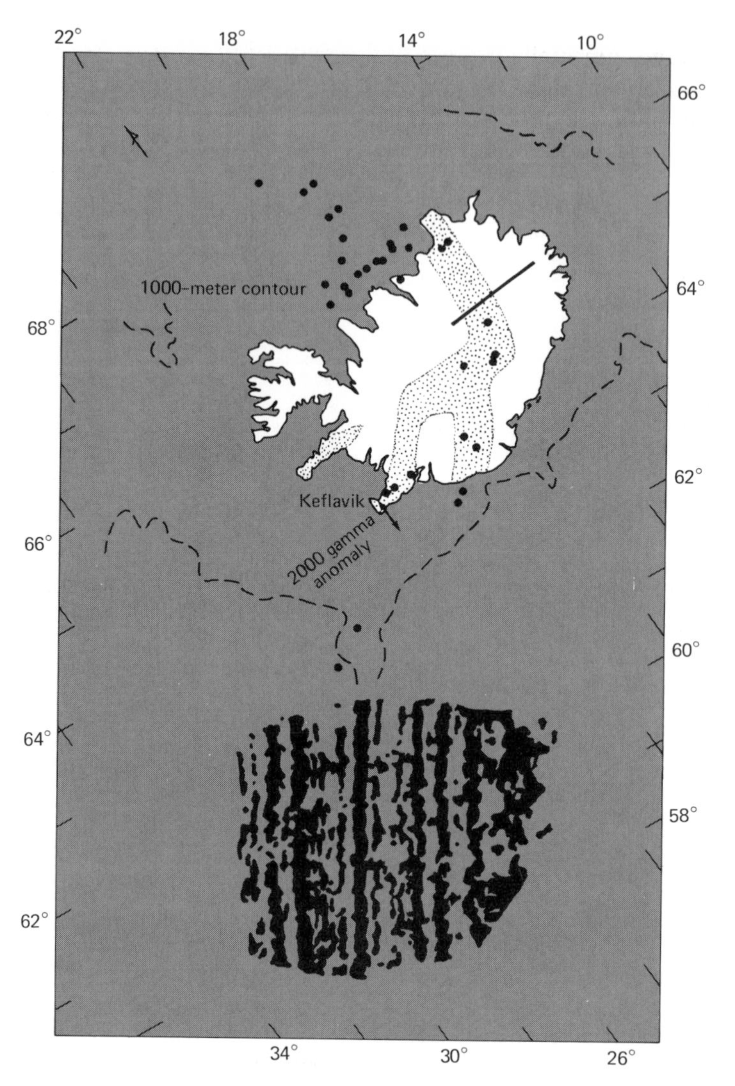

Figure 5-12. Magnetic anomalies over the Reykjanes Ridge, an extension of the Mid-Atlantic Ridge. Black strips indicate positive anomalies; white areas indicate negative anomalies. Similar so-called "zebra strips" were previously recorded on either side of the Mid-Atlantic Ridge and led to the concept of sea-floor spreading. (*Source:* J. R. Heirtzler *et al.*, 1966, *Mid-Oceanic Anomalies over the Reykjanes Ridge*, New York, Pergamon.)

the rate of 1.5 centimeters per year and confirmed that the seafloor was indeed *mobile*. This rate suggests that the continents on either side of the Atlantic Ocean were joined during the Mesozoic era. Finally, the paucity of sediments older than 70 million years can be explained by the conveyor-belt mechanism of seafloor spread. That is, periodically fresh material from the earth's interior would spread on either side of a midoceanic ridge, gradually sweeping the floor of the ocean. Younger material would push aside previously deposited older material, in this way maintaining the relatively younger material throughout the ocean floor.

As a result of cumulative evidence of systematic magnetic reversals, of radioactive-dating patterns, and of the rate of flow of material on either side of the Mid-Atlantic Ridge, the seafloor-spreading hypothesis was overwhelmingly accepted before the end of the 1960s and has been incorporated into a unified theory of global plate tectonics.

5-6 PLATE TECTONICS

Plate tectonics *The concept that the crust and upper mantle of the earth are divided into segments or plates that are always in friction with each other, consequently generating earthquakes, mountain ranges, and so on.*

The modern theory of global plate tectonics considers that the earth's crust is broken into several large and small segments—or plates. The plates are over 150 kilometers thick and occupy most of the earth's crust and the uppermost portion of the mantle. The plates constitute the lithosphere. The lithosphere is riding on the relatively soft layer of asthenosphere (see Chapter 2), thereby causing the plate movement. Boundaries between these plates are characterized by frequent volcanic and earthquake activities. Major plates are identified and named for their prominent geographic landmasses: Eurasian, American, African, Pacific, Indian, and Antarctic (Figure 5-13).

The Eurasian plate includes the continents of Europe and Asia and their peripheral submerged areas. The American plate includes North and South America and the western half of the Atlantic Ocean. The African plate is comprised of the entire continent of Africa, the eastern half of the Atlantic Ocean, and the western part of the Indian Ocean (west of the Mid-Indian ridge). The Pacific plate is predominantly oceanic. The Indian plate occupies much of the Indian Ocean (east of the Mid-Indian ridge), the subcontinent of India, and western Australia. The Antarctic plate occupies most of the Antarctic continent and the Antarctic Ocean.

The plates are strong and rigid and are in constant motion (Figures 5-14, 5-15). The plates sail on and are buoyed up by the hot, molten, plastic asthenosphere. The plates can collide or can slide past each other. If collision takes place within an oceanic basin, elongate trenches, such as Tonga Trench in the southwestern Pacific, will form. If the collision of plates occurs along a relatively steep coast, original oceanic deposits may be squeezed (or in fact destroyed) and folded into a mountain chain, such as the Andes in South America. When two plates slide past each other, major zones of faulting result, such as the San Andreas Fault in California. In this case the Pacific plate is moving northwest with respect to the American plate.

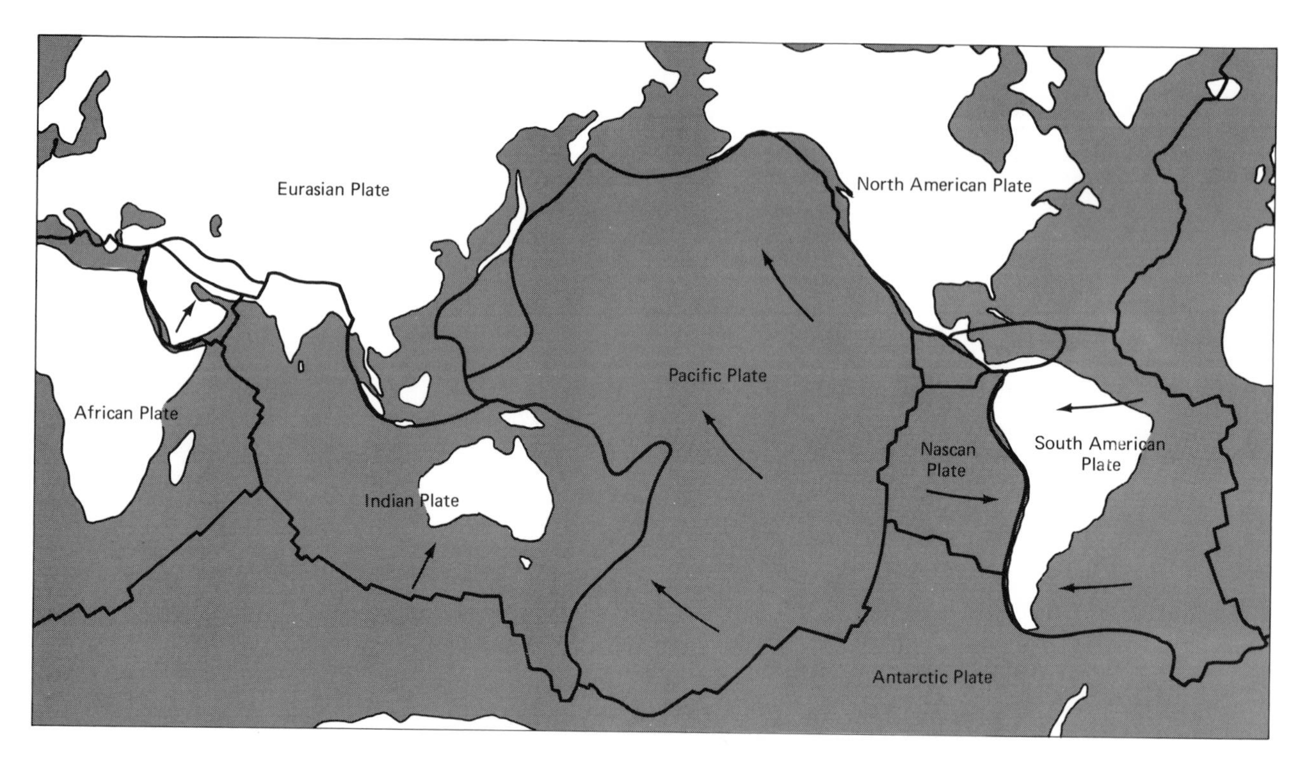

Figure 5-13. Major plates of the earth. Arrows show direction of plate movement.

During the course of their growth plates not only collide or slide past one another; their crustal material is also destroyed. The sea-floor spreading explains how this material is created at the midoceanic ridges. The crustal material of plates is most effectively destroyed along elongated deep trenches (see Chapter 4) of the deep ocean. Through them, the material finds passage into the hotter parts of the mantle where melting and assimilation take place. Often, the assimilated fresh material may reemerge at midoceanic ridges. This overall process is known as *subduction*, and the sites at which destruction of crustal material takes place are called *subduction zones*.

Major conclusions of the theory of plate tectonics theory are that: (1) modern ocean basins are relatively young (less than 200 million years old); (2) ocean floors have been kept younger by the process of the seafloor spreading; (3) the interaction between plates is well documented by such geologic features as mountain chains, island arcs, oceanic deeps and trenches, midoceanic ridges, major fault zones, and the distribution of earthquake and volcanic zones; and (4) ocean floors and continents are mobile.

The general sequence of events in the formation of the ocean basin may now be summarized. Soon after the earth's formation, its surface

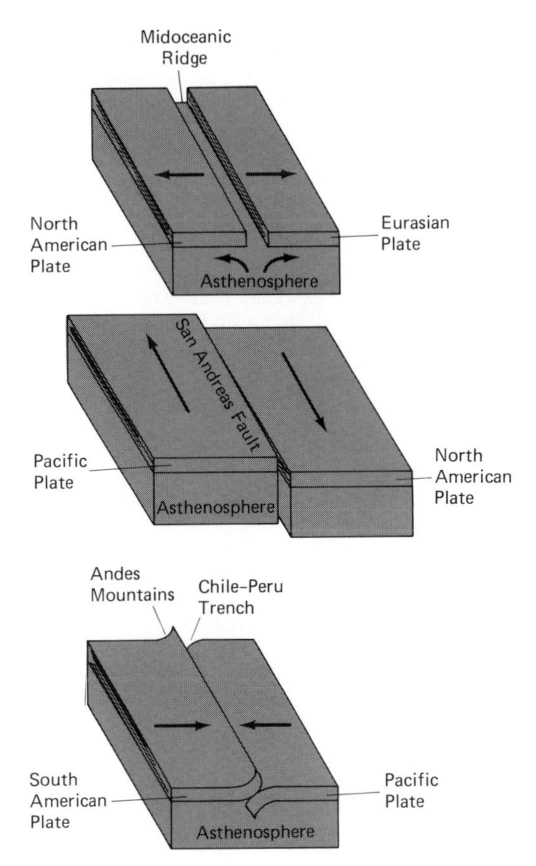

Figure 5-14. Plates are strong and rigid and are in constant motion. As a result, new mountains, ridges, oceans, and trenches are formed. (a) Two plates, North American and Eurasian, are drifting from each other because of sea-floor spreading at the midoceanic point in the Atlantic Ocean. Another modern example of plate movement of this type is evident in the Red Sea and the Gulf of Aden (see Figure 5-15). (b) Two plates, Pacific and North American, are sliding past each other along the San Andreas Fault. The Pacific Plate is moving to the northwest. As these plates move, strain builds up in the rocks. When it reaches its threshold, energy is released similar to that in a tightly coiled spring. As a result, earthquakes occur. (c) When two plates collide, one may bend to form a trench, whereas the other plate that is raised gives rise to mountains. The Andes Mountains and adjacent Peru-Chile Trench are a classic example. Another example is that of the Himalayas and the Indo-Gangetic Trench.

cooled, forming the crust. The crust was formed by a stratification of heavier material (iron, magnesium, etc.) sinking toward the core and lighter material (silica, and aluminum) moving toward the surface. The primordial crust formed swells where the lighter material (sial) had concentrated and became continents. Where the lighter material was relatively less concentrated, low relief areas were formed. The low relief areas became suitable sites for the collection of sea water derived from the earth's interior (see Chapter 3).

Figure 5-15. The Red Sea and the Gulf of Aden show how modern seaways are formed by the process of sea-floor spreading. Note the remarkable fit between the landmasses. (Photo courtesy of NASA.)

Ocean basin *The part of the floor of the ocean that is deeper than 2000 meters below sea level.*

In subsequent geologic time, the ocean basins originated and later widened along the axis of the midoceanic ridges. The widening of the basins was caused by upwelling convection currents (divergent and emergent) produced by radioactivity. The restless state of the plate-carrying continents and the geologically rapid seafloor spreading suggest that continental break-ups and their subsequent drifting and births and deaths of ocean basins were more common throughout the earth's history. Sea-floor spreading on either side of a midoceanic ridge is between less than 1 to 16 centimeters per year. These rates are least in the Mid-Atlantic Ridge and highest in the East Pacific Rise. According to Sir Edward Bullard, at a 16 centimeters per year rate of spreading, the entire floor of thePacific Ocean (15,000 kilometers wide) would be produced in only 100 million years. This is a rapid speed of spreading in terms of geological time. The Atlantic Ocean formed less than 150 million years ago when the American, Eurasian, and African plates split apart. Since that time, the Atlantic Ocean has grown in size. The Indian Ocean also formed in recent geologic time when the Indian plate drifted away from Gondwana, and ultimately collided with the Eurasian plate in the north as mentioned earlier. At present, because of the plates in motion, the Pacific basin is shrinking at the expense of the Atlantic and Indian oceans. The Pacific Ocean may possibly disappear sometime in the future as an unknown number of oceans have in the geologic past. An ancient ocean, the *Uralian,* once existed between Siberia and western Russia. This ocean disappeared because of a collision between the Russian and Siberian plates, forming the Ural mountains.

Plate movements and sea-floor spreadings have influenced the size and configuration of the ocean basin in the past and continue to do so today. Future oceans in the making are now identified in the Gulf of Aden and the Red Sea (Figure 5-15).

SUMMARY

1. Geological oceanography entails the study of various aspects of the ocean basin structure and includes the study of the earth's interior, of continental drift, of seafloor spreading, and of polar wandering. Plate tectonics and paleomagnetism are branches of geological oceanography.
2. The hypothesis of continental drift was first proposed by Alfred Wegener. According to him, all continents were joined into a single land mass called Pangaea; later, Pangaea broke into two continents—Laurasia and Gondwana, separated by the ancient Tethys sea.
3. The concept of seafloor spreading relates to the welling up of fresh material on either side of a midoceanic ridge.
4. Plate tectonics holds that the earth's crust and outer mantle are broken into large and small segments or plates. These plates can collide with each other or pass each other. Mutual interaction among plates produces earthquake belts, mountain ranges, ocean trenches, and ridges.
5. Paleomagnetism is the fossilized magnetism of iron-bearing rocks, especially basalt. Soon after the emergence of iron-rich lava at the surface, the iron particles become magnetized. The orientation of such atoms also occurs when iron-rich particles settle in water.
6. Polar wandering refers to the drifting of continents and ocean floors which have imperceptibly changed their positions throughout time.

7. Ocean basins originated and subsequently widened along the axes of the midoceanic ridges. The widening of basins is caused by upwelling convection currents generated by radioactivity in the earth's interior.
8. Based upon the generally restless state of plates and the relatively rapid spreading of the ocean floor, we may postulate that oceans have formed and disappeared many times in the earth's geologic history.
9. Future oceans in the making are identified in the Gulf of Aden, the Red Sea, and near the San Andreas Fault.

Suggestions for Further Reading

Anderson, D. L. 1971. "The San Andreas Fault." *Scientific American, 225* (5), 58–68.

Bullard, Edward. 1969. "The Origin of the Oceans." *Scientific American, 221,* 66–75.

Dewey, J. F. 1972. "Plate Tectonics." *Scientific American, 227* (5), 57–68.

Hallam, A. 1975. "Alfred Wegener and The Hypothesis of Continental Drift." *Scientific American, 232* (2), 88–97.

Heezen, B. C., and I. D. MacGregor. 1973. "The Evolution of the Pacific." *Scientific American, 229* (5), 102–112.

Heirtzler, J. R., and W. B. Bryan. 1975. "The Floor of the Mid-Atlantic Rift." *Scientific American, 233* (2), 79–89.

McKenzie, D. P. 1972. "Plate Tectonics and Sea-Floor Spreading." *American Scientist, 60,* 425–435.

McKenzie, D. P., and J. G. Sclater. 1973. "Evolution of the Indian Ocean." *Scientific American, 228* (5), 63–72.

Sullivan, Walter. 1974. *Continents in Motion: The New Earth Debate.* New York: McGraw-Hill.

Sullivan, Walter. 1976. "Experts Discuss Mysteries of Shifting Ocean Floors." *The New York Times,* April 4, pages 13–14.

The Physical Properties of Seawater

6-1 INTRODUCTION

THE PHYSICAL AND CHEMICAL PROPERTIES OF SEAWATER ARE a direct consequence of the atomic structure of water (Figure 6-1). Water is the union of hydrogen and oxygen linked with covalent bonds. A *covalent bond* exists when elements share their electrons to form a compound. In water, hydrogen and oxygen are linked through bond angles of 105°. Each atom of hydrogen and oxygen has electrons that are distributed unequally in such a way that each hydrogen atom has a partial positive charge and each oxygen atom has a partial negative charge. This unusual simultaneous positive and negative behavior of water gives it a *dipolar* molecular structure. Each positive center (or H atom) is attracted to, and forms a weak link with, a negative center (or O atom) in another molecule. The link of hydrogen to oxygen atoms is called the *hydrogen bond*. The strength of this bond is 10 percent of a covalent hydrogen-oxygen bond.

Covalent bond *The linkage between two atoms in a molecule resulting from the sharing of electrons.*

Because it is an aggregation liquid, water behaves as if it had a much larger molecule than would be indicated by simple formula of H_2O. Consequently, typical properties of water appear abnormal when compared with those of a nonpolar substance such as methane (CH_4) or hydrogen sulfide (H_2S). Because of the hydrogen bond, water has a boiling point (100°C) higher than would be expected. If hydrogen-bonding suddenly ceased, the oceans would vaporize instantly.

6-2 CONDUCTIVITY

Conductivity refers to the capacity of seawater to transmit electrical currents and depends on the concentration of ions and their speed.

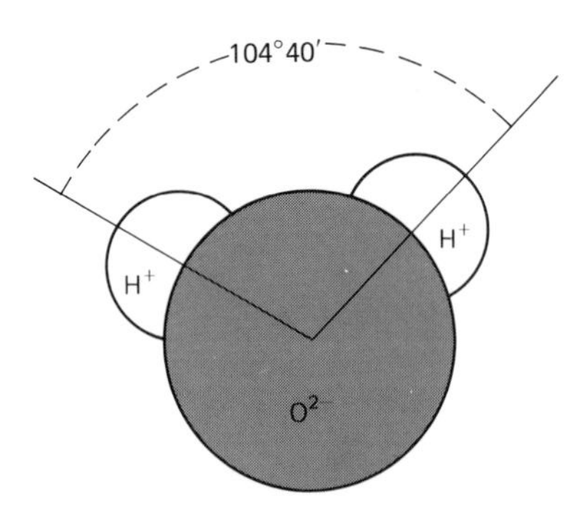

Figure 6-1. Diagrammatic sketch of a water molecule. The hydrogen atoms make an angle of 104° 40′ with the oxygen atom. Because of the lack of symmetry, one side of the water molecule has a net positive charge, and the other side has a net negative charge.

Charged atoms are called *ions*. The more ions per unit volume of water, the greater will be its conductivity. The chemical theory of conductivity is as follows: When salt (sodium chloride or NaCl) dissolves in water, it separates as Na^+ and Cl^- ions, which are engulfed by water molecules. Positive sodium ions are drawn toward the negative oxygen of the water molecules; negative chloride ions are attracted to the positive hydrogen part of the water molecule. In this way, the ions of salt (Na^+ and Cl^-) are freed from each other and brought into contact with water molecules. If positive and negative electrodes are immersed in a container of water, the positive sodium and the negative chloride ions will be attracted to the oppositely charged electrodes. As these ions continue to move through the surrounding water molecules to the electrodes, they produce electric currents. The conductivity of seawater enables us to determine its salinity.

6-3 SALINITY

Salinity *Index of the amount of total dissolved solids in seawater. It is expressed in parts per thousand by weight in 1 kilogram of seawater.*

Salinity is the saltiness of the ocean and is generally computed as the number of grams of dissolved salts in 1000 grams of seawater. Oceanographers, from their extensive analysis of seawater samples collected from the open ocean, have learned that in 1 kilogram of seawater there are 35 grams of dissolved salts. This concentration is commonly expressed as 35 parts per thousand, or 35‰. The salinity of the ocean varies from 33‰ to 38‰ with an average of 35‰. The salinity of seawater is largely dependent on the difference between evaporation and precipitation and, to a lesser extent, on the stream runoff, and freezing and melting of ice. In areas of high evaporation, such as in the Red Sea, salinity approaches 40‰, but near the mouth of a river, it may be as low as 20‰. In general, higher salinity is common in arid equatorial zones.

Halocline *A layer of water in which salinity changes drastically.*

The salinity of seawater varies with depth. The greatest change in salinity occurs between 100 and 1000 meters. This zone of rapid variations in salinity is called *halocline* layer. This rapid change in salinity corresponds to temperature (see Section 6-4) and dissolved oxygen (Figure 6-2). The surface distributions of salinity and temperature in the oceans are given in Figure 6-3.

Because the total amount of dissolved salts is variable and relative proportions of the major elements (Na^+, Cl^-, Mg^{2+}, Ca^{2+}, K^+, and SO^{2-}) are constant, it is possible to determine salinity by manipulating any of these elements. The common practice is to use the chloride ion (Cl^-) since it has the highest dissolved concentration in seawater. The salinity of seawater using chloride content or chlorinity (Cl‰) as a basis can be determined by the equation

$$\underset{\text{salinity}}{S^0/_{00}} = 1.8 \times \underset{\text{chlorinity}}{Cl^0/_{00}}$$

Most salinity determinations based upon this equation are accurate. The salinity of seawater can be determined by measuring the electrical conductivity of seawater. The underlying principle is simply that a salt solution does conduct electric currents. At a known temperature, the higher the salinity, the greater the conductivity of that solution.

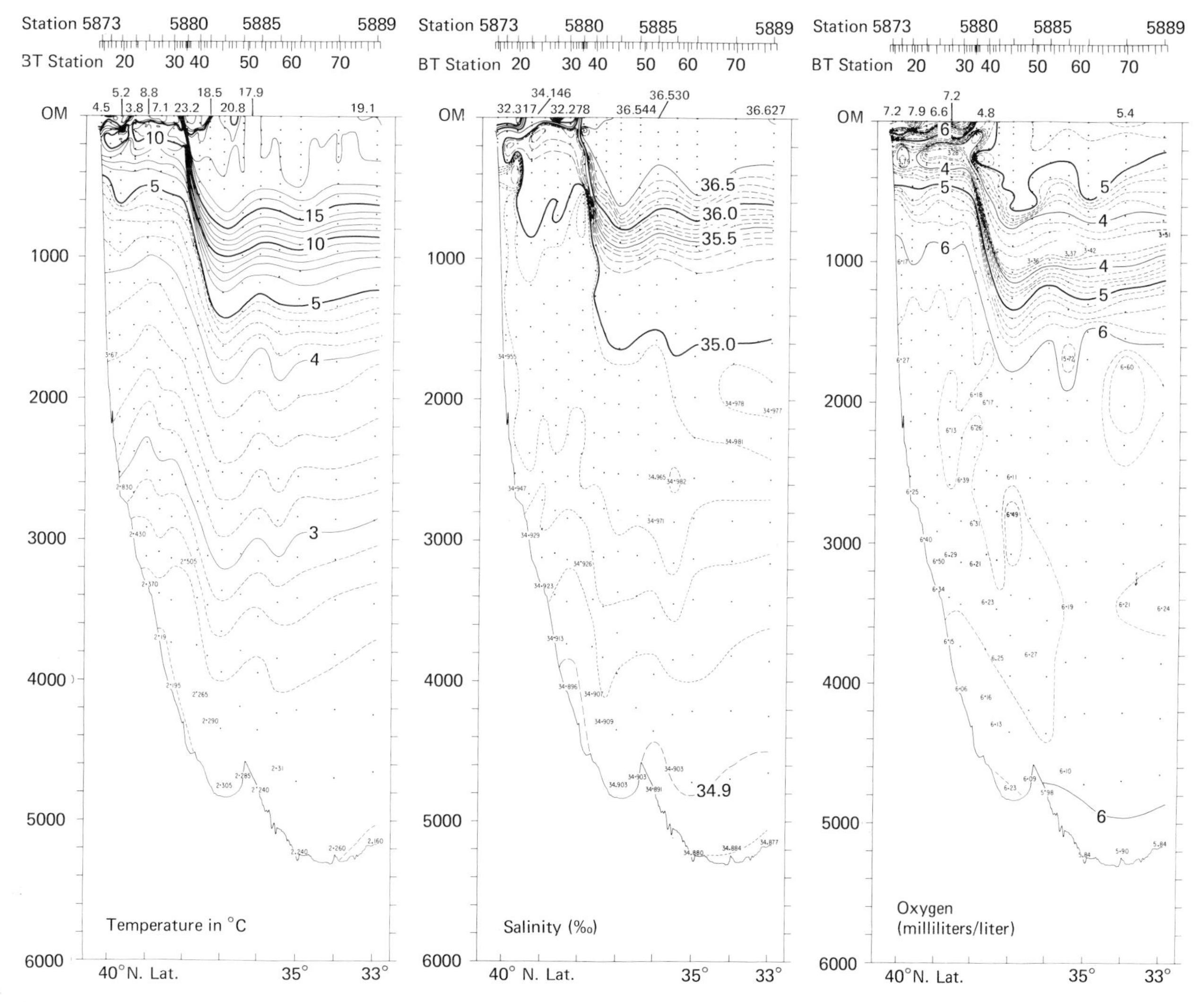

Figure 6-2. Typical distribution of temperature, salinity, and dissolved oxygen in the ocean with depth. (Photo courtesy of Woods Hole Oceanographic Institution.)

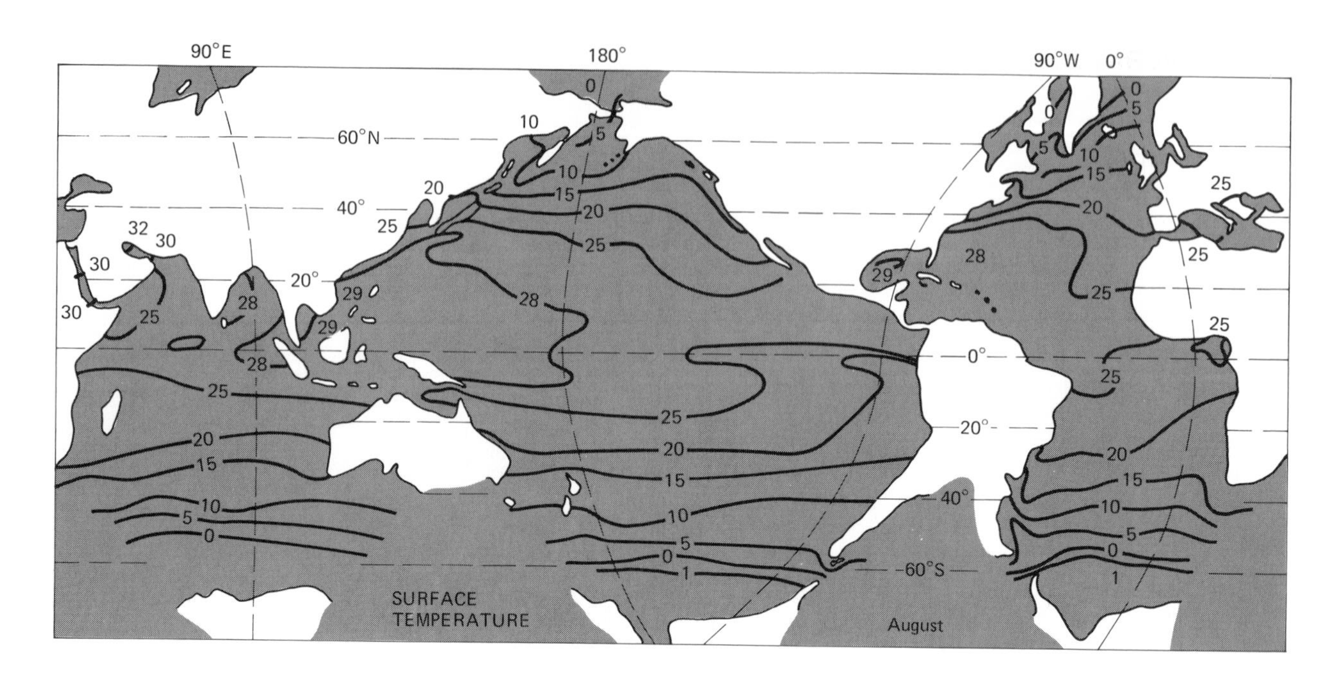

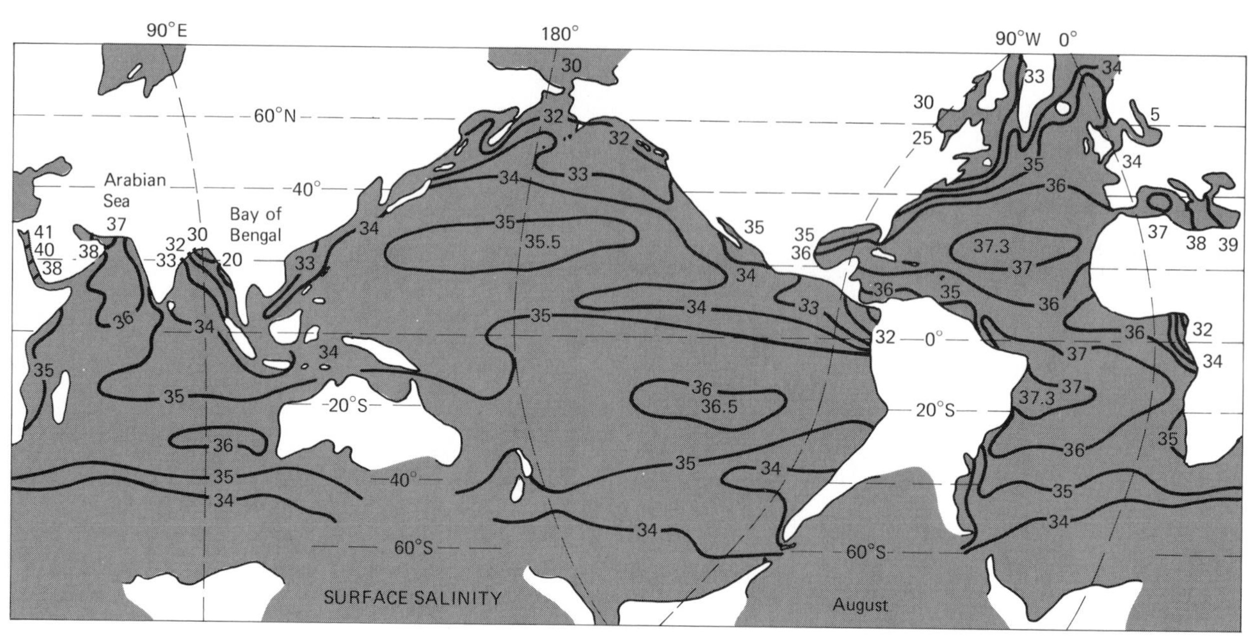

Figure 6-3. Surface temperature *(top)* and salinity *(bottom)* distribution in oceans in August. (*Source:* G. L. Pickard, 1963, *Physical Descriptive Oceanography*, Oxford, Pergamon.)

6-4 TEMPERATURE

The temperature is an important physical property of seawater. The temperatures of surface waters vary considerably throughout the world. Temperatures below the surface vary with depth, air circulation, turbulence, geographic location, and distance from heat-generating sources such as volcanoes. In general, temperatures in seawater vary from below −5°C to over 33°C (Figure 6-3). The freezing point of salt water is 1.9°C.

Oceans are a giant pump that transfers heat from the equator to the poles. Heat from the sun moves from lower latitudes to upper latitudes, where it is released to the atmosphere. This transfer is effected in the surface waters of the ocean by strong currents (for example, the Gulf Stream) that move warm tropical waters to the polar region (see Chapter 7). Deep waters (7500 meters) originate in upper latitudes. Temperatures in the oceans fall into three zones: (1) a surface (or mixed) layer that reflects the average temperature of that latitude; (2) a deep (or bottom) layer that reflects the origin of water in upper latitudes; and (3) a thermocline between 100 and 1500 meters deep in which temperature gradually decreases uniformly from high surface value to low deep value (see Figure 6-2).

The *thermocline* indicates a transfer of heat vertically from surface waters to deep waters as well as horizontially across waters. Although some of this transfer occurs by molecular heat diffusion, much of it is accomplished by small eddy currents that transport water vertically, thus mixing salinities as well as temperatures.

Thermocline *A layer of water in which rapid changes in temperature occur in the vertical column.*

6-5 DENSITY

Density *(D)* is defined as mass per unit volume. Mass (*M*) and volume (*V*) are expressed in grams and grams per cubic centimeters. The density of sea water varies from 1.02 to 1.07 grams per cubic centimeter and is dependent on temperature, salinity, and pressure or depth. The density of seawater is higher at lower temperatures, higher salinity, and greater depth. Colder deep waters, such as those of the Antarctic Ocean, are denser. Changes in the density of seawater are brought about largely by evaporation.

Density *Mass per unit volume of a substance. In the metric system expressed as grams per cubic centimeter (g/cc).*

Seawater forms layers, according to differing densities, in response to forces of gravity and buoyancy. In the top 100 meters, the density of seawater is governed largely by winds and waves; consequently it tends to be uniform. Below this depth, density is largely influenced by

Pycnocline *That portion of the ocean zone where water density increases rapidly in response to changes in temperature and salinity.*

temperature and salinity and is slightly different in the top 100 meters. Corresponding changes in temperature (thermocline), salinity (halocline), and density (pycnocline) are almost absent in deep waters. The density of seawater and its corresponding relation to temperature, salinity, and depth are shown in Figure 6-4.

6-6 LIGHT PENETRATION

Euphotic zone *The surface layer of the ocean where enough light is received to support photosynthesis. This zone is usually 80 to 100 meters below sea level.*

Aphotic zone *Deeper parts of the ocean where there is not enough light to permit photosynthesis by plants.*

Solar light can penetrate the oceans to a superficial depth of 100 to 200 meters. The well-lighted region of the *euphotic zone* exists between the surface layer and 80 meters; the poorly lighted region of the *disphotic zone* ranges from 80 to 200 meters; below 200 meters is the *aphotic zone,* the zone of perpetual darkness. Within the aphotic zone, a few luminescent fishes provide their own occasional light.

6-7 THE COLOR OF SEAWATER

Sea water seems to be bluish, because, within the spectrum of incident light, blue is the color scattered most by water molecules (Figure

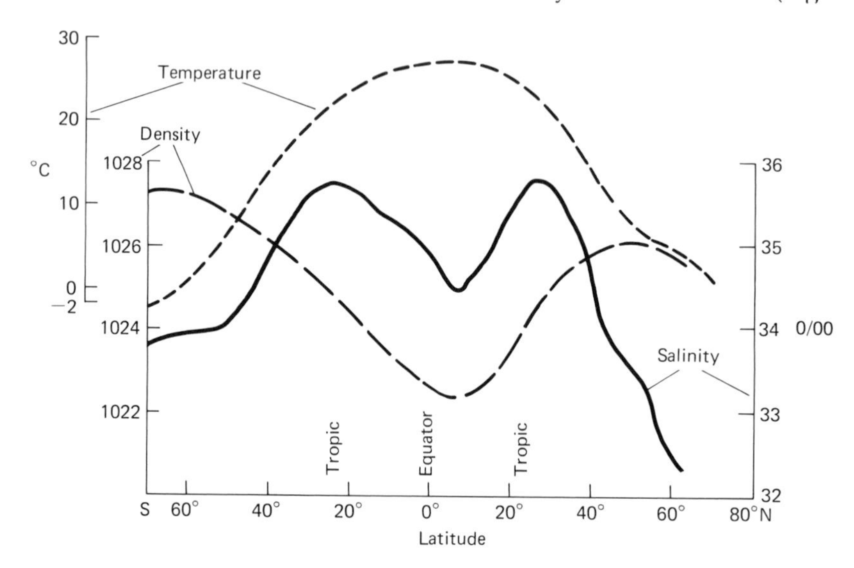

Figure 6-4. Average variation of salinity, temperature, and density at the sea surface. (*Source:* G. L. Pickard, 1963, *Physical Descriptive Oceanography*, Oxford, Pergamon.)

6-5). Consequently, more blue than other components of light is radiated from water under the surface. This phenomenon is called *selective scattering*. Furthermore, blue is the least absorbed color in pure seawater and thus is most readily reflected. Oceans are characterized by their deep blue appearance, particularly in tropical, subtropical, and middle-latitude regions. Oceans in upper latitudes are not quite so blue. When seas carry considerable organic material, such as living products and waste, a green color results. Sometimes seas may be densely populated with colored organisms. For example, the algae *Trichodesmium* imparts a red color to the Red Sea and gives this body of water its name.

6-8 SURFACE TENSION

The cohesion of molecules in liquids produces *surface tension*. Surface tension results when, because of molecular attraction, a thin film

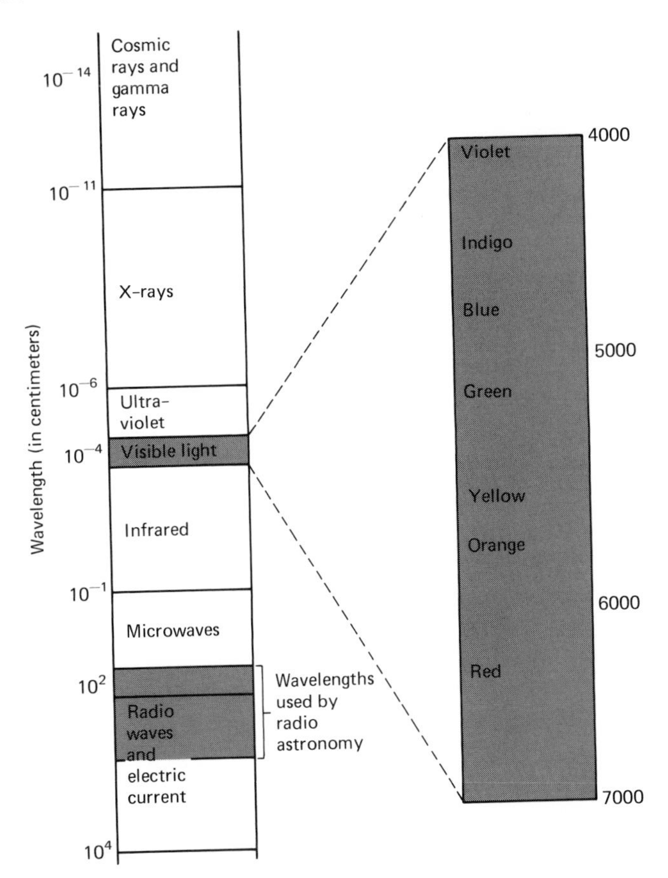

Figure 6-5. Electromagnetic radiation spectrum.

forms at the surface of water and behaves like a membrane. This membrane is always under tension and is always seeking to contract. Surface tension can be observed by floating a needle on water. If a steel needle is greased and carefully placed on the surface of water, it will float even though the density of the needle is greater than that of water. The needle will not break the surface because it is unable to pull apart the water molecules. Instead, it is depressed until the upward-moving buoyant force is equal in magnitude to the weight of the needle.

The surface tension of seawater is slightly greater than that of freshwater under the same temperature. The surface tension of freshwater at 20°C is 73 dynes per centimeter (Table 6-1).

6-9 VISCOSITY

Viscosity *A measure of a fluid's resistance to flow.*

If water and oil are permitted, under equal pressures, to flow through separate pipes, they will have different rates of flow. This difference is attributable to internal fluid resistance or *viscosity*. The viscosity of a fluid depends greatly on its chemical makeup and temperature and is less significantly influenced by salinity.

Viscosity is an important physical property of seawater because of its considerable influence on marine life and on the motions of water.

6-10 THERMAL PROPERTIES OF SEAWATER

Heat capacity *The amount of heat required to raise the temperature of a substance.*

Water, unlike most compounds, is unique for its ability to avoid rapid temperature changes. The ocean can absorb great amounts of solar heat during the day without undergoing drastic changes in temperature. Oceans have, in other words, high heat capacity.

TABLE 6-1 Principal Physical Properties of Freshwater

Property	H_2O_n
Molecular Weight	$(18)_n$
Density	1.0 g/cc
Boiling point	100°C
Melting point	0°C
Specific heat	1.0
Heat of evaporation	540 cal/g
Melting heat	79cal/g
Surface tension at 20°C	73 dyne/cm
Viscosity at 20°C	0.01 poise

As a result of the high latent heat energy produced by evaporation, oceans can supply an enormous quantity of thermal energy to the atmosphere in the form of vapor. On cooling, water vapor condenses as rain in an endless cycle.

Because of its high latent energy of fusion, seawater is never completely frozen except in some polar areas. Deep water for example, can maintain a flow of currents carrying nutrients for marine organisms during winter.

6-11 MEASUREMENTS OF SALINITY, TEMPERATURE, AND DEPTH

Physical oceanographers use Nansen bottles to obtain seawater samples for determining various physical properties as discussed in Figure 6-6 (see Chapter 10). The most commonly used instrument for the determination of salinity, temperature, and depth is the STD probe (Figure 6-7). This tool contains a sensitive unit equipped with an induction salinometer, an electronic thermometer, and a pressure transducer. The probe, once immersed, transmits continuous electronic pulses (through a cable linking it with the ship), which are recorded on a graph recorder. Oceanographers can thus obtain simultaneous knowledge of salinity, temperature, and depth. Temperature distribution in the ocean is also studied by bathythermographs, Nansen bottles (containing reverse thermometers and a salinometer), and electrical conductivity. Combined readings of salinity, temperature, and density are often obtained by using sophisticated instruments such as the *in situ* monitoring system (Figure 6-8). Simultaneous readings of salinity, temperature, and density are also taken by a network of buoys equipped with sensitive sensors (Figure 6-9).

Bathythermograph *An instrument that measures temperature at various depths in the ocean.*

In situ *In place or in position. From the Latin.*

SUMMARY

1. The physical and chemical properties of water are attributable to its structure. Water is composed of two atoms of hydrogen and one atom of oxygen, bonded covalently.
2. Conductivity is the capacity of seawater to transmit electric currents. Conductivity measurements help to determine salinity.

Figure 6-6. Nansen bottles are routinely used by oceanographers to obtain seawater samples for physical and chemical studies. (Photo courtesy of Woods Hole Oceanographic Institution.)

3. Salinity is defined as the number of grams of dissolved salts in 1000 grams of water. The salinity of seawater ranges from 20‰ near a mouth of a river to 40‰ in the Red Sea. The average salinity of seawater is 35‰.
4. Temperature varies widely in the ocean both vertically and horizontally. Temperatures in the sea range from –5°C to 33°C. The freezing point of salt water is not at 0°C but at 1.9°C.
5. A thermocline between 100 and 1500 meters is characterized by generally decreasing temperatures.
6. Density is defined as mass per unit volume. The average density of sea water is between 1.02 and 1.07 grams per cubic centimeter.
7. A blue color predominates in oceans, because it is scattered more effectively than other colors by water molecules.

Figure 6-7. STD (salinity-temperature-depth) probe. (Photo courtesy of Woods Hole Oceanographic Institution.)

8. Surface tension involves cohesion of molecules, which form a thin film at the surface of water and behave much like a membrane.
9. Viscosity involves differential rate of flow of a liquid because of internal force of resistance.

Suggestions for Further Reading

Defant, A. 1972. *Physical Oceanography,* 2 volumes. New York: Pergamon.

Pickard, G. L. 1975. *Descriptive Physical Oceanography,* second edition. London: Pergamon.

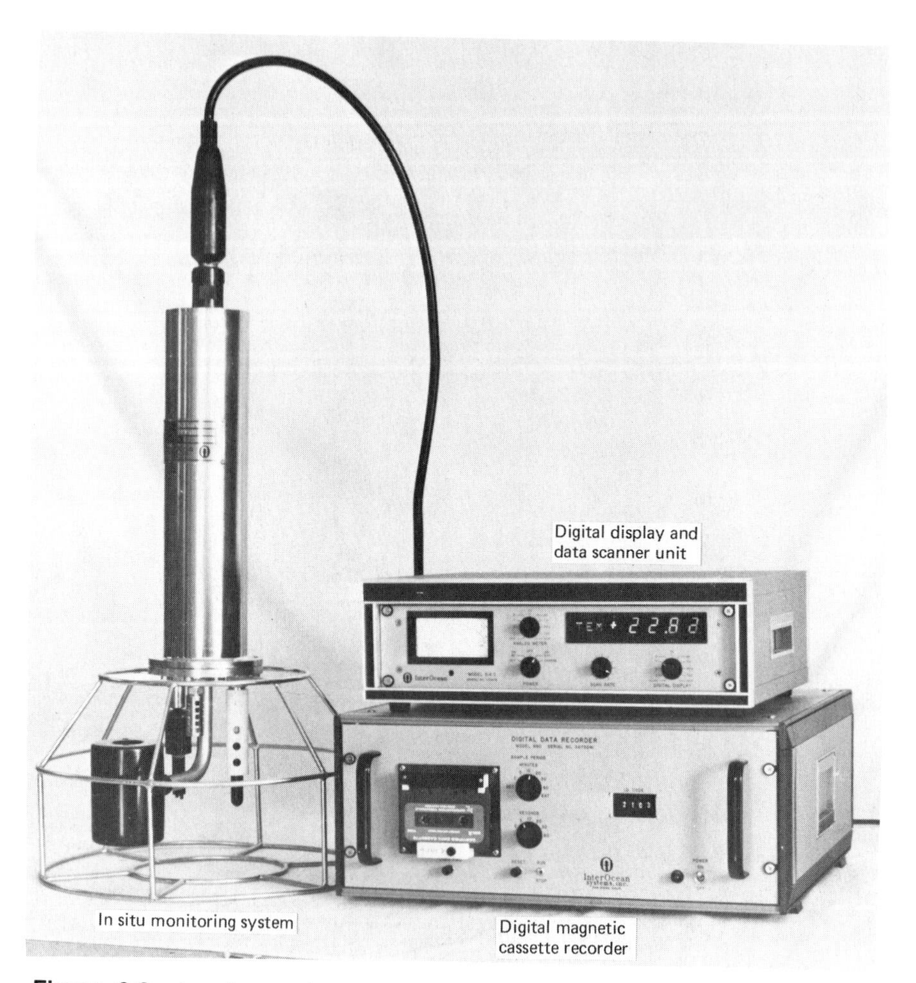

Figure 6-8. *In situ* monitoring system for oceanic environmental analyses. (Photo courtesy of Inter-Ocean Systems, Inc.)

Spar, J. 1973. *Earth, Sea and Air: A Survey of the Geophysical Sciences.* Reading, Massachusetts: Addison-Wesley.

Sverdrup, H. U., M. W. Johnson, and R. K. Fleming. 1942. *Oceans: Their Physics, Chemistry, and General Biology.* Englewood Cliffs, New Jersey: Prentice-Hall.

Von Arx, W. S. 1975. *An Introduction to Physical Oceanography.* Reading, Massachusetts: Addison-Wesley.

Williams, Jerome. 1973. *Oceanographic Instrumentation.* Annapolis, Maryland: Naval Institute Press.

Figure 6-9. A buoy collects and transmits oceanographic data concerning: temperature, pressure, and salinity at various depths; waves; surface current; and surface winds and related parameters. (Photo courtesy of Woods Hole Oceanographic Institution.)

7
Ocean Circulation

7-1 INTRODUCTION

LARGE-SCALE MOVEMENTS OF CURRENTS IN THE OCEANS are termed *oceanic circulation.* Oceanic circulation is largely induced by the sun's heat that is received on the earth's surface. Solar heat interacts with surface water to generate two major types of circulation: (1) *wind-driven circulation,* in which the atmosphere plays a vital role and which is restricted to the upper 100 meters, and (2) *thermohaline-driven circulation,* for which the ocean, less efficient than the atmosphere, acts as a heat engine, and which is operative at a greater depth. Oceanic circulation is also affected by the earth's rotation, by gravitational attraction of lunar and solar forces, and by differences in pressure levels.

7-2 THE CORIOLIS EFFECT

In 1884, Gaspard Gustave de Coriolis, a French scientist, observed that because of the earth's rotation any moving fluid is deflected to the right in the Northern Hemisphere and to the left in the Southern Hemisphere. This phenomenon is the *Coriolis effect* (Figure 7-1). As equatorial air rises from the surface of the ocean and moves toward the poles, it exhibits the Coriolis effect. No Coriolis effect is present when an air mass follows its path along the equator.

Coriolis force *A force produced as a consequence of earth's rotation. It causes objects in motion to be deflected to the right in the Northern Hemisphere and to the left in the Southern Hemisphere.*

Although the Coriolis effect is best observed in the atmospheric pattern over the rotating earth, other moving objects such as fired bullets, running automobiles, or trains in motion display it to a lesser degree.

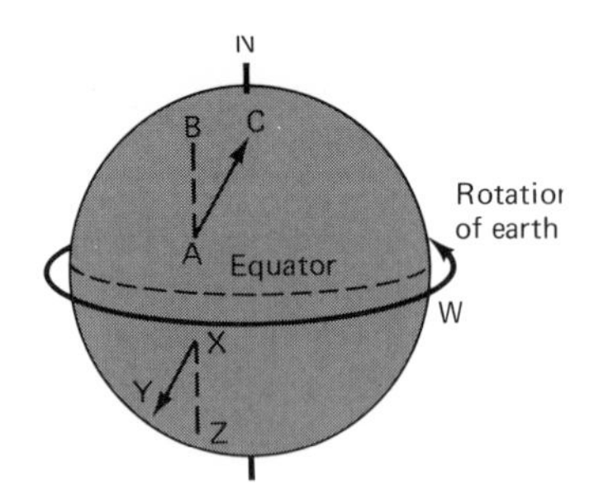

Figure 7-1. Coriolis force is produced as a result of the earth's rotation. A moving object is deflected to the right in the Northern Hemisphere (AC) and to the left in the Southern Hemisphere (XY). On a stationary earth these objects would move without deflecting their respective paths (AB and XZ).

7-3 THE EKMAN SPIRAL AND EKMAN TRANSPORT

V. Walfrid Ekman, a Swedish scientist, in 1902 demonstrated mathematically that under ideal ocean conditions a systematic decrease in current speed and a change in its direction at increasing depths occurred. The resulting spiral is the *Ekman spiral* (Figure 7-2). According to Ekman, under ideal conditions, surface water moves 45 degrees to the right of the wind in the Northern Hemisphere, and 45 degrees in the opposite direction in the Southern Hemisphere.

Under natural conditions, however, the Ekman spiral does not operate in its predicted theoretical fashion. Oceans are not in uniform states, and a single wind impact is not so prolonged. The net motion of the entire mass of moving water flowing at right angles to wind direction is called *Ekman transport*. The *Ekman layer* marks a depth of 100 meters; there current and frictional forces defining the Ekman spiral are quite active. Below this depth these forces are virtually nonexistent.

7-4 GEOSTROPHIC CURRENTS

Geostrophic current *A current that emerges from the earth's rotation and is the product of a balance between Coriolis effect and gravitational force.*

Wind-generated surface currents of the ocean are not only subject to the Ekman spiral and to the Coriolis force but to the force of gravity as

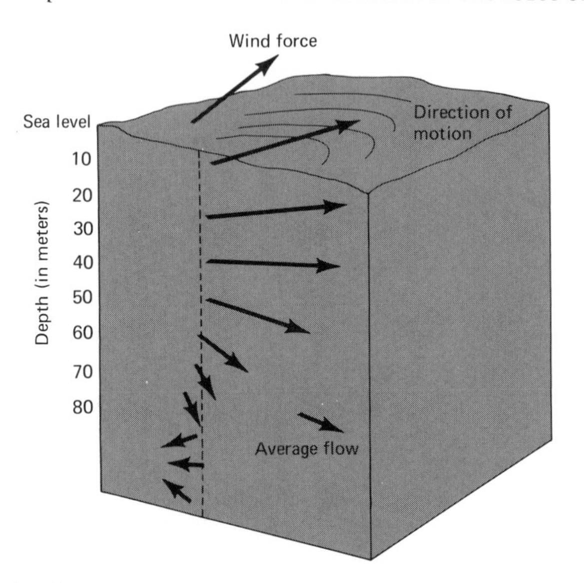

Figure 7-2. An Ekman spiral is formed when a steady wind blows over an ocean of deep depth and of uniform viscosity. As a result, change occurs in the direction of flow and the decrease in speed with increased depth.

Ekman spiral *A theoretical model to explain the influence of a steady wind dragging over an ocean of unlimited depth and breadth and of uniform viscosity. As a consequence, a surface flow at 45° to the right of the wind is formed in the Northern Hemisphere. Water at increasing depth will drift in directions increasingly to the right until, at about 100 meters depth, the water is moving in a direction opposite to that of the wind. The net water transport is 90° to the wind, and velocity decreases with depth.*

well. When a current is produced as a result of the near balancing of the Coriolis effect and gravitational forces, it is called a *geostrophic current* (Figure 7-3). Major surface currents of the ocean are now regarded by oceanographers to be geostrophic.

Essentially, winds blowing over surface waters will push the water, piling it up in the direction that the wind is moving. This forms so-called "hills." The hills set up high-pressure areas near their highest points and low-pressure areas away from their peaks. The pressure difference so developed generates a gravitational force that tends to move water from the high- to the low-pressure areas. However, as the water moves, it is deflected by the Coriolis force. As a result, the moving water shifts its direction until the two forces—Coriolis and gravitational—balance each other.

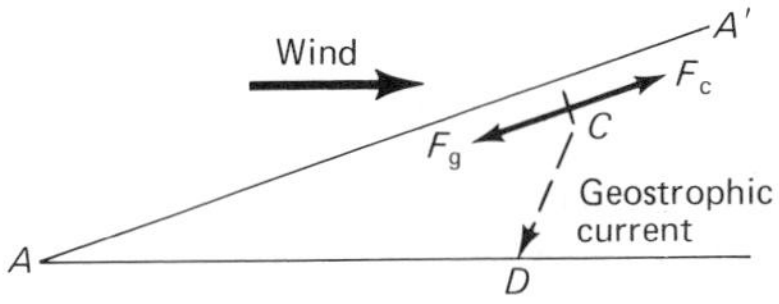

Figure 7-3. Schematic representation of the geostrophic current for the Northern Hemisphere. The gently sloping surface of the ocean is produced in the direction of prevailing winds (AA′). "Piled-up" water flows downhill in response to gravity (F_g), but the direction of flow is deflected by the Coriolis effect (F_c). The direction of flow continues to change until these forces (F_g and F_c) are balanced. The resultant flow of water is called a geostrophic current (CD).

7-5 LANGMUIR CIRCULATION

Wind drag is the chief driving force of currents at the upper layer of the ocean. Irving Langmuir first observed that strong winds moving at high speed often produce convention cells with alternate right- and left-hand circulation patterns known as *Langmuir cells* (Figure 7-4). The long axes of these cells run parallel to the wind direction. Langmuir cells are significant with respect to transportation of heat, dissolved gases, and nutrients. They also play an important role in intermixing of surface waters.

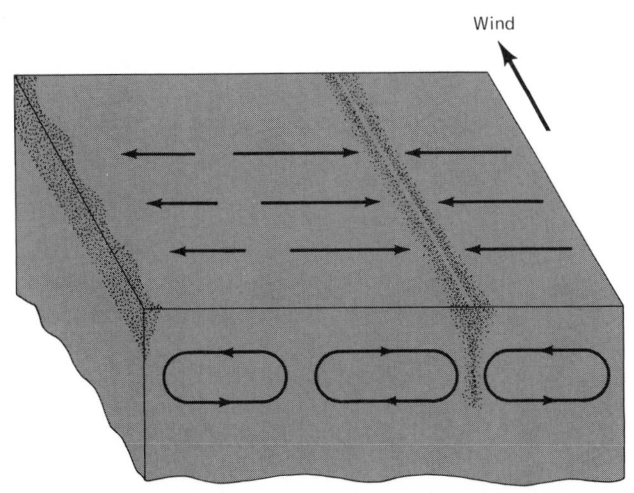

Figure 7-4. Schematic representation of Langmuir cells. Steady winds blowing across the ocean surface produce a cellular pattern with alternate right- and left-hand circulation.

Sargassum *Drifting seaweed found in the Sargasso Sea.*

Langmuir cells have alternate convergences and divergences. Convergence can be noted aerially. For example, floating seaweed such as *Sargassum* in the Sargasso Sea easily show a convergent sea pattern. Divergent cells are sandwiched between two convergences.

The Sargasso Sea, located in the North Atlantic Ocean, is a unique feature in the sea world. In the past it was regarded as a realm of navigational hazards, because it is a large eddy. Its boundary is demarcated by major current systems such as the Gulf Stream and the Canary currents. The ring of these currents combine to develop a powerful eddy that occupies approximately 5 million square kilometers and rotates clockwise in response to earth's motion. Oceanographers call this great eddy of surface water the Sargasso Sea.

7-6 WIND-DRIVEN CIRCULATION

The mechanics of wind-driven circulation may best be described by considering atmospheric circulation patterns. First envision a simple model of atmospheric circulation in which there are no land barriers and no rotation of the earth. The sun's heat is received unevenly on the earth's surface and is relatively greater per unit area at the equator than at the poles. This unbalanced heat budget tends to adjust itself by passing additional heat from equatorial to polar regions. The transfer of heat from the equator leaves a low-pressure area behind. Relatively cold air in the vicinity of the equator immediately flows in to fill the gap. But this air mass also expands and rises and is once again uplifted. In contrast, at the poles, as the air is cooled, it subsequently contracts and sinks, forming a high-pressure area. Polar air then flows back toward the equator. In this fashion, an ideal atmospheric (wind) system between the poles and the equator is set up (Figure 7-5).

Gyre *A circular spiral motion of water.*

In reality, however, the wind system is more complex, mainly because of land barriers and the rotation of the earth. Because of both obstacles, wind-driven circulation gives rise to large circular patterns in the oceans called *gyres* (or rings) (Figure 7-6). Gyres correspond to major wind patterns; for example, there is a counterclockwise gyre in the subtropical zone south of the equator. But because much of the Southern Hemisphere, particularly in upper latitudes, is predominantly a water hemisphere, the currents there do not flow in gyres but simply circle the globe. The Antarctic Circumpolar Current is the striking example of this motion.

Major gyres are now recognized by oceanographers. Each gyre has its own characteristic range of temperature, salinity, dissolved oxygen, phosphate content, and organic activity (Figure 7-7).

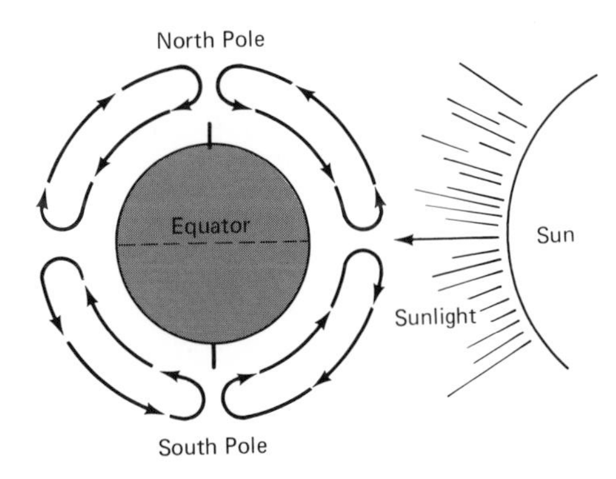

Figure 7-5. Idealized air circulation pattern if the earth had no land barriers and no rotation. The Equator receives the greatest heating; polar regions the least heating.

Equatorial Drift

Trade winds drive surface waters slowly toward the equator and at the same time westward in both hemispheres. This generates the famous *equatorial drift* of westward-moving water. In the middle latitudes of both hemispheres, westerly winds produce a similar eastward drift. The wind-driven currents include the Gulf Stream in the North Atlantic, the Kuroshio in the North Pacific, the Brazil Current in the South Atlantic, and the Agulhas in the Indian. The Antarctic circumpolar Current, moved by gusty west winds around 60°S, flows around the globe, forming a ring. As a result, a continuous westwind drift girdles the Antarctic Ocean (Figure 7-8).

The Gulf Stream

"There is a river in the sea—the Gulf Stream," so declared Lt. Matthew Fontaine Maury in his 1855 classic, *Physical Geography of the Sea*.

The Gulf Stream currents are made of three principal currents, namely the Florida Current, the Gulf Stream Current, and the North Atlantic Current. The Florida Current is pushed by northern equatorial currents in a clockwise direction and passes through the narrow straits between Cuba and Florida. These currents moved at about 9 kilometers per hour and transport 20 to 40 million cubic meters of water every second. When the Florida Current leaves the Gulf of Mexico and moves northward along the continental shelf, it is called the Gulf Stream Current. After the Gulf Stream Current passes through the Grand Banks near Newfoundland, it is known as the North Atlantic Current. As the North Atlantic Current approaches Europe, it branches; one part flows north to become the Norwegian Current, and the other moves south

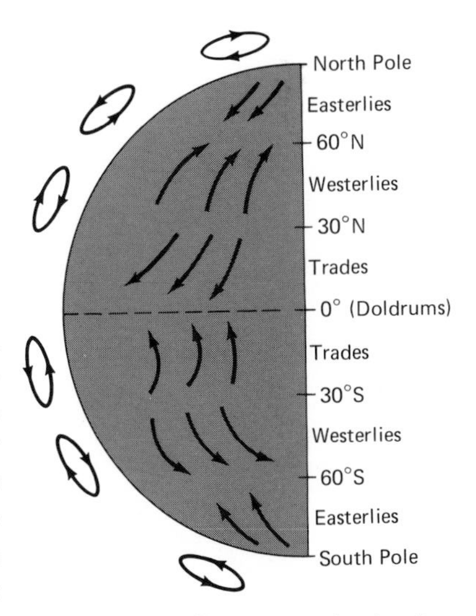

Figure 7-6. Planetary air circulation and winds.

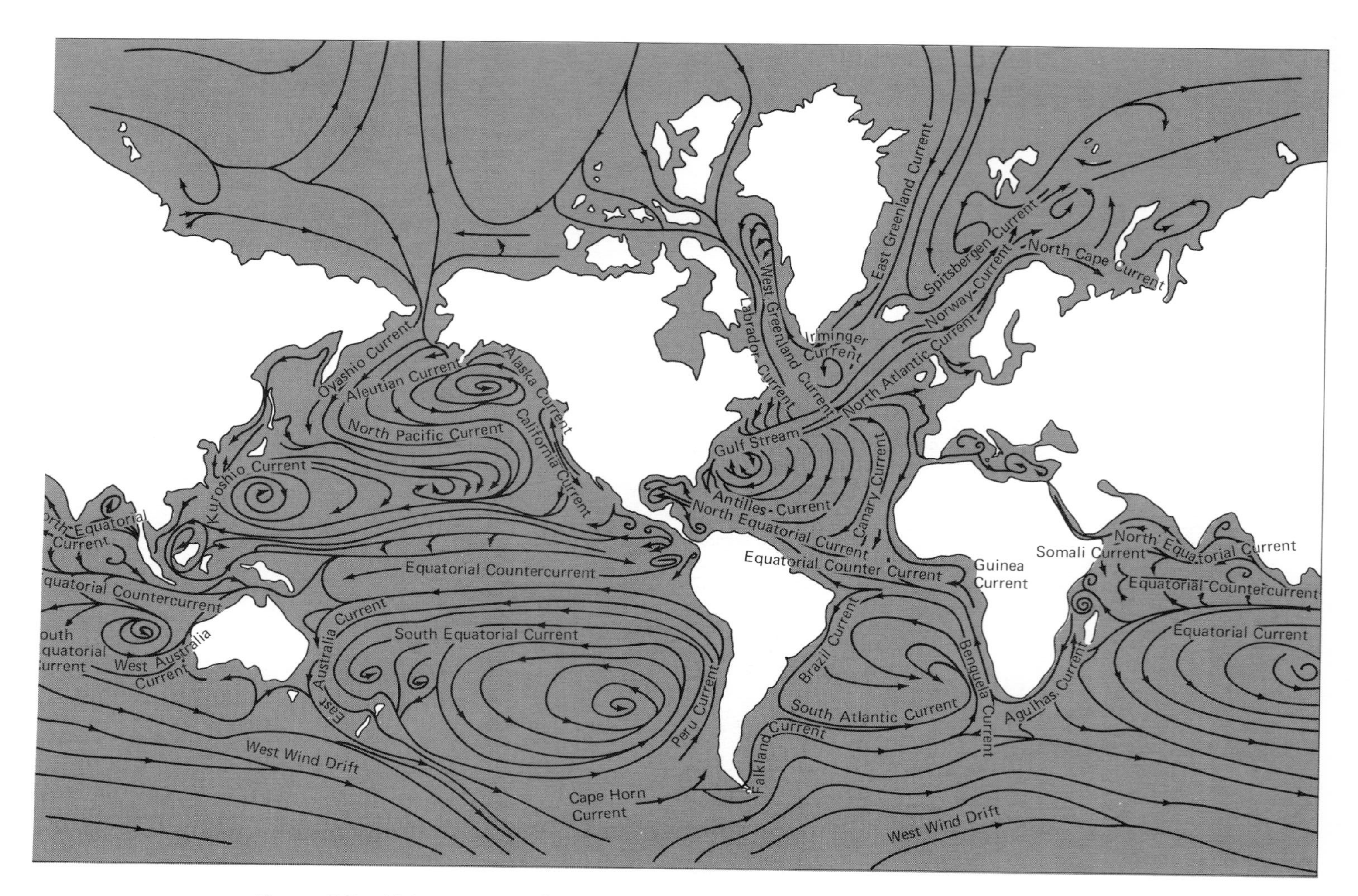

Figure 7-7. Major ocean surface currents. Currents flow in clockwise direction in the Northern Hemisphere and vice versa in the Southern Hemisphere (see Figure 7-1).

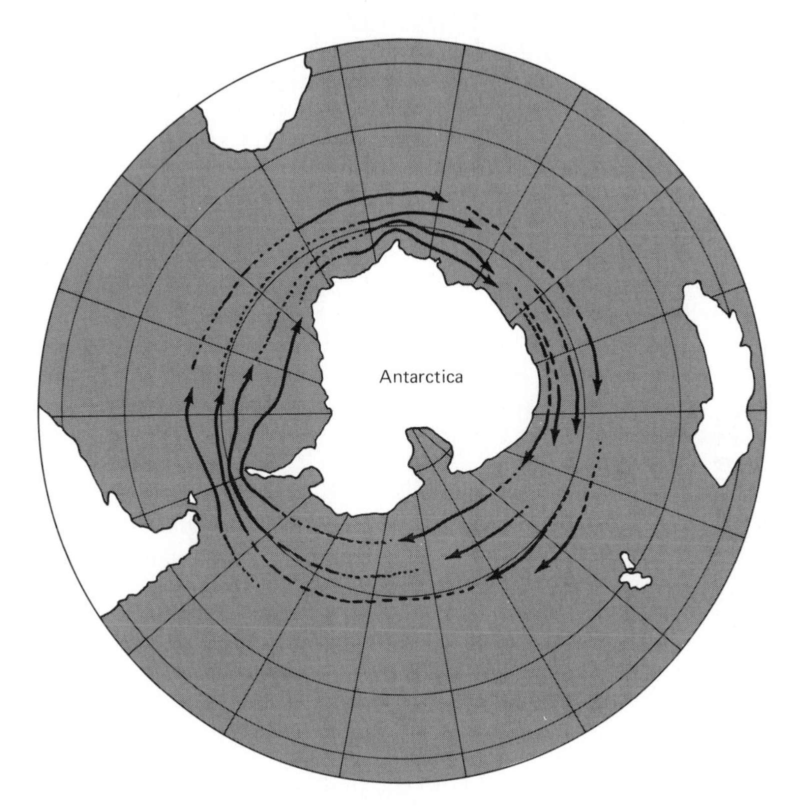

Figure 7-8. The Antarctic circumpolar current.

along the coast of Great Britain. There warm water exerts considerable influence on the climate of the European coast, particularly on that of Britain and of Scandinavia. The southern-flowing current is now known as the Canary Current; it moves along the European coast to the tradewind belt and is then driven back to the Atlantic by equatorial drift. In this way, a clockwise gyre is completed.

7-7 OCEANWATER MASSES

A body of water with a particular range of temperature and salinity is called a *water type* and occupies different depths according to its distinctive temperature, salt content, and content of dissolved gases such as oxygen and carbon dioxide. Changes in temperature and salinity occur at the boundary of a water body both through mixing with surrounding waters and through the earth's generation of heat at the bottom of the ocean. Two or more water types compose a larger body of water with distinctive temperatures and salinities. This is a water mass.

A *water mass* is a large, uniform body of water with a definite range of temperature and salinity values. The major water masses of oceans, for example, include Antarctic bottom water (salinity, 34.66‰; temperature, −0.4°C), North Atlantic intermediate water (salinity, 34.73‰; temperature, 2.2°C to 3.5°C) and Indian central water (salinity, 34.60‰ to 35.50‰; temperature, 8° to 15°C).

Water masses in oceans are classified according to two factors: the depth at which they reach vertical equilibrium and the geographical source region. In general, water masses are classified as surface, central, intermediate, deep, and bottom. The upper 100 meters of the ocean is covered by the *surface water mass;* from 100 meters to the base of the main thermocline (200 meters) is the *central water mass;* from below the central water mass to about 3000 meters is the *intermediate water mass; deep* and *bottom water masses* occupy the greatest depths of the ocean basins (Figure 7-9).

The water mass is in continuous motion. A radioactive tracer study of water masses shows that 750 years have passed since the Antarctic bottom water in the Atlantic was at the surface, but the same water mass is determined to be 1500 years old in the Pacific.

7-8 THERMOHALINE CIRCULATION

Thermohaline circulation *Vertical circulation of ocean water brought about by density differences resulting from temperature and salinity variations.*

Thermohaline circulation is brought about by cold, dense water. That is, it is a flow induced mainly by the varying temperature and salt content of seawater. The dense cold water developed in the Arctic and Antarctic Seas sinks to great depths. Variations in temperature and salinity are brought about by chilling, evaporation, freezing, precipita-

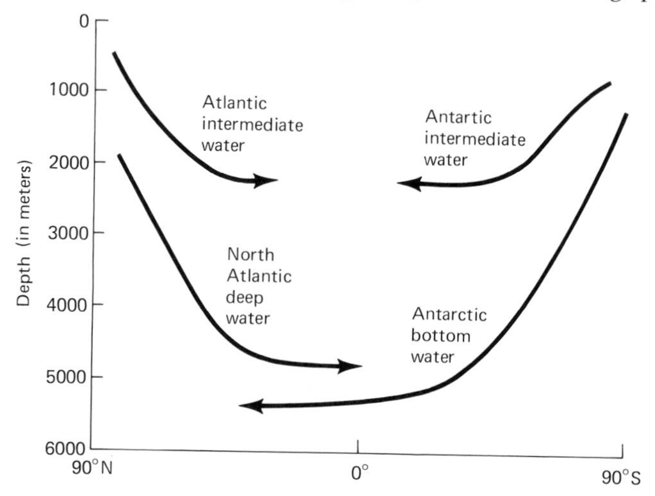

Figure 7-9. Principal subsurface water masses in the Atlantic Ocean.

tion, and melting. Major types of thermohaline-driven currents follow (also see Table 7-1).

Antarctic Bottom Water

In the vicinity of Antarctica, especially in the Weddell Sea, waters are exposed to great cold during the winter and consequently yield low-temperature and high-salinity water masses. High salinity is a result of ice formation.* Cold, saline waters sink and flow along the ocean bottom toward the equator. These are Antarctic bottom waters, which are measured as far north as 45°N. The Antarctic bottom water mass is also subject to eastward flow around the Antarctic continent because of the depth impact of the west-wind surface drift. As this drift mixes with the bottom-water mass, it gives rise to a new homogeneous water mass, the Antarctic circumpolar water. The deeper reaches of the Antarctic circumpolar water mass furnish deep water to the Indian and South Pacific oceans, and some of it circumnavigates Antarctica.

TABLE 7-1 Characteristics of Some Water Masses

Mass	*Ocean of Origin*	*Depth (in meters)*	*Salinity (⁰/₀₀)*	*Temperature (°C)*
Antarctic bottom	South Atlantic (Weddell Sea)	4000 to bottom	34.66	−0.4
Antarctic circumpolar	South Atlantic	100–4000	34.68–34.70	0.5
Antarctic intermediate	South Atlantic	500–1000	33.8	2.2
South Atlantic central	South Atlantic	100–300	34.65–36.00	6–18
Arctic deep and bottom	North Atlantic	1300–4000 (deep) 1300–bottom (bottom)	34.90–34.97	2.2–3.5
North Atlantic intermediate	North Atlantic	300–1000	34.73	4–8
North Atlantic central	North Atlantic	100–500	35.10–36.70	8–19
Mediterranean	—	1400–1600	37.75	13
Pacific equatorial	Central Pacific	200–1000	34.60–35.15	8–15
Indian central	Indian	100–500	34.60–35.50	8–15
Red Sea	—	2900–3100	40.00–41.00	18
Black Sea	—	0–200	16.00	various

SOURCE: J. Williams et al., 1968, *Air and Sea: The Naval Environment*, Annapolis, Naval Institution Press.

*Formation of ice on the icebergs causes the elimination of their salt. As a result, icebergs can float, and, at the same time, the residual salt increases the salinity of the surrounding waters.

Arctic Deep and Bottom Waters

Atlantic deep water and Atlantic bottom water masses are produced off the coast of Greenland. These waters are also called the North Atlantic deep and bottom waters. Relatively less dense than Antarctic bottom water; they flow over the latter into the South Atlantic as far as 60°S.

The Mediterranean Density Current

The Mediterranean density current arises because of the difference in the density of waters between the Mediterranean Sea and the adjacent Atlantic Ocean (Figure 7-10). The Mediterranean is a shallow and selectively enclosed sea which is exposed to hot and dry climatic conditions. As a result, the salinity of its surface water increases to about 38.6‰, and it thus becomes denser. Density differences in waters cause them to intermingle. At the Strait of Gibralter, over the rock sill, dense Mediterranean water (salinity over 38‰) flows into the Atlantic as bottom water. The waters of the Atlantic (having salinity less than 36‰) move into the Mediterranean as surface water. The Mediterranean density current carries over 2 million cubic meters per second into the Atlantic.

The Mediterranean current continues its journey down the continental slope to a depth of about 2000 meters. At this depth, it en-

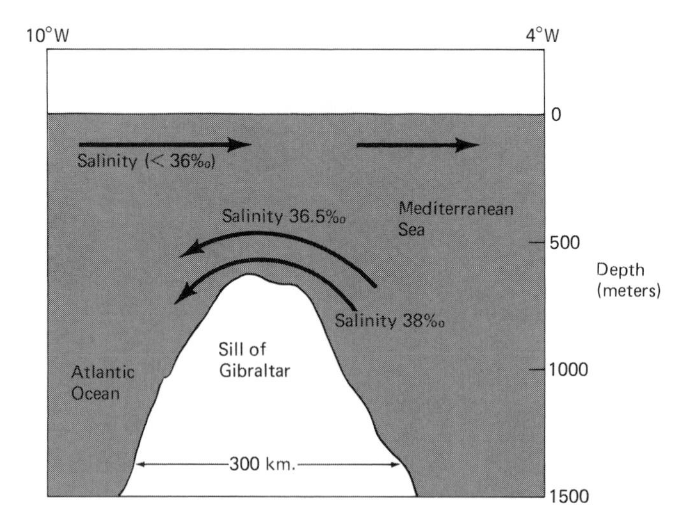

Figure 7-10. Vertical distribution of salinity and temperature in west-east cross section through the Strait of Gibraltar at 36°N. Arrows indicate main direction of spreading of the water. Vertical scale exaggerated 200X. (*Source:* Dietrich, Günter, 1963, *General Oceanography*, New York, Wiley.)

counters Atlantic deep water of relatively higher density. The Mediterranean current thus floats *on* the Atlantic water until it becomes well intermixed.

A similar exchange of water occurs on a smaller scale between Mediterranean Sea and the Black Sea. The high-density waters of the Mediterranean flow as bottom currents into the Black Sea; there they sink to constitute a stagnant water body nearly devoid of oxygen.

7-9 MEASUREMENTS OF CURRENTS

Speed and direction of ocean currents are basic data in which oceanographers are constantly interested. In the past, information on currents was obtained through small bottles that were usually either tagged or contained a card indicating when and where they were placed in the water. Finders were asked to fill out and return the cards with information regarding when and where they picked up the bottles (Figure 7-11).

The shallow float, a sealed aluminum tube, is used to measure currents beneath the surface. Often the swallow float is equipped with a buoy and attached to a sounding device. In this manner movements at given depths can be monitored. The buoy provides a reference point to the moving ship and to the connected swallow float. The float can detect feeble movements of deep-ocean currents, which may be a few centimeters per second. Two other devices commonly used for the measurement of currents are the Ekman current meter and the Savonius rotor current meter. A current meter prepared for launch as part of bouy group mooring is shown in Figure 7-12.

SUMMARY

1. The movement of ocean currents is called the oceanic circulation and is caused by wind, density differences of seawater, the earth's rotation, and gravitational forces.
2. The Coriolis force is induced by the earth's rotation. According to the Coriolis effect, any moving fluid tends in the Northern Hemisphere to be deflected to the right and in the Southern Hemisphere to the left.
3. The Ekman spiral is a theoretical consideration that predicts a systematic decrease in current speed and change in its direction at increasing depth.

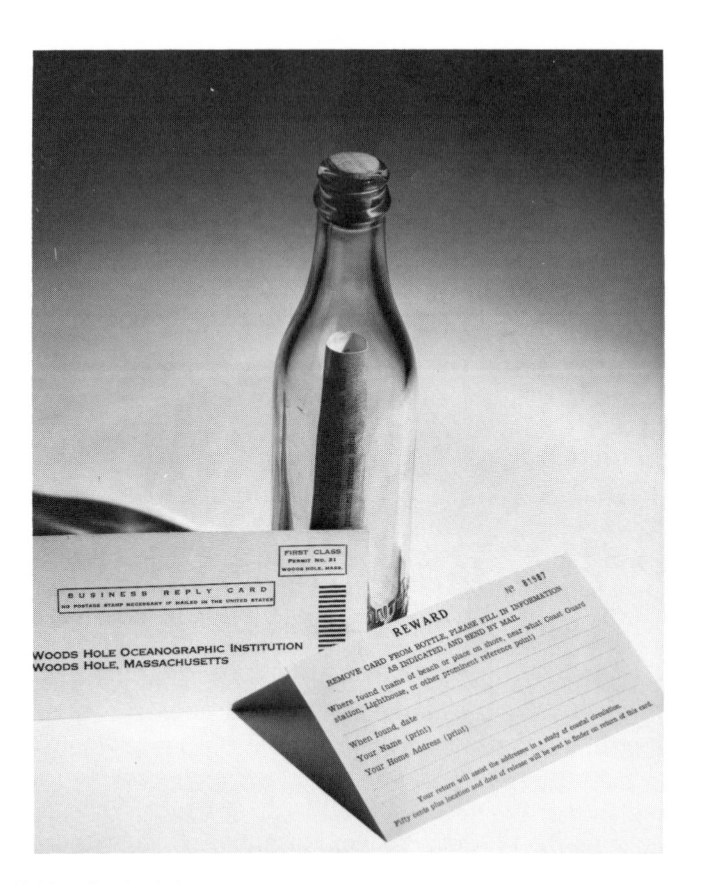

Figure 7-11. Basic information on ocean currents is sometimes obtained by dropping tagged bottles into oceans. Finders are asked to return the bottle (with necessary information as shown). (Photo courtesy of Woods Hole Oceanographic Institution.)

4. Geostrophic currents are generated as a consequence to the balancing between the horizontal pressure gradient and the Coriolis effect. Most ocean currents are geostrophic.
5. Langmuir circulation is cellular circulation set in motion by strong and prolonged winds that blow in one direction. It is characterized by vertical spirals running parallel to the wind direction in alternately clockwise and counterclockwise directions.
6. Water masses have separate ranges of temperature, salinity, and chemical nature.
7. Wind-driven currents generated as a result of differences in water pressure in response to winds, density variations in the sea, and changes in the slope of the sea surface induced by winds.
8. Wind-produced currents are common in the surface water of the ocean. Examples of the surface currents include the Gulf Stream in the Atlantic, the Kuroshio Current in the Pacific, and the Agulhas

Figure 7-12. Launching a telemetering buoy from the *R/V Chain*. This buoy is used to trace currents and to return data via radio to a surface ship. (Photo courtesy of Woods Hole Oceanographic Institution.)

Current in the Indian. The west-wind drift is a classic example of surface currents in the Antarctic Ocean.

9. Thermohaline currents are produced primarily by the differences in the densities of water in the ocean.
10. Oceanographers measure currents by using many methods and instruments such as small floating bottles, the swallow float, and current meters.

Suggestions for Further Reading

Gaskell, T. F. 1972. *The Gulf Steam*. London: Cassell.

Kinsman, Blair. 1965. *Wind Waves: Their Generation and Propagation on the Ocean Surface*. Englewood Cliffs, New Jersey: Prentice-Hall.

Munk, W. 1955. "The Circulation of the Oceans." *Scientific American* (September).

Smith, F. G. W. 1973. *The Sea in Motion*. New York: Crowell.

Turekian, K. K. 1976. *Oceans*, second edition., Englewood Cliffs, New Jersey.

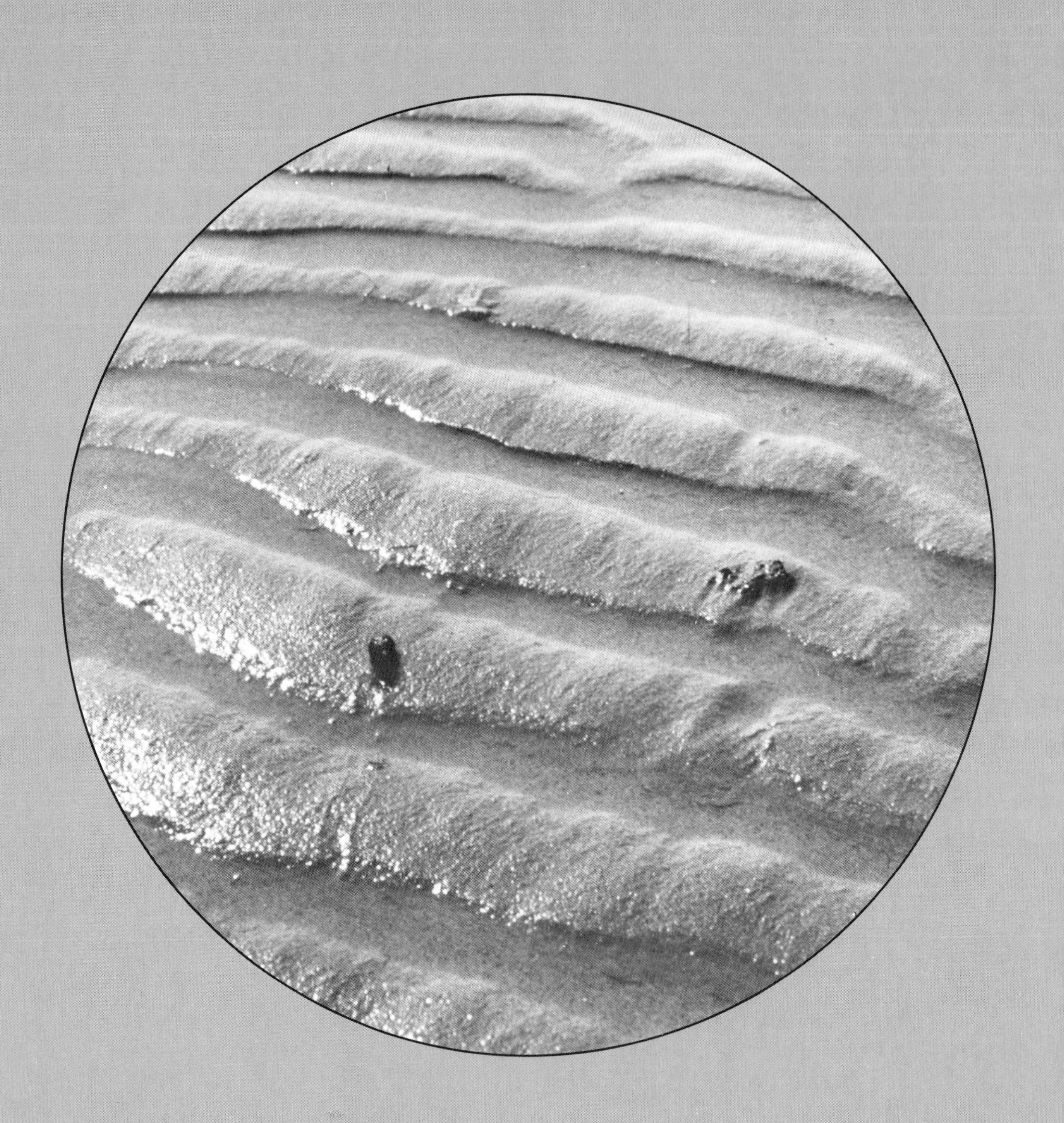

Ocean Waves and Tides

8-1 INTRODUCTION

WAVES ARE A MOST CONSPICUOUS FEATURE OF THE planet Ocean. Their sheer size and vigor have always impressed artists, poets, and wave-watchers. The scientific study of the waves began in the early nineteenth century when Franz Gerstner, a German scientist, proposed to explain the phenomenon of waves. According to him, water particles in a wave move in circular orbits. In 1825, Ernst and Wilhelm Weber, in making experimental observations of a wave tank, concluded that waves are reflected without loss of energy. In the twentieth century, oceanographers such as Harold U. Sverdrup and Walter Munk of Scripps Institution of Oceanography undertook a serious and detailed study of waves in order to predict wave and surf movements for naval operations during World War II. This sort of knowledge is necessary for offshore petroleum explorations, marine mining, marine engineering, and the planning and development of coastal areas.

Wave *An oscillatory movement in a body of water manifested by an alternate rise and fall of the surface.*

8-2 WAVE PARAMETERS

An ideal wave train can be described in terms of its (1) *period* (the time it takes two successive crests to pass a fixed point), (2) *wavelength* (the distance between two consecutive crests), and (3) *height* (the vertical distance between a trough and a crest). The speed of a moving wave can be determined as follows:

Wave period *The time for a wave crest to traverse a distance equal to one wavelength.*

$$\text{speed of wave } (C) = \frac{\text{wavelength } (L)}{\text{period } (T)}$$

Major components of a typical wave are depicted in (Figure 8-1).

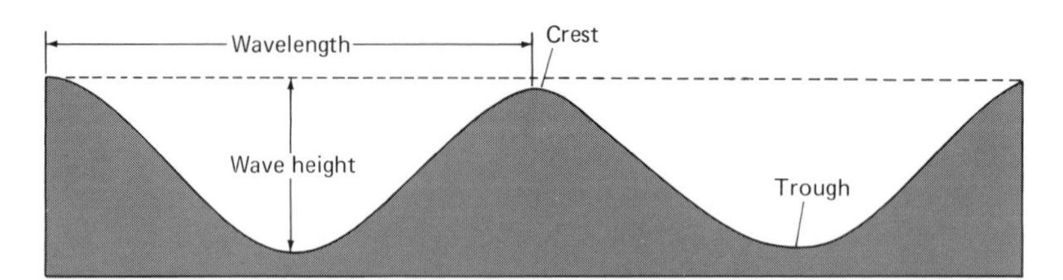

Figure 8-1. Wave parameters.

8-3 MECHANISM OF WAVE FORMATION

The exact mechanism of wave formation is complex, and much of the knowledge of wave formation is theoretical. When blowing winds impart their energy to the water in the form of friction and pressure on the smooth surface of the sea, waves are formed. The effectiveness of wind in generating waves depends on three factors: (1) its average speed, which determines its force, (2) its duration; and (3) the extent of open water across which it blows (the *fetch*). When gusty winds blow for a long time and cover large extents of the open water, waves of great height (sometimes up to 20 meters) can result.

Waves are measured by such electronic recording devices as pressure transducers. A *pressure transducer* is a pressure-sensing device equipped with a sensitive strain gauge (or potentiometer) that records on a metal diaphragm the slightest change in pressure caused by wave energy and which subsequently transmits it as an electronic pulse. The intensity of waves is reflected by the strength of these electrical pulses.

The distinction between the motion of *wave form* and the motion of *water mass* is important. Waves are carriers of energy imparted to them by wind; water masses are not. In deep water, wave forms continue to move forward; but water masses (or the water particles) are, except for a slight amount of forward movement, essentially stationary. When a wave is in deep water, the motion of individual particles at the surface follows a circular orbital pattern and the orbital radius falls off quickly with depth. For example, at a depth equal to one-half the wavelength, the orbital radius is reduced to 4 percent of its surface value. As a result, the water motion gyrates to and fro instead of circularly, and the speed of the water particles decreases rapidly with depth (Figure 8-2).

Beaufort Scale *A numerical scale for the estimation of wind force, based on its effect on common objects.*

This mechanism can be illustrated by placing a tennis ball on a water surface. When a wind-produced wave passes by, the ball will follow a circular orbital movement, bouncing up and down without moving forward. Another ball just below the surface of the water will behave in the same manner but will have a smaller radius to its circular orbit. As a measure of the influence of wind on the general state of the sea, oceanographers use the *Beaufort Scale* (Table 8–1). The Beaufort

Scale is an arbitary numerical scale involving the general estimation of wind force based on its effects on the intensity and heights of the waves produced.

8-4 WIND-PRODUCED WAVES

Three types of wind-generated waves are: sea, swell, and surf. Oceans are characterized by the simultaneous occurrence of several trains of *sea waves* of differing wavelengths and directional movements, which result overall in an irregular and chaotic wave pattern (Figure 8-3). Because seas are so complex and variable, unusual statistical methods must be employed to describe their parameters. Height, wavelengths, and periods can be quantitatively measured. Sea waves of over 12 meters are commonly observed near an area immediately under wind impact.

Sea *Generally chaotic waves produced by wind.*

As waves move away from the winds that ruffle them, they assume a uniform pattern and begin to move in trains of equivalent period and height. These trains are called *swells* (Figure 8-4). Swells can travel thousands of miles. As swells approach shore, their pattern is modified by shallower water. Consequently they are shortened and steepened, and as their particle velocity in the crest increases, they move more swiftly. This modified form of swell is called *surf* (Figure 8-5). Surf is quite different from sea and swell waves in one important respect: it does not have the circular motion of water particles. Thus wave trains are broken

Swell *Long-period waves (as opposed to short-period waves that are characteristic of a storm).*

Surf *The breaking waves in a coastal region.*

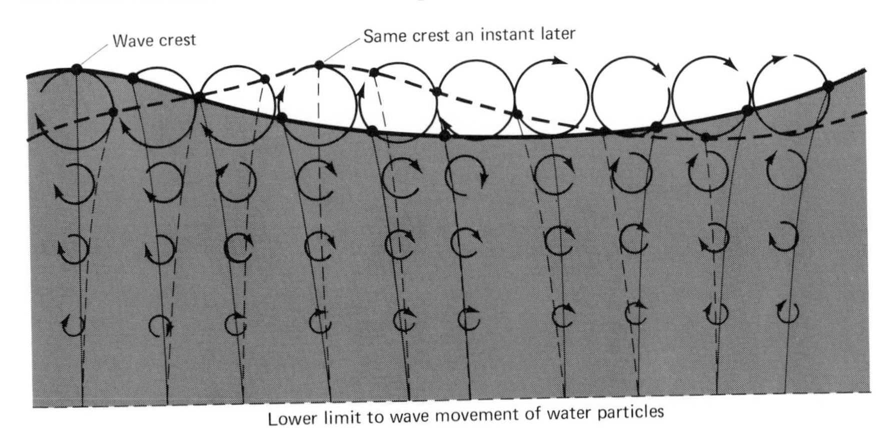

Figure 8-2. Cross section of a seawave in motion. Circles represent the orbit of water particles in the wave. At the surface the orbital diameter described by a water particle is equal to the wave height; at a depth of .5 wavelength the orbit of a water particle is only .04 that of a particle at the surface.

TABLE 8-1 Beaufort Wind Scale and Sea State Chart

Sea State	*Beaufort Number*	*Terminology*	*Wind Speed (miles per hour)*	*Average Wave Height (feet)*	*Description*
0	0	Calm	0–1	0	Sea like a mirror.
1	1	Light air	1–3	0.05	Ripples with appearance of scales; no crests.
2	2	Light breeze	4–7	0.18	Small wavelets; crests have glossy appearance but do not break.
3	3	Gentle breeze	8–12	0.19–1	Large wavelets, crests begin to break; foam of glossy appearance; a few white horses.
	4	Moderate breeze	13–18	2–3	Small waves becoming longer, fairly frequent white horses.
4	5	Fresh breeze	19–24	4–5	Moderate waves taking longer form; many white horses formed; some spray.
	6	Strong breeze	25–31	6–10	Large waves begin to form: many white crests; more spray.
5	7	Moderate gale	32–38	11–16	Sea heaps up; white foam from breaking waves begins to be blown in streaks.
	8	Fresh gale	39–46	17–28	Moderately high waves of greater length; edges of crests begin to break into spindrift; foam blown in well marked streaks; spray affects visibility.
6	9	Strong gale (storm)	47–54	29–40	High waves; dense streaks of foam along wind direction; sea begins to roll; visibility affected.
7	10	Whole gale (heavy storm)	55–63	41–59	Very high waves with long overhanging crests; sea takes white appearance; rolling heavy; visibility affected.

TABLE 8-1 Beaufort Wind Scale and Sea State Chart (Cont.)

Sea State	*Beaufort Number*	*Terminology*	*Wind Speed (miles per hour)*	*Average Wave Height (feet)*	*Description*
8	11	Violent storm	64–72	60–73	Exceptionally high waves; small and medium-sized ships lost from view; crests blown into froth.
9	12	Hurricane	73–82	74–80	Air filled with foam and spray; sea completely white with driving spray; visibility very seriously affected.

SOURCE: From NOAA, *Diving Manual*, 1975.

up to become a foam line, an agitated body of aerated water (Figure 8-6). As waves approach shore, they begin to draw nearer the bottom. At that point the depth of water is about one-half the wavelength [Figure 8-6 (bottom)]. Moreover, the length and height of the waves increase, and their speed is reduced. That is, the shallow bottom exerts significant influence upon the waves. In particular, the wavefronts bend or refract according to the bottom water topography (see Section 8-5). When the water becomes extremely shallow, these approaching but modified waves break.

Figure 8-3. Seas (or seawaves) in the open ocean. (Photo courtesy of R. B. Theroux, National Marine Fisheries Service.)

Figure 8-4. Swells are uniform waves that can travel thousands of miles from one ocean to another. (Photo courtesy of R. B. Theroux, National Marine Fisheries Service.)

Longshore current *A nearshore current that flows parallel to the shore.*

Once waves break, water is moved into the surf and ultimately to the beach at an angle parallel to it. This parallel movement of water constitutes a *longshore current* between breakers and the shoreline (Figure 8-7). When breaking waves approach a straight shoreline, water flows parallel to the shore. But in the case of an irregular shoreline, water moves perpendicularly seaward (particularly at the mouth of a bay) and produces a narrow, swift current known as a *rip current* (Figure 8-7). Rip currents may be dangerous for swimmers. If a swimmer is caught in a rip current zone, he should swim across it for a short distance instead of against it. The presence of rip currents along a shore is detected by streaks of water flowing seaward. The rip currents are different from *undertow*. Undertows are waves returning to the sea primarily as a bottom flow, and they are relatively less efficient in returning water seaward than are the rip currents.

8-5 WAVE REFRACTION

Wave refraction *The process by which the direction of a train of waves moving in shallow water at an angle to the contours is changed. The part of the wave train advancing in shallower water moves more slowly than that part still advancing in deeper water, causing the wave crests to bend toward alignment with the water contours.*

Waves approaching shore no longer move perpendicularly to it but instead move in a parallel direction. This happens because the shoreward face of the wave moves more slowly than the seaward side, eventually causing *refraction* (Figure 8-8). Wave refraction has been studied in great detail because of its influence on the distribution of wave energy. Knowledge of wave energy is vital for coastal engineering

Figure 8-5. Breaking waves (surf) approaching shore. Note angle of approach. (Photo courtesy of S. McGinn.)

projects and architectural planning, including the design of buildings and highways along the shore areas.

The distribution of wave energy in the nearshore region can be computed by using *orthogonal* technique. *Orthogonals** (or perpendicular waves) converge on headlands to concentrate wave energy and spread

*Orthogonals are arbitary lines drawn at right angles to wave crests, enabling one to visualize refraction or bending clearly.

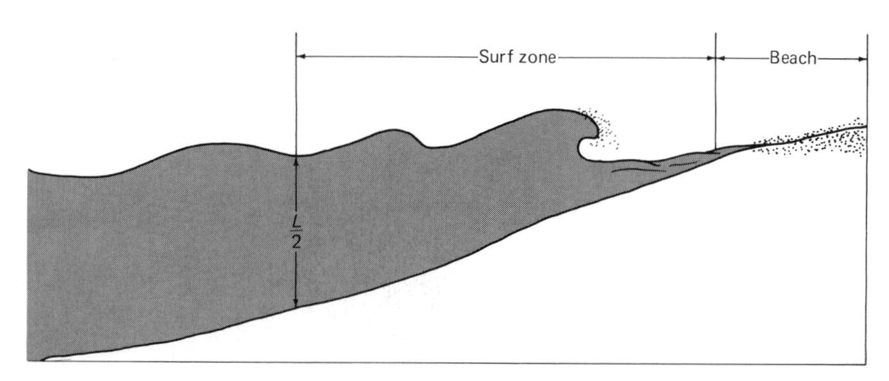

Figure 8-6. *Top:* Surf forming foamline, an agitated body of aerated water near shore. (Photo courtesy of Woods Hole Oceanographic Institution.) *Bottom:* Diagrammatic representation.

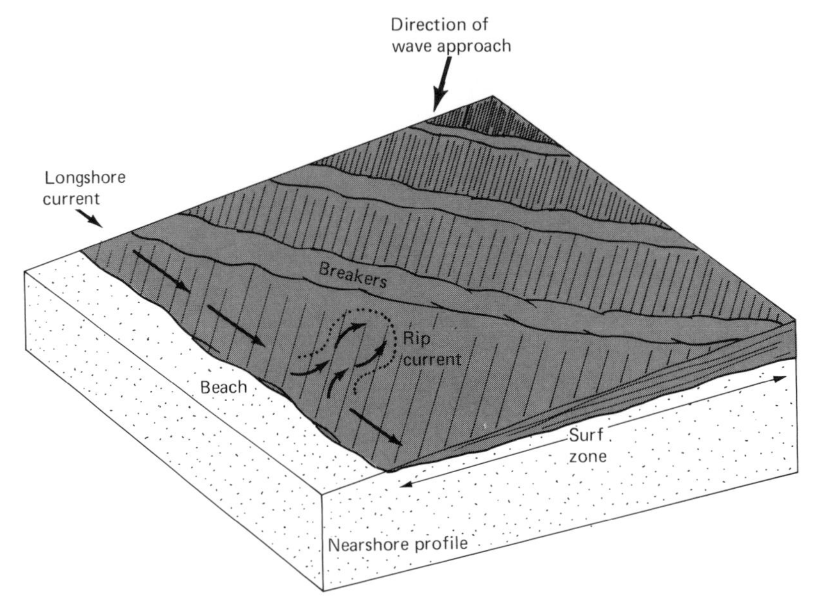

Figure 8-7. Longshore currents run parallel to the shore. Rip currents bounce back to the sea.

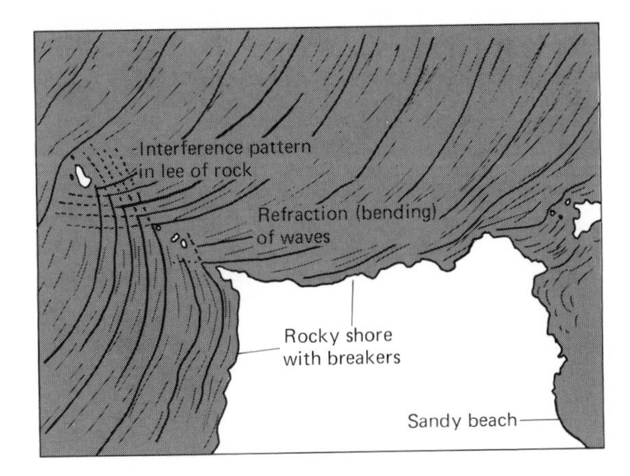

Figure 8-8. Wave refraction.

in embayments and bays to dilute energy. When convergence occurs in headlands, the rate of destruction by waves is quite effective. When divergence takes place, such as in bays, destruction by waves is much less extensive, but they do receive eroded material from headlands (Figure 8-9).

8-6 TYPES OF SEAWAVES

Catastrophic Waves

Catastrophic waves are produced with the sporadic occurrence of volcanoes, earthquakes, or landslides in the ocean. These waves are

Catastrophic waves *Sudden violent and temporary waves caused by earthquakes, volcanic activity, etc.*

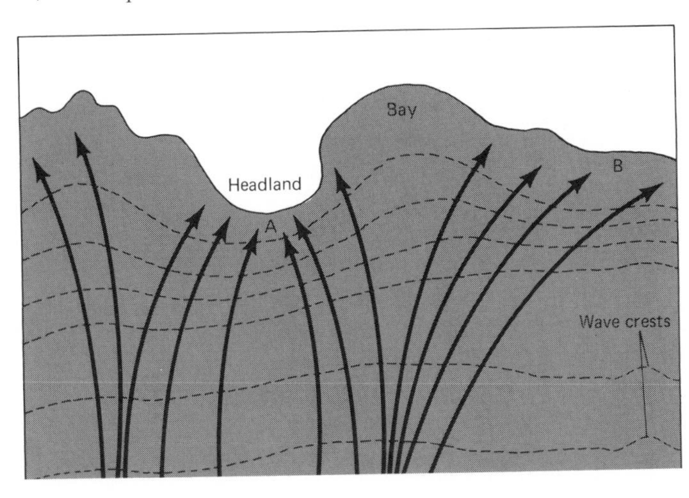

Figure 8-9. Wave-energy distribution concentrates in the headlands and diversifies in bays.

generated less frequently than are wind-generated waves, but they can have a far more damaging impact on coastal areas, especially in densely populated regions. Three common types of catastrophic waves are: tsunamis, landslide surges, and storm surges.

Tsunamis *Long-period sea waves produced by an earthquake or volcanic eruption.*

Tsunamis (seismic sea waves) are triggered by earthquakes or volcanic eruptions beneath the ocean floor. These waves have wavelengths of approximately 160 kilometers and a speed of 600 to 700 kilometers per hour. In the open ocean, they have heights of 3 to 6 meters but on approaching shore become even higher. Their high speed gives them great momentum and large destructive power. Most tsunamis are in the Pacific Ocean and are not so common in the Atlantic and Indian oceans. The Hawaiian Islands experience a tsunami every 25 years. The Japanese name *tsunami* was adopted by oceanographers in favor of the earlier misnomer *tidal wave*.

When Krakatoa volcano erupted in the Pacific in 1883, it triggered tsunamis that moved at about 500 kilometers per hour and reached peaks of 30 meters. The coastal regions of Java and Sumatra were severely devastated by these waves. In 1896, a Japanese tsunami killed over 25,000 people and destroyed 10,000 homes. During 1964, tsunamis that resulted from the Alaska earthquake engulfed several coastal areas, particularly Kodiak Island in Alaska, the western coasts of Canada, and coastal areas as far away as Japan.

A tsunami often affects an area with a radius of more than 150 kilometers; in exceptional circumstances it can have a wavelength as high as 1000 kilometers. Because a tsunami is really a series of waves, usually separated by intervals of 15 minutes to an hour or more, when a tsunami hits a coast, it comes in waves after waves for hours and hours. In most cases, the third to eighth waves are the most dangerous; they are the largest. In recent years, a network of early warning systems has made it possible to reduce the destructive impact of tsunamis.

Storm Waves

Storm surge *Unusually high water waves resulting from strong wind action.*

Storm waves, or surges, are generated following a combination of abnormal meteorological and oceanic conditions. These high-water level waves are produced, for example, when gusty winds accompany hurricanes, high tides, and exceptionally low atmospheric pressure. Storm waves can cause great damage to coastal areas. In 1900 a storm wave of 7 meters struck the city of Galveston, Texas, and drowned several thousand people.

Internal Waves

Internal wave *A wave that forms at the boundary of two water layers having different densities.*

Internal waves are found below the surface of the sea (Figure 8-10). Although these are similar to ordinary wind-generated surface waves,

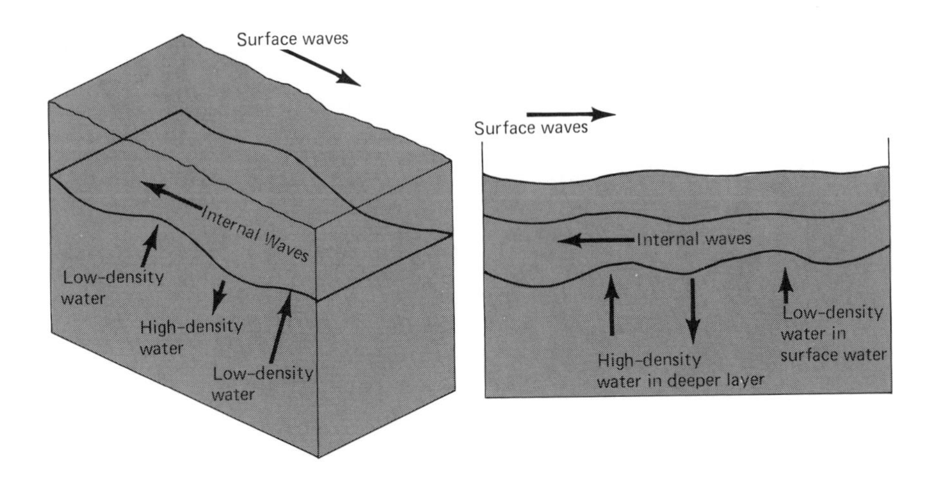

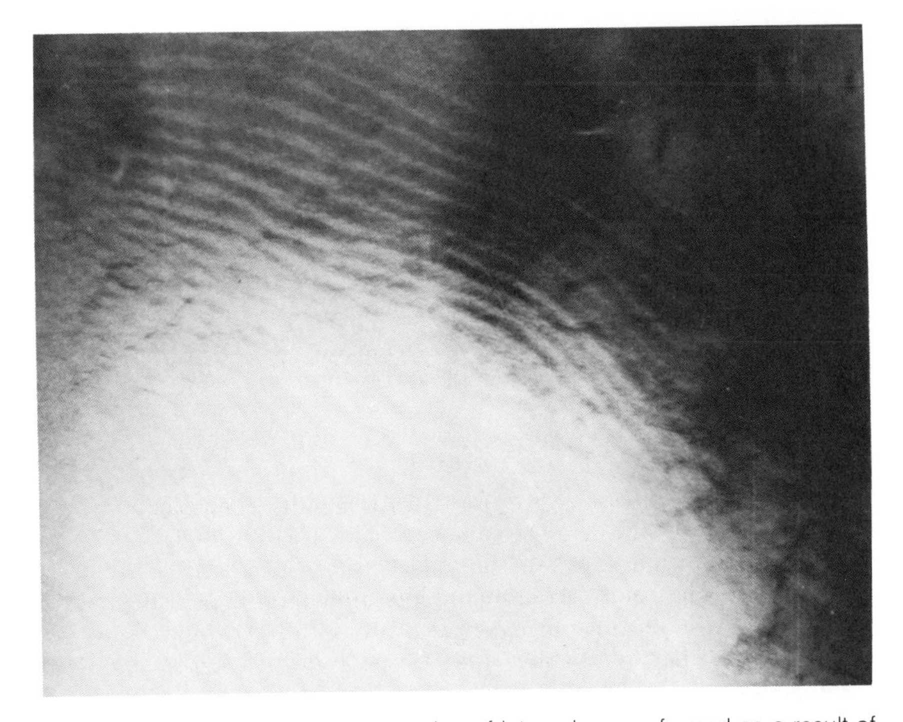

Figure 8-10. Schematic representation of internal waves formed as a result of differences in water densities. Arrows indicate wave direction. *Top left:* Block diagram. *Top right:* Cross section. *Bottom:* Surface slicks caused by internal waves appear in the reflection of the sun in this outstanding U-2 photograph taken at 21,000 meters at the head of Hudson Canyon. (Photo courtesy of National Oceanographic Data Center.)

they differ in certain respects. For example, internal waves move so slowly that they are visibly undetectable, and they commonly occur only where density variation exists. These waves have a greater height than surface waves. Because of their very slow movement and long extent, internal waves can be detected only by mechanical observation, for example, by sensitive temperature recordings. When dense internal waves rise, they form *slicks* on the ocean surface. A slick can be distinguished from surrounding waters by its abundance of fine sediments or plankton.

Stationary Waves or Seiches

Seiche *A stationary wave.*

Stationary or *standing waves* are characterized by the absence of forward movement; instead the water surface moves up and down. These are also called *seiches.* Seiches are common in enclosed water bodies such as bays and lakes. These waves are produced by storms that are associated with drastic atmospheric conditions. Stationary waves can be devastating in terms of loss of life and property damage.

8-7 TIDES

Tide *The periodic rise and fall of the planetary ocean level in response to the gravitational interaction of the earth, moon, and sun.*

The periodic rise and fall of sea level that results from the gravitational attraction exerted on the earth by the moon, and to a lesser extent by the gravitational pull of the sun, is the *tide.* Tides are complex because of: (1) the movements of the moon in relation to the earth's equator, (2) changes in positions of the moon and the sun with respect to the earth, (3) uneven distribution of the water on the earth's surface, and (4) irregularities in the configuration of ocean basins. Because of these variables, tides vary. For example, in the Bay of Fundy, between New Brunswick and Nova Scotia, they are over 12 meters high, but along the coast of the Mediterranean Sea, they are virtually unnoticeable.

Measurements of tides can be made by using an ordinary graduated pole anchored to the ocean bottom (Figures 8-11 and 8-12). Periodically at, say, one-hour intervals, readings of the water level on the pole are recorded. The height of the water at given interval will determine the construction of the *tidal curve.* At some U.S. tide stations, sophisticated tidal gauges are employed to record continuous tidal data (Figure 8-13). Most tidal curves record two high tides and two low tides per *tidal day,* occurring every 24 hours and 50 minutes (Figure 8-14). The time, 12 hours and 25 minutes, between high or low tides is called the *tidal period.* The combination of two high and two low tides is called semidaily or *semidiurnal* tides. When there is only one high tide and one low tide daily, they are called daily or *diurnal* tides. Often, the tidal curves record two

Diurnal tides *Tides with one high and one low water mark during a tidal day.*

Figure 8-11. High- and low-water levels recorded by a tide-gauging station. (Photo courtesy of National Oceanographic Data Center.)

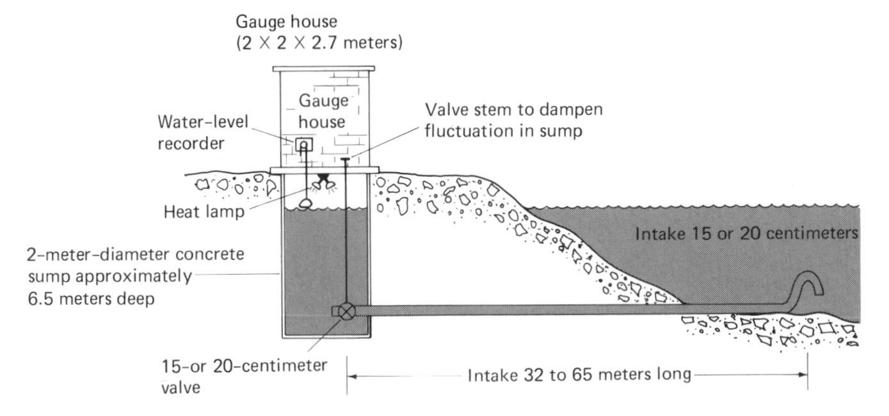

Figure 8-12. Diagrammatic representation of a typical water-level gauging station. (*Source:* National Oceanographic Data Center.)

Figure 8-13. A tide gauge is used to monitor tides in lake, estuarine, harbor, and ocean environments. The instrument can be attached to existing structures or can be lowered directly to the bottom. Its accuracy is independent of water depth. (Photo courtesy of Bass Engineering, Inc.)

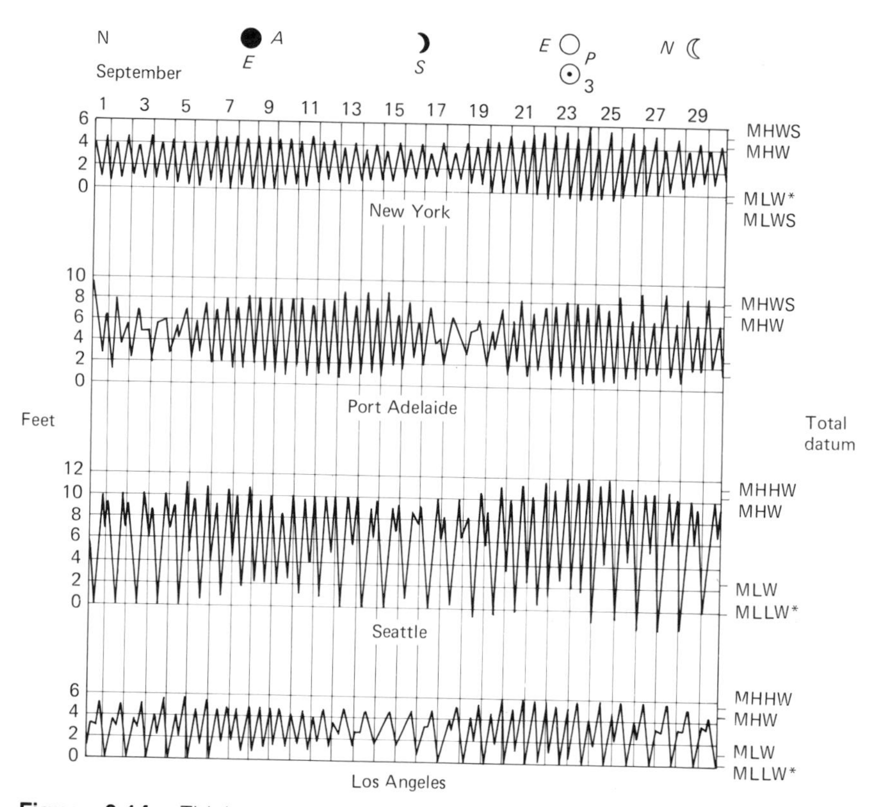

Figure 8-14. Tidal curves for selected coastal cities. (*Source:* Nathaniel Bowditch, *American Practical Navigation,* rev. ed., H. O. Publ. 9, Washington, D.C.)

high tides and two low tides per tidal day, but if there are differences between two high tides and between two low tides, *mixed tides* result. Mixed tides are abbreviated HHW (higher high water) and LHW (lower high water). Similarly, low and lower marks are expressed as LLW (lower low water) and HLW (higher low water). Tide-predicting machines are used at many locations in the United States for tidal forecast purposes (Figure 8-15). This information is of great value to fishermen, sailors, marine miners, and divers.

Tides are at their maximum when the moon and the sun are in the same plane as the earth. These *spring tides* occur every 14 days, at new and full moons. When the moon and the sun are at right angles to each other, low tides occur. These *neap tides* occur every 14 days, always at half moon (Figure 8-16).

Diurnal and semidiurnal tides occur because of the gravitational attractions of the earth, the moon, and the sun. The moon, because of its closeness to the earth, is the strongest influence on tides. The tide-producing force of the moon is twice as strong as that of the sun. The moon takes 29.53 earth days to complete one revolution around the earth. During the course of this revolution, the earth and the moon are gravita-

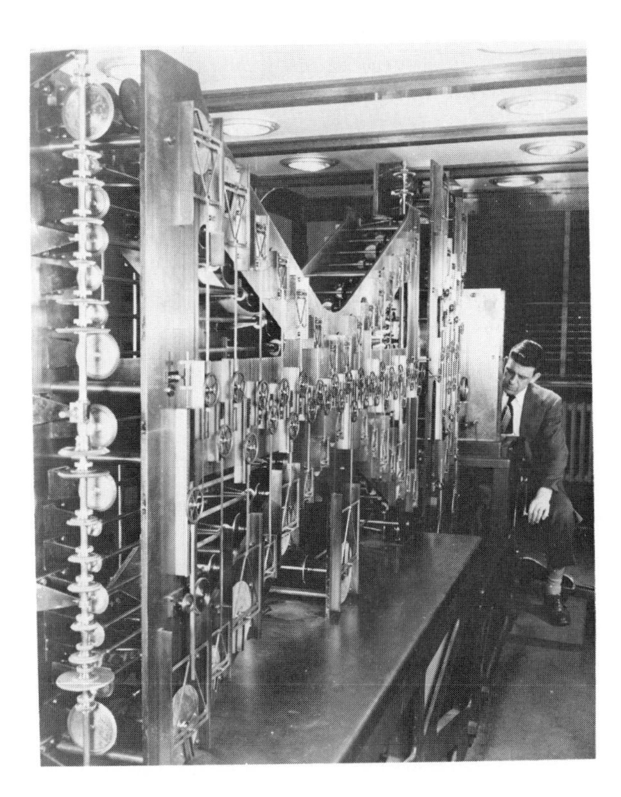

Figure 8-15. Tide-predicting machine. (Photo courtesy of NOAA.)

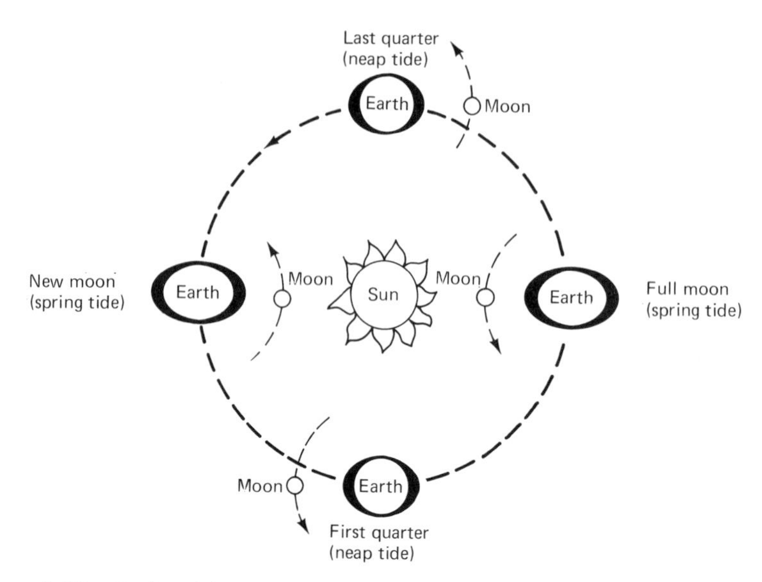

Figure 8-16. Spring tides occur with greatest amplitude when the sun and moon are lined up at the time of new and full moon. Neap tides occur with lowest amplitude when the moon is at first and last quarter.

tionally attracted. But this gravitational attraction is balanced by *centrifugal force,* which originates from their orbital motion about each another (Figure 8-17). Interaction between gravitational and centrifugal forces is the primary source of tides.

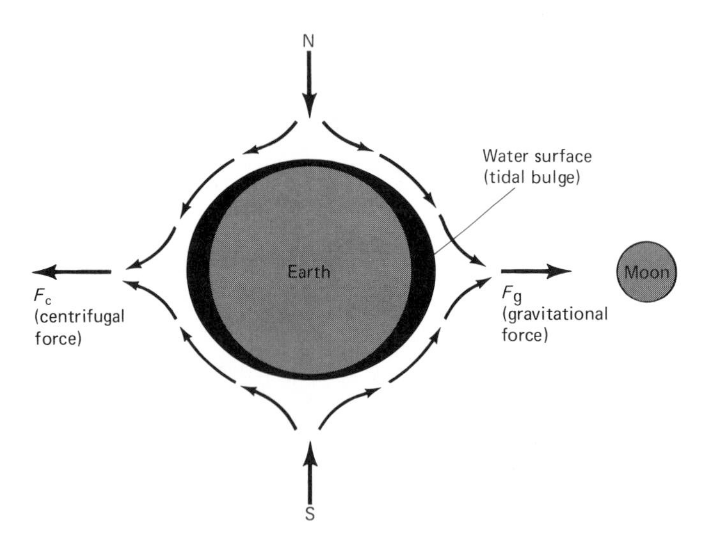

Figure 8-17. The primary cause of tides is the gravitational pull of the moon. Although its pull is only 1/300,000 of the earth's gravitational force, it is sufficient to raise the sea level on the moon-facing portion of the earth; centrifugal force balances the opposite side.

Tidal Bores

Often under favorable conditions in some tidal rivers, incoming high tides reverse their flows as they move upstream in breaking waves called *tidal bores* (Figure 8-18). The favorable conditions for tidal bores include strength of the incoming tidal wave, slope and depth of the channel, and river flow.

Tidal bore *A steep-nosed tide crest rushing (along with a high tide) upstream.*

Tidal bores have been reported to rush between 25 and 30 kilometers per hour and have heights of up to 10 meters. Notable examples of tidal bores include the Tsientang Kiang River at Hangschou Bay in China (which has a 3.5-meter height and moves at about 1 meter per second) and the Amazon River (which has a 5-meter height and moves at more than 0.5 meter per second). In North American, the Petitcodiac River flow reverses periodically as a result of the tidal bore phenomenon (see Figure 8-18).

8-8 POWER FROM TIDES

At present, tidal power plants are in operation only in France and Russia. But more than 160 locations in many parts of the world have been identified as potential sites for tidal power plants (Figure 8-19). In the United States, Passamaquoddy, Maine, is being considered for a tidal power plant. In Great Britain, the Severn River near the Bristol Channel is another site under serious consideration.

The Rance River power plant, near St. Malo, France, is the world's first commercially successful tidal power plant (Figure 8-20). The tidal amplitude in the area ranges from 9 to 14 meters. The plant was begun in 1961 and was completed in 1967 at a cost of about 100 million dollars. Across the Rance River, a dam was constructed. In the central part of the dam, 24 reversible turbines were mounted to operate during flood and ebb tides. Each turbine has a capacity to generate 10,000 kilowatts of power, for a total of 240,000 kilowatts. Because of its reversible operation, the plant can capture tidal power from waters both when they advance at high tide and when they recede. Turbines thus generate power as the reservoirs empty and as they fill. The flow of the water is about 18,000 cubic meters per second. The power from the St. Malo plant is channelled to the national electric grid of France.

Figure 8-18. Tidal bore at Petitcodiac River at the head of the Bay of Fundy. (Photo courtesy of NOAA.)

SUMMARY

1. Waves are formed in response to prevailing winds. Waves are described in terms of height, wavelength, and period.
2. The mechanism of wave formation is complex and is not yet fully understood.
3. Major wind-generated waves include seas, swells, and surf.
4. Wave refraction is a phenomenon that waves display as they approach shore.
5. Longshore currents and rip currents are important types of wave currents.
6. Catastrophic waves include tsunamis, which are produced either by volcanic or by earthquake activity.
7. Storm waves or surges are produced by abnormal meteorological and oceanic conditions.
8. Internal waves are found below surface waters. When they emerge at the surface, they are identifiable as slicks and carry fine sediments and plankton.
9. Tides are a daily phenomenon produced by the gravitational force of the moon, and to lesser extent, by the sun.
10. Tides are harnessed to manufacture electricity, for example, in France. There are 160 locations throughout the world under consideration to obtain energy from tides.

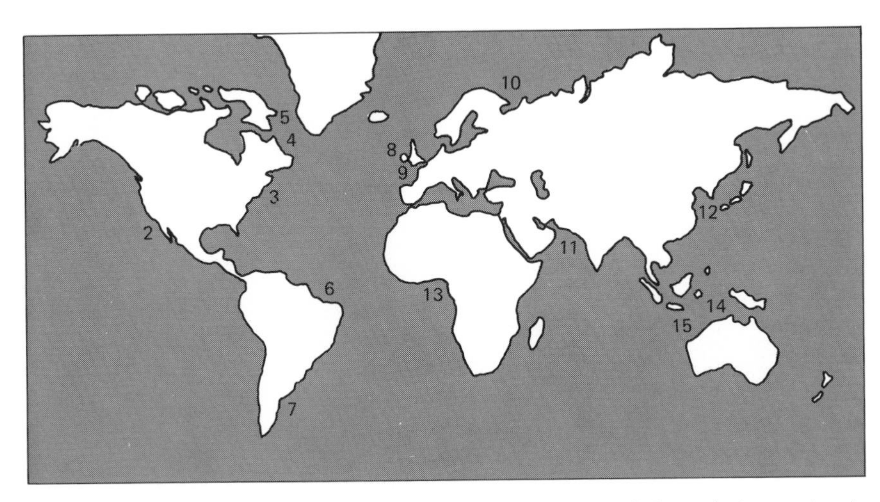

Figure 8-19. World's potential tidal power-plant sites. Selected sites only: 1. Cook Inlet (Alaska); 2. Baja California; 3. Passamaquoddy (Maine); 4. Bay of Fundy; 5. Frobisher Bay (Canada); 6. Maranho (Venezuela); 7. San Jose Gulf (Argentina); 8. Severn River (Great Britain); 9. Rance River (France); 10. Kislaya (USSR); 11. Cambay River (India); 12. Seoul River (South Korea); 13. Abidjan (Africa); 14. Darwin (Australia); 15. Kimberleys (Australia).

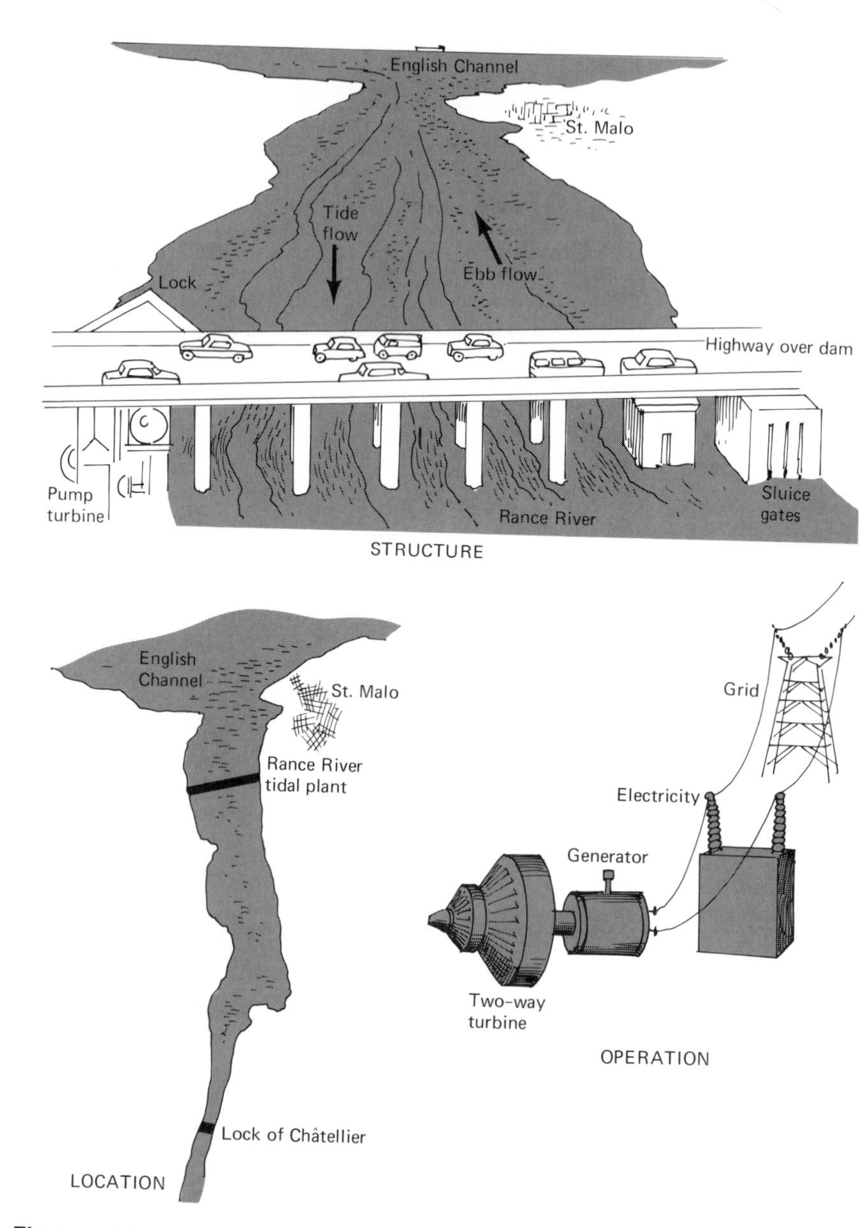

Figure 8-20. Tidal energy is obtained at Rance River at St. Malo, France. This power plant has a system of two-way turbines, which is triggered by incoming and outgoing tides.

Suggestions for Further Reading

Bascom, W. 1964. *Waves and Beaches: The Dynamics of the Ocean Surface.* Garden City, N.Y.: Doubleday.

Goldreich, Peter. 1972. "Tides and the Earth–Moon System." *Scientific American* (May).

Groen, P. 1967. *The Waters of the Sea.* New York: Van Nostrand Reinhold.

U.S. Coast and Geodetic Survey. 1965. *Tsunami, The Story of the Seismic Sea-Wave Warning System.*

Von Arx, W. S. 1975. *Introduction to Physical Oceanography,* second edition. Reading, Massachusetts: Addison-Wesley.

Shoreline Processes and Sediments

9-1 SHORE AND SHORELINES

THE REGION EXTENDING FROM THE POINT OF LOW tide to the landward limit of effective wave action is called the *shore*. The *shoreline* is a flexible boundary that marks the position of sea level between low tide and high tide. The terms *shore* and *shoreline* are often used interchangeably. The term *coast* is different and denotes an area that extends landward from a shore. Shore features change constantly because of the variable nature of waves and currents. Thus, many topographic features of the shore are not permanent, and consequently not all features are constant on all shore areas of the world at one time.

Shoreline *The boundary between sea and land.*

The shore or beach zone marks the area between the point of low tide to the foot of the sea cliff. The shore is divided into three zones: (1) The *backshore,* which occupies the area in front of the sea cliff, is usually associated with one or two *berms* (small, terracelike, low ridges built by storm waves). (2) The *foreshore* area extends seaward from berms to the low-tide point. (3) The *offshore* zone extends seaward from the point of low-tide (Figure 9-1).

Two types of shorelines; emergent and submergent, exist. On an *emergent shore* sea level falls in relation to land; many of its features surface. The shore area of California is a good example. On a *submergent shore* sea level rises in relation to the land; shore features are drowned. The shoreline of New England is a good example.

9-2 SHORELINE PROCESSES

The overall processes along the shore are very complex and dependent upon many variables, such as the intensity and direction of

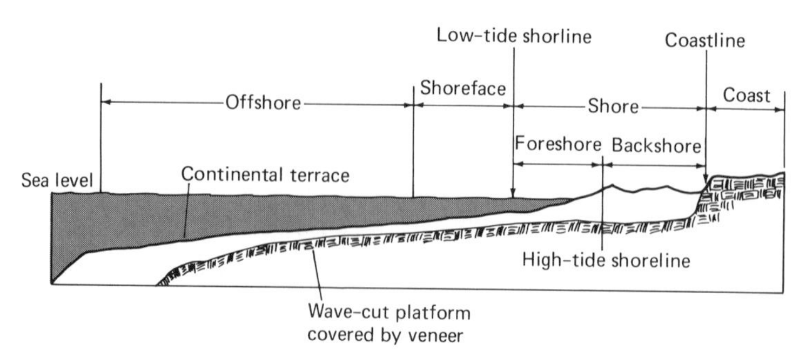

Figure 9-1. Principal divisions of shore and shoreline features.

Pleistocene *Subdivision of the Quarternary Period of geologic time. It began one million years ago and terminated 11,000 years ago. During its span, widespread continental glaciation occurred periodically.*

waves, the kind and durability of the rock along the shore, the openness of a coast to attack by waves, the depth of water offshore, and the nature of man-made obstacles along the shore. Moreover, all shores in the world have been influenced by glacial and postglacial sea level changes in recent geologic time (since the Pleistocene or the Ice Age). At times of low sea level, much of the shallow seabed of the continental shelves was exposed to subaerial erosion, which formed numerous shallow-sea canyons, channels, and valleys. Glacial material such as gravel, pebbles, and boulders was deposited in huge quantities during these times. The postglacial rise of sea levels has formed many new coastlines throughout the world.

Changes in Sea Level

We are accustomed to daily sea level changes along many shores of the world caused primarily by tides. These changes may be determined by checking tide tables for a given area. However, past sea level changes may be determined only from nature's logbook, as recorded in marine sediments. During the peak of the Ice Age (Pleistocene Epoch), extensive areas of the earth were under glaciation. The sea level fell about 100 meters, and about 5 percent of the total water from the ocean was withdrawn to support extensive glaciation.

A change in sea level is caused by the upward or downward movement of either the ocean or the land or by their joint movement. *Eustatic* sea level changes are caused by the movement of the ocean. One such change was a worldwide phenomenon during the Ice Age, when freezing and thawing of glaciers were common. When a change in sea level is caused primarily by crustal movements, it is *tectonic*. Tectonic movements are local and spasmodic. Any sea level change is difficult to distinguish under most circumstances.

Shore Erosion

Shore erosion by forces of waves and currents can be observed at work. In general, waves attack the shore in three principal ways: (1) direct impact, (2) hydraulic pressure and solution, and (3) by corrosion or abrasion. Owing to their consistency and direct impact on the shore, waves and currents are most devastating where the shore is composed of weak and unconsolidated material. The breakers in their attack against the shore are restricted by their upper and lower limit. The upper limit of their effectiveness is the maximum height that waves achieve at high tide. Often waves are most effective along shore where rocks are cracked and jointed. The power of waves drives water into these openings and compresses the air in them. This in turn further accelerates the wave force. When water retreats from these openings, there is a sudden expansion of the air, releasing enormous energy called *hydraulic pressure*. The hydraulic pressure mechanically breaks down rock material with time. If the rocks under attack are soluble, such as limestone, the hydraulic pressure can bring about their total destruction. Wave corrosion (or abrasion) involves the grinding action of cobbles, gravel, and sand when hurled against cliffs or rolled and dragged across the shore. Under the right circumstance, abrasion by waves can dislodge large blocks from cliffs, thereby modifying a shoreline in time. This action is also described as *artillery action*. When shorelines are irregular (that is, when they have considerable indentations), erosion by waves and currents is most effective, particularly in the headlands where the wave energy is highly concentrated (see Section 8-5). Waves effectively erode the headlands, in this way forming seacliffs. The eroded sediments from headlands are usually spread along the shore by waves and currents. This depositional activity is sometimes aided by rivers discharging sediments into bays.

Sediment Transport Along The Shores

When waves break at an angle against the shore, swash will run up the beach in the direction of wave propagation. Owing to the force of gravity, its backwash will move down seaward. The swash and backwash move sand particles in a sawtooth-like manner down the beach. Movement in this manner constitutes *beach drifting* (Figure 9-2). When beach drifting and longshore currents move in the same direction, they can efficiently transport both sand and gravel. However, when their directions are opposite, the coarser sediment (gravel) will follow the beach drifting, and the finer sediment (sand) will travel with the longshore currents. The combined transportation of sand and gravel by beach drifting and longshore current is known as *longshore drifting* (see Figure 9-2).

Man-Made Interference with Shoreline Processes

Knowledge of longshore and beach drifting is vital in the location, design, and maintainance of harbors. Therefore basic understanding of these dynamic driftings is essential (Figure 9-3). Longshore drift can daily carry several hundreds to thousands of kilograms of sand and gravel in suspension along many shores of the world. When longshore drifting is obstructed by artificial features such as harbors and jetties, however, it fills up these sites with sand causing serious dredging and maintenance problems and erosion of beaches further down from the sites. Moreover, these modifications often affect bathing beaches and waterfront towns. To protect these sites, ocean engineers have installed groins, jetties, and breakwaters. *Groins* are gentle walls extending from the high tide area seaward (Figure 9-4). Although groins have managed to slow sand removal by longshore drift, the beaches further down become starved for sand. They are scoured by wave action and are altered from predominantly sandy beaches to gravel beaches, as is happening to the famous Waikiki Beach in Hawaii.

Jetties, unlike groins, are longer structures built at right angles to the shore (Figure 9-5). Jetties are built to keep longshore drift sediments away from entrances to harbors or rivers. They have, however, caused problems similar to those created by groins. *Breakwaters* are rock or concrete structures built parallel to the shore to prevent the erosive effect of high energy waves (Figure 9-6). Although these structures are effective, they allow longshore drifts to deposit sand behind them, once again, as in the case of groins and jetties, causing dredging problems and the erosion of beaches further down.

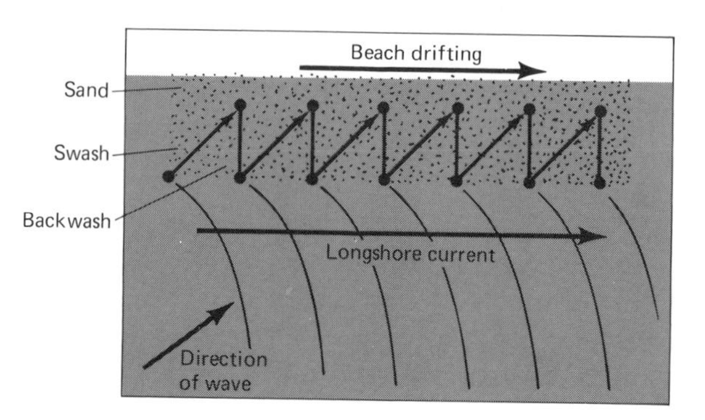

Figure 9-2. Longshore drifting and beach drifting along the shores.

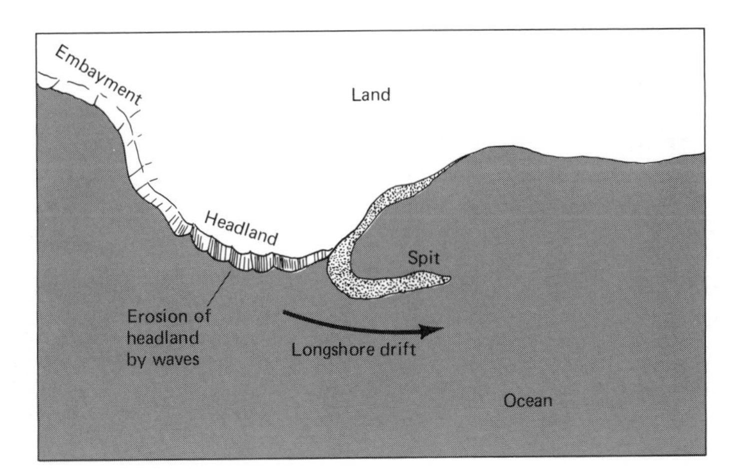

Figure 9-3. Longshore drift forming a spit or sandspit. Some of the eroded material at the headland is carried away by the drift to build the spit. Arrow indicates direction of the longshore drift.

9-3 SHORELINE FEATURES

Most shoreline features are the result of the erosional and depositional activities of waves and currents. Principal erosional features (Figure 9-7) include: (1) *sea cliffs,* found commonly along the coast and

Figure 9-4. A groin at Santa Monica, California. Note that it is relatively smaller than a jetty. (Photo courtesy of U.S. Army Coastal Engineering Research Center.)

Figure 9-5. Jetties are built to prevent shoaling of a channel by longshore drift at Ballona Creek, California. (Photo courtesy of U.S. Army Coastal Engineering Research Center.)

usually marked by a nick or scarp resulting from wave erosion (Figure 9-8). Running toward the sea from the base of a cliff is a (2) *terrace* or *wavecut terrace,* a steplike feature produced by wave action. Continual wave action against a sea cliff over time may produce *cavities* or *sea caves.* Prolonged action by sea waves through soft rocks of a headland may form a (3) *sea arch.* If waves continue to act against a sea cliff, the sea arch may be totally destroyed, leaving behind only an isolated remnant of rock in front of its cliff. This feature is referred to as the (4) *stack* (Figure 9-9). The erosional activity of sea waves also forms broad indentations, such as (5) *coves, bays,* or *bights,* on coasts.

Figure 9-6. A breakwater protecting a shore area at Santa Monica, California. (Photo courtesy of U.S. Army Coastal Engineering Research Center.)

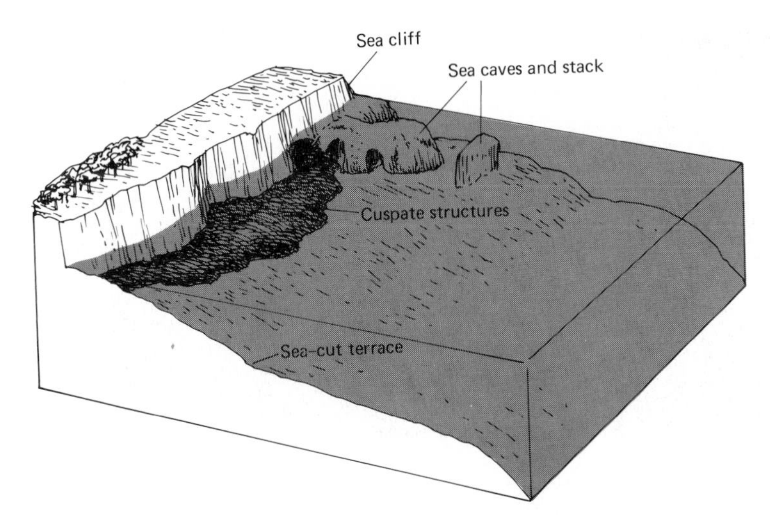

Figure 9-7. Principal erosional features of shorelines.

The most common shoreline features formed by the depositional activity of sea waves include beaches, bars, spits, and tombolos (Figure 9-10). A *beach* is an ephemeral depositional feature characterized by accumulation of rock debris, usually sand, and sometimes pebbles and

Beach *The deposit of sand and gravel along a shore that is usually the result of active transport by waves and currents.*

Figure 9-8. A typical steep sea cliff at Punta Este, Isla Mona, Puerto Rico. (Photo courtesy of C. A. Kaye, U.S. Geological Survey.)

Figure 9-9. Elephant Rock on the Oregon Coast typifies such erosional features as sea stacks and a sea cave. (Photo courtesy of J. S. Diller, U.S. Geological Survey.)

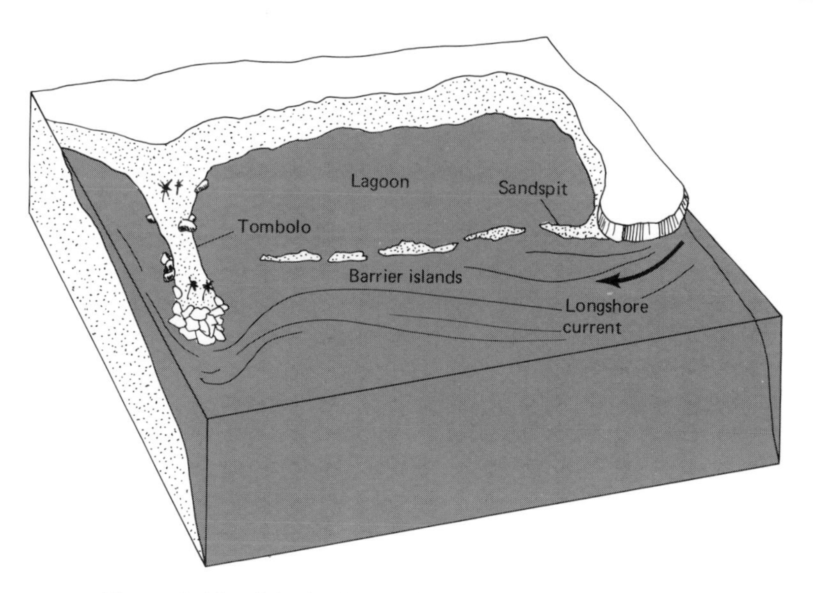

Figure 9-10. Principal depositional features of shorelines.

gravel along a shore (Figure 9-11, 9-12). Rock debris is provided by eroded material from headlands and rivers and occasionally by landslides, tsunamis, and hurricanes. Beaches may run hundreds of ki-

Figure 9-11. A rocky beach. (Photo courtesy of C. D. Walcott, U.S. Geological Survey.)

Figure 9-12. A beach of limestone gravel. (Photo courtesy of I. C. Russell, U.S. Geological Survey.)

lometers along a shore, such as, for example, the southeast coast of the United States.

Systematic studies of California beaches show that the widths of those beaches vary seasonally, probably because waves are much stronger in winter than in summer. In summer, the relatively low-energy waves are unable to transport beach material to deep water. As a result, wide beach areas develop. In winter, high-energy waves remove much beach material and leave a relatively narrow beach behind.

Sand or gravel deposits built on the seafloor by waves and currents are generally referred to as a *bar*. The most common bar deposit is a *spit*. A spit is a bar with one end linked to land and the other terminating at sea. The formation of spits is particularly favored by longshore drifting. A *tombolo* is a sand or gravel depositional shore feature that links an island with the mainland or that joins two islands (Figure 9-13).

Tombolo *A sand or gravel deposit that links an island with the mainland or joins two islands.*

Cuspate deposits of beach material built by wave action along the foreshore are called *beach cusps*. In beach cusp, sand, gravel, or cobbles are heaped together in more or less regularly spaced ridges that bend at right angles to the sea margin, thinning out to a point near the water's edge.

9-4 MARINE SEDIMENTS

Because of its gigantic volume of water and its depth, oceans provide an ultimate sink for all terrestrially derived eroded material, including fragments of rocks of varying shapes and sizes. Eroded ma-

Figure 9-13. A tombolo connects an island with the mainland. (Photo courtesy of I. C. Russell, U.S. Geological Survey.)

terial is transported by running water, wind, glaciers, and groundwater (as an example, see Table 9-1). Through geologic time, loose sediment deposited on the ocean floor is solidified (lithified) into sedimentary rocks.

TABLE 9-1 Modes of Entry and Rates of Accumulation of Sediments in the Atlantic Ocean

Zone	*Mode of Entry*	*Rate of Accumulation (millimeters per thousand years)*
50°N	Glacial from Arctic polar easterlies Jet stream	—
15°N–50°N	Jet stream St. Lawrence River	0.3–0.5[a] 2–7[b]
15°S–15°N	Trade winds South American and African rivers	0.7–0.8[a] 1–8[b]
50°S–15°S	Jet stream	0.2–0.4[a] 2–6[b]
50°S	Glacial from Antarctic polar easterlies Jet stream	2

SOURCE: J. J. Griffen, H. Windom, and E. O. Goldberg, 1968, *Deep Sea Research, 15,* 433.

[a]Mid-Atlantic Ridge valleys.

[b]Continents to flanks of ridge.

These rocks are formed by inorganic and organic activities. Inorganic sedimentary rocks are formed, for example, when intense evaporation of seawater, such as in the Persian Gulf, leads to precipitation of various evaporates such as rock salt (halite) and gyprock (gypsum). Organic sedimentary rocks are formed when, for example, corals secrete calcium carbonate shells. When these organisms die, they provide significant organic material to form coral limestones (see Chapter 12). The most common sedimentary rocks are sandstone, limestone, and shale. Sedimentary rocks are important sources of oil, natural gas, coal, and many ore minerals such as iron and manganese (See Chapter 14).

Nomenclature

Sediment *General term applied to loose solid material made available from the breakdown of pre-existing rocks and organic matter.*

Sedimentary rock *Rocks formed of sediment.*

Clastic *Unconsolidated fragments of pre-existing rocks.*

In order to understand the origins and distribution of marine sediments, we will introduce common terminology. A *sediment* is a loose deposit of solid debris on the earth's surface, originally contained in air, water, or ice under usual conditions. A *sedimentary rock* is consolidated (lithified) sediment. *Sedimentation* is the process that involves weathering (breakdown of rocks), transportation, and deposition (Figure 9-14). *Texture* is the mutual relation between particles (grains) in rock and is generally observed under a microscope. *Composition* refers to the mineralogical and chemical structure of a rock. A *clastic* rock is composed primarily of detrital (loose) material (examples include sandstone and shale). A *nonclastic* rock is composed predominantly of organic (biological) and chemical material (examples include reef limestone, saltrock, and dolomite).

Wentworth's grade scale provides a systematic classification of clastic rocks in terms of their grain size (Table 9-2). According to this scheme, each grade size varies from its previous one by the ration of $^1/_2$, and each has a designated name. For example, sizes between 2 and $^1/_{16}$ millimeters are sand, between $^1/_{16}$ and $^1/_{256}$ millimeter silt, and below $^1/_{256}$ millimeter clay. Wentworth's grade scale of sediment is important because it provides a standardized terminology. The information concerning various grade sizes of sediments can be obtained by *sieving*, which involves measuring the grain (or particle) size by passing a sample of known dried weight through a nest of sieves (with diminishing mesh diameter). The sieving method groups the aggregate samples into different size classes (each class with a specific grain diameter in millimeters). Subsequently these are weighed separately, thereby obtaining graphic or numerical data concerning the individual grain sizes in a sediment. These data are then statistically analyzed. These data are very useful in identifying and distinguishing former sand deposits of a beach, dune, or river.

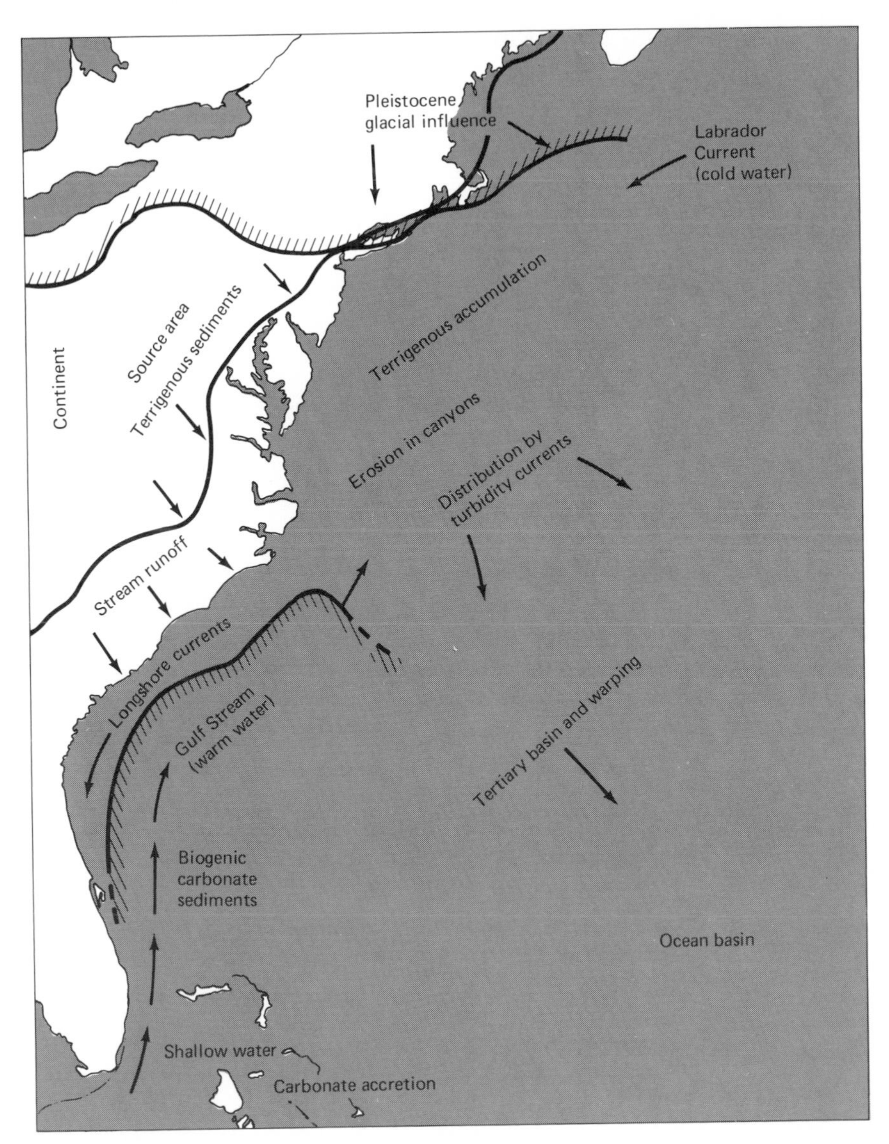

Figure 9-14. Dynamics of marine sedimentational processes are observed in waters off the East Coast of the United States.

Sources of Marine Sediment

Marine sediment is derived from various sources: from air (cosmic or airborne), water, and land (both exterior and interior). The most prominent source of clastic rock is land-derived sediment carried by rivers that annually empty a billion tons into the sea (Table 9-3). The

TABLE 9-2 Wentworth's Particle-Size Classification

Grade Limits (diameters in millimeters)	Particle	Grade Limits (diameters in millimeters)	Particle
over 256	Boulder	$\frac{1}{2}-\frac{1}{4}$	Medium sand
256–128	Large cobble	$\frac{1}{4}-\frac{1}{8}$	Fine sand
128–64	Small cobble	$\frac{1}{8}-\frac{1}{16}$	Very fine sand
64–32	Very large pebble	$\frac{1}{16}-\frac{1}{32}$	Coarse silt
32–16	Large pebble	$\frac{1}{32}-\frac{1}{64}$	Medium silt
16–8	Medium pebble	$\frac{1}{64}-\frac{1}{128}$	Fine silt
8–4	Small pebble	$\frac{1}{128}-\frac{1}{256}$	Very fine silt
4–2	Granule	$\frac{1}{256}-\frac{1}{512}$	Coarse clay
2–1	Very coarse sand	$\frac{1}{512}-\frac{1}{1024}$	Medium clay
1–1/2	Coarse sand	$\frac{1}{1024}-\frac{1}{2048}$	Fine clay

supply of airborne material might be much greater than in rivers, particularly in the southern Atlantic off the Sahara desert or in waters off active volcanic belts. When an organic population dominates, such as in a reef area, sediments of biogenic origin will be abundant.

As soon as suspended clastic sediments are delivered to the sea, they are immediately buffered by continual wave agitation, which causes clastic debris to separate. Coarse particles are deposited near shore and clay-size particles are carried farther into deeper waters. Sometimes, sporadic currents such as turbidity currents accelerate the movement of coarse and fine particles and carry them together into relatively deep waters. As soon as dissolved sediments arrive in the sea environment, silica and calcium are withdrawn by certain organisms that use them to build their siliceous or limy shells. However sediment is

TABLE 9-3 Ten Top Rivers of the World Ranked by Sediment Yield

River	Drainage Basin (10^3 square kilometers)	Average Suspended Load per Year: Metric Tons ($x10^6$)	Average Suspended Load per Year: Metric Tons (per square kilometers)	Average Discharge at Mouth (10^3 cubic feet per second)
Amazon	5776	363	63	6400
Yangtze	1942	499	257	770
Mississippi	3222	312	97	630
Brahmaputra	666	726	1090	430
Ganges	956	1451	1518	415
Mekong	795	170	214	390
Indus	969	435	449	196
Red	119	130	1092	138
Nile	2978	111	37	100
Missouri	1376	218	159	69

SOURCE: Simplified after J. N. Holeman, 1968, *Water Resources Research*, *4*, 4. Copyright American Geophysical Union.

brought to the seas, it is subject to consolidation over time, and it eventually forms sedimentary deposits.

9-5 THE DISTRIBUTION OF MARINE SEDIMENTS

Marine sediments are distributed in two ways: (1) in shallow water as deposits at the continental margin (notably on shelves and slopes) and (2) as deposits in deep oceans.

Continental margin *Seaward extended portion of the landmass.*

Shallow-Water Sediments

In general, glacial continental shelves are characterized by sediments composed predominantly of a mixture of sand, gravel, and cobbles. On nonglacial continental shelves, deposits consist usually of riverborne mud. In other areas of shelves, where wave and current action are strong, and on the seafloor, coarse rock material and gravel are usually deposited (Figure 9-15). A few lime deposits have built up on the

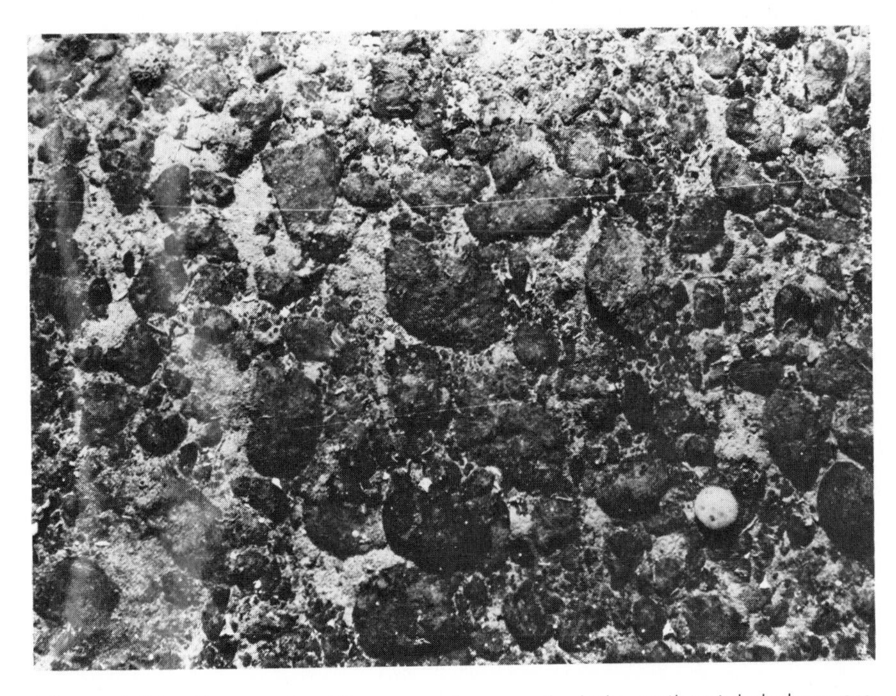

Figure 9-15. Many parts of the ocean floor, particularly continental shelves, are covered by glacial boulders, gravel, and sand. (Photo courtesy of Woods Hole Oceanographic Institution.)

continental shelves of the oceans. The bulk of these deposits are restricted to warm tropical and subtropical waters (from 30°N to 30°S), where coral reefs flourish (see Chapter 12).

Most continental slopes are steep, and sediment is not deposited in great thickness there. Slope regions are characterized by bedrock at the surface and are overlaid with a thin layer of fine silt and mud. Sometimes barren rocks are covered by gravel and sand. In the waters off Spain, the continental slope is altogether devoid of loose sediment.

Deep-Sea Sediments

Deep-sea sediment covers 85 percent of the ocean floor. Two types of sediment dominate: (1) terrigenous (land-derived material) and (2) pelagic (ocean-derived material) (Figure 9-16).

Terrigenous sediments *Land-derived material deposited in the ocean.*

Terrigenous sediment comes from various sources, including windborne volcanic ash that has settled into the deeper parts of oceans over long periods of time. The bulk of terrigenous material is present as

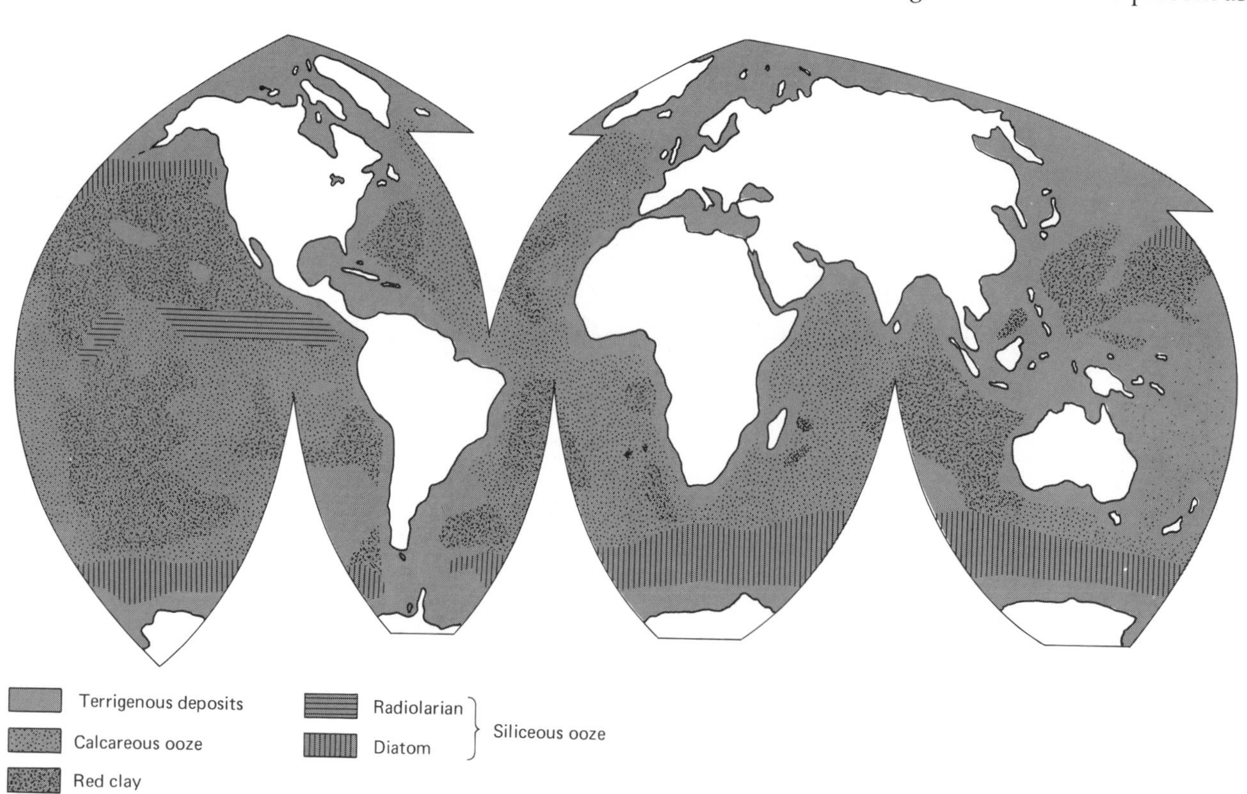

Figure 9-16. Distribution of marine sediments in oceans.

silt, as sand from glacial ice, and as suspended sediment in the running water of sea currents. In the Antarctic and Arctic oceans are extensive terrigenous deposits of glacial silt. Turbidity currents transport enormous quantities of mixed gravel, sand, silt, and clay to the deeper parts of the ocean, consequently forming *turbidite* deposits (see Chapter 4). Terrigenous deposits may also contain desert sand and mud. A significant portion of the ocean floor, mainly in the Pacific, is extensively covered with a very-fine-grained deposit known as *brown clay*. The brown clay originated on land and subsequently drifted into the ocean.

Pelagic deposits cover about 85 percent of the ocean floor and consist mainly of *ooze* deposits. At least 30 percent of an ooze deposit consists of the hard parts of microscopic organisms. Following the death of these organisms, the shells, or debris, sink to the seafloor to form calcareous or siliceous ooze (Figures 9-17, 9-18).

Pelagic sediment *Refers to sediments of the deep ocean in contrast to those derived from the land.*

Ooze *Fine-grained pelagic deposits containing a minimum of 30 percent organic material. Oozes may be classified according to their chemical makeup– as siliceous (silica-containing) or calcareous (limy)–or often according to the organism whose remains dominate the deposit–such as radiolarian ooze and foraminiferal ooze.*

Calcareous ooze is the most extensive deposit on the ocean floor. Calcareous ooze consists of calcium carbonate shells of planktonic organisms, notably foraminiferans and coccoliths (calcareous algae; see Chapter 11), as well as lesser amounts of clay and associated debris. Two marine deposits, *Globigerina* ooze and pteropod ooze, are each dominated by a particular organism. In *Globigerina ooze,* remains of limy shells of minute one-celled organisms called *Globigerina* are found. *Globigerina* ooze covers most of the floor of the western Indian Ocean, the mid-Atlantic Ocean, and the equatorial and southern Pacific Ocean. In

Figure 9-17. Foraminiferal assemblage constitutes a principal raw material in forming lime-ooze deposits on the ocean floor. (Photo courtesy of Joseph Morley, Lamont-Doherty Geological Observatory of Columbia University.)

Figure 9-18. Pteropod assemblage constitutes a principal raw material in forming siliceous-ooze deposits on the ocean floor. (Photo courtesy of Joseph Morley, Lamont-Doherty Geological Observatory of Columbia University.)

pteropod ooze, the remains of small snails are abundant. *Pteropod ooze* is highly concentrated linearly from north to south in the Atlantic Ocean.

Siliceous ooze contains the remains of silica-rich material of organisms such as diatoms and radiolarians. *Radiolarian ooze* consists mainly of minute single-celled organisms that are highly concentrated near the equatorial regions of the Pacific Ocean. *Diatomaceous ooze,* also composed of single-celled organisms, is concentrated in the upper latitudes of the oceans, especially in the northern Pacific and the northern Antarctic oceans.

The preferential distribution of *radiolarian* and *diatomaceous* oozes is determined by the supply of nutrients delivered by upwelling currents. Radiolarians grow in great numbers in the equatorial upwelling, which is rich in phosphorus and other elements obtained from the decay of organisms. Radiolarians multiply so rapidly that their growth surpasses that of any other organism. In colder oceans, especially the Antarctic, upwelling currents supply nutrients to diatoms.

Pelagic sediment formed from direct precipitation from seawater is called *authigenic sediment.* The silicate mineral *phillipsite* is found in high concentration in much of the Pacific Ocean. Manganese nodules are another striking example of authigenic deposits (see Chapter 14).

Pelagic deposits also include cosmic material from meteoritic dust and fragments. Every day 10,000 to 100,000 tons of cosmic material is delivered into the oceans. A considerable amount of airborne volcanic material also enters the ocean intermittently following volcanic eruptions.

9-6 RATES OF SEDIMENTATION

The exact rate at which sediment is deposited in the ocean is not accurate, because many variables influence sedimentation. For example, coarse sediments sink faster than clay-sized sediments; deposits in deltaic or reef regions are relatively greater than those in the deep sea. Rates of sedimentation varies considerably from a few millimeters to 100 or more centimeters per thousand years (Table 9-4). It is realized, however, that average rates of sedimentation are slower on ocean floors than on continental shelves and slopes. Rates are determined by detailed studies of core sediments obtained from various parts of the ocean and by employing radioactive isotopes, such as carbon-14, which provide useful information about the rate of sedimentation since the last Ice Age.

In the Pacific and Indian oceans, rates of sedimentation range between .1 and 1 centimeter per thousand years, whereas in the Atlantic Ocean the range is between 1 and 10 centimeters for the same period of time. On continental shelves and on upper slopes, rates range between 20 and 30 centimeters per thousand years. Off the coast of southern California, the rates of sedimentation range between 5 to 40 centimeters per thousand years. A summary of sedimentation rates of pelagic clay is given in Table 9–4.

TABLE 9-4 Rates of Pelagic Clay Accumulation

Area	*Rate (in millimeters per thousand years)*
North Pacific	4–7 (Recent)
Pacific off Mexico	11 (Recent)
Atlantic Continental Rise off Florida	50–500 (Recent)
Caribbean southwest of Hispaniola	28 (Pleistocene and Recent)
Argentine Basin	17–34 (Recent)
Argentine Basin	60–110 (Late Glacial)
Southeastern Pacific	80

SOURCE: James Gilluly, Aaron C. Water, and A. O. Woodford, 1968, *Principles of Geology*, third edition. W.H. Freeman, Copyright © 1968.

9-7 AGE OF SEDIMENTS IN THE OCEAN

The thickness of abyssal deposits, as determined by seismic explorations, is about 300 meters. Therefore, the total thickness of sediment on the ocean floor is quite trivial in terms of geologic time. In terms of time it represents only about 5 percent of the 4.5-billion-year-old earth. The oldest rocks on the ocean floor are of Jurassic age (that is, 200 million years or so). This realization, of course, led to the concept of seafloor spreading. In the absence of such a phenomenon, and assuming constant deposition, the thickness of sediment in, for example, the Pacific basin, would be more than 3 kilometers.

Figure 9-19. Geological oceanographers examine a core sample containing marine sediments. (Photo courtesy of Woods Hole Oceanographic Institution.)

9-8 COLLECTIONS OF SEDIMENTS

Geological oceanographers obtain sediments from the ocean in a variety of ways in order to understand their nature and geologic ages (Figure 9-19). When ocean bottoms are characterized by hard rock, dredges are used (Figure 9-20). A dredge is a sturdy sediment sampler that is dragged along the bottom of the sea. Dredges can collect sediments from any depth (Figure 9-21). For collection of sediments, particularly in their order of deposition through geologic time, a gravity or a piston corer is used. A gravity corer is a hollow tube that forces its way into soft sediments by the force of gravity (Figure 9-22). A piston corer is more complex in design and is efficient in collecting sediments

Figure 9-21. Smith-McIntyre grab retrieval. (Photo courtesy of National Marine Fisheries Service.)

Figure 9-20. Oceanographers use a rock dredge to collect sediment samples from the ocean floor. (Photo courtesy of Woods Hole Oceanographic Institution.)

Figure 9-22. Gravity corer is lowered to obtain sediment samples from the ocean floor. (Photo courtesy of National Marine Fisheries Service.)

from the ocean bottom (Figure 9-23). For undisturbed mud samples from the ocean floor, a specially designed near-surface box sediment sampler is employed (Figure 9-24). But deep-sea drilling is the most efficient way of obtaining sediments from the deepest parts of the ocean (see Chapter 5).

Figure 9-23. Piston corer is lowered to obtain sediment samples from the ocean floor. (Photo courtesy of Woods Hole Oceanographic Institution.)

Figure 9-24. A near-surface box sediment sampler is often used by oceanographers to obtain an undisturbed mud sample from the surface of the ocean floor. (Photo courtesy of Woods Hole Oceanographic Institution.)

SUMMARY

1. The shoreline marks the position of the sea level between high and low tide. The shore is divided into back shore, foreshore, and offshore zones. Two types of shorelines are emergent and submergent.

2. Erosional shoreline features include sea cliffs, terraces, sea arches, stacks, coves, bays, and bights.
3. Depositional shoreline features include beaches, bars, spits, and tombolos.
4. A beach is a wave-washed temporary deposit of sand, gravels, and cobbles on or near the shore. Common beach features include beach cusps.
5. Marine sediment is loose material brought into oceans from land, outer space, dead organisms, and water. Loose sediment may be inorganic, organic, or chemical in origin. When sediment is compacted through sedimentation, it becomes sedimentary rocks. Sedimentary rocks such as sandstone and limestone often contain significant amounts of oil, natural gas, coal, and metallic ores.
6. Marine sediments are widely distributed in the ocean, both vertically and horizontally.
7. Major deep-sea deposits cover 40 percent of the ocean floor and include turbidites, brown clay, calcareous ooze, siliceous ooze, and small amounts of cosmic deposits.
8. Authigenic deposits are formed from direct precipitation from seawater.
9. Rates of sedimentation in the ocean vary from place to place. Radioactive dating of oceans gives rates of from .1 to 40 centimeters per thousand years.
10. The age of sediment in oceans is relatively less than that of sediment on land. The oldest rocks on the ocean floor are approximately 200 million years old.

Suggestions for Further Reading

Inman, D. L., and B. M. Brush. 1973. "The Coastal Challenge." *Science, 181,* 20–32.

King, C. A. M. 1972. *Beaches and Coasts,* second edition. London: Edward Arnold.

Komar, P. D. 1976. *Beach Processes and Sedimentation.* Englewood Cliffs, New Jersey: Prentice-Hall.

Pettijohn, F. J. 1975. *Sedimentary Rocks,* third edition. New York: Harper & Row.

Pettijohn, F. J., P.E. Potter, and R. Siever. 1972. *Geology of Sand and Sandstone.* New York: Springer.

Shepard, F. P. 1974. *Submarine Geology,* third edition. New York: Harper & Row.
Thomsen, D. E. 1972. "As the Seashore Shifts." *Science News, 101,* 396–397.

10
Chemical Oceanography

10-1 INTRODUCTION

CHEMICAL OCEANOGRAPHY IS CONCERNED WITH THE STUDY OF the oceans as chemical systems. It studies the principal influences of various chemical processes of life and on seawater (Figure 10-1).

The science of chemical oceanography began in the latter half of the seventeenth century when Robert Boyle first examined the chemical makeup of seawater. Later Antoine Lavoisier discovered that water was a mixture of hydrogen and oxygen. He also successfully identified the major elements in seawater, including chloride, sulfate, sodium, potassium, calcium, and magnesium.

As was mentioned in Chapter 1, Forchammer developed the law of relative proportions, which stated that, regardless of the absolute concentration of total dissolved substances in a given amount of seawater, the ratios between the major elements remain virtually constant. That is, in seawater, chloride ions always comprise 55.25 percent of dissolved solids. If the salinity were 35‰, the total chloride in solution would be 0.5525 × 35‰ or 19.35 grams per kilogram. This value will be virtually constant regardless of sample site. Forchammer's principle was an important development in chemical oceanography, because calculations of salinity, chlorinity, and density relations are determined through its application.

William Dittmar, who verified Forchammer's principle, based his conclusions on an analysis of 77 water samples representative of all oceans. Following World War II, the scope of chemical oceanography enlarged far beyond mere identification of elements and determination of salinity and chlorinity. At present, chemical oceanographers are involved in isotopic studies that measure ages of and rates at which marine sediments are deposited, that determine paleoclimatic conditions, and that determine the complex chemical interaction between the atmosphere and oceans. Many chemical oceanographers are currently involved with problems of ocean pollution.

Figure 10-1. Oceanographers retrieve a large seawater sample for further chemical analyses. (Photo courtesy of J. G. Schilling, Graduate School of Oceanography, University of Rhode Island.)

10-2 CHEMICAL METHODS

Seawater samples are commonly obtained by a specially designed cylindrical metal tube called the Nansen bottle (Figure 10-2). The Nansen bottle has valves at either end; inside it is coated with a chemically resistant plastic to ensure against contamination from seawater. The bottle contains thermometers that measure temperature at the time of collection.

The operation of Nansen bottles is relatively simple. Several bottles are attached at regular intervals on a hydrographic wires. Each bottle when lowered into the sea at a desired depth undergoes variation in its position before tripping (Figure 10-3). It is vertical on immersion and inclined during actual tripping; after the messenger hits the bottle, it overturns, collecting the water sample. When the bottle is returned to the ship, it is stored in a refrigerator to prevent further chemical altera-

Figure 10-2. An oceanographer obtains seawater from a Nansen bottle for further chemical analyses. (Photo courtesy of Woods Hole Oceanographic Institution.)

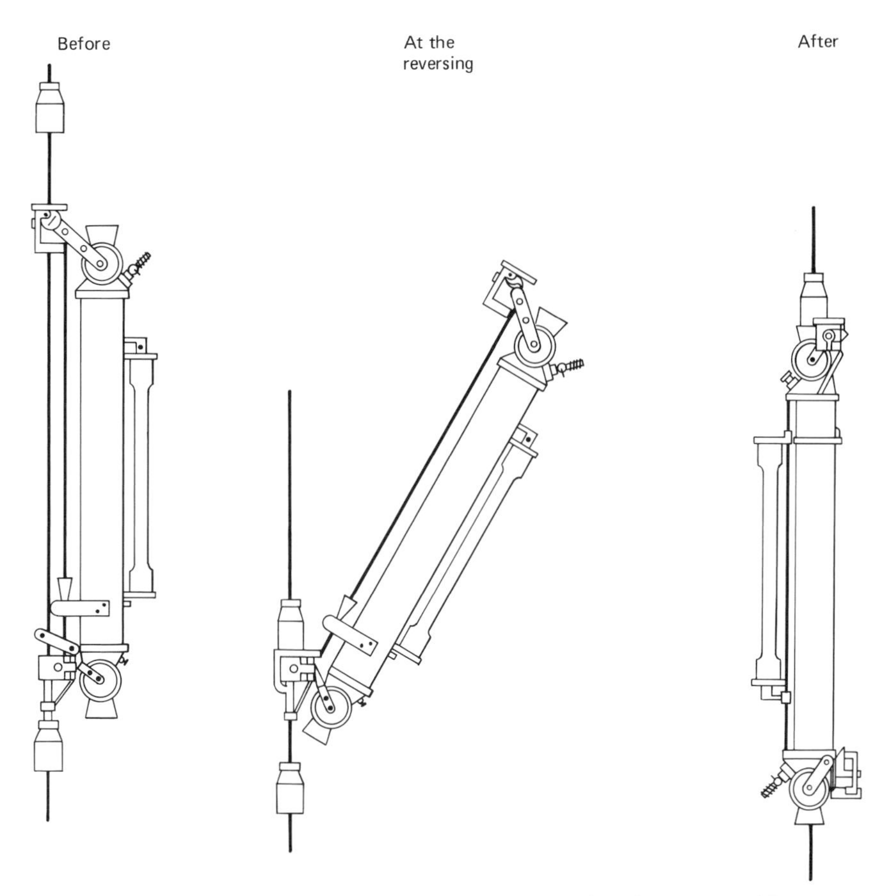

Figure 10-3. Three positions of a Nansen bottle during the course of its operation to collect seawater samples.

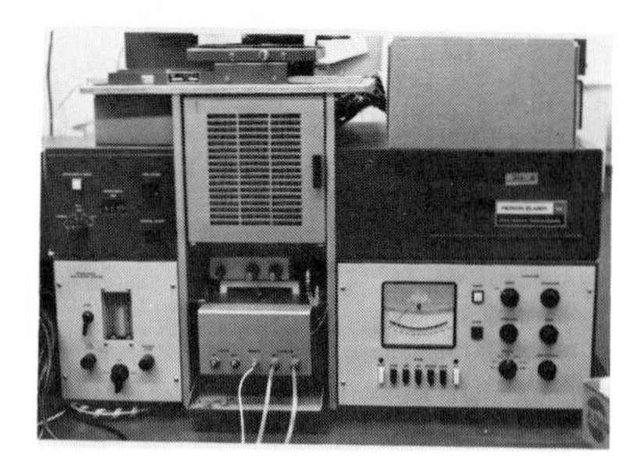

Figure 10-5. The atomic absorption spectrophotometer (AAS) is a sophisticated analytical instrument used in the determination of trace elements in seawater. (Photo courtesy of R. Vallari.)

tions of seawater. Small amounts of sea water are drawn periodically from the master stock for further chemical analysis.

A Bowen-Bodman water sampler is used to collect large volumes of seawater to study dissolved elements (Figure 10-4). The chemical determination of seawater includes elemental composition, gaseous content (such as carbon dioxide, oxygen, and other gases), and nutrient content (such as phosphates, nitrates, and silicates). Elemental identification is generally carried out by using a sophisticated atomic absorption spectrophotograph (Figure 10-5). Gas chromatographs are

Figure 10-4. A Bowen-Bodman water sampler is used to collect large volumes of seawater in order to study trace metals present. (Photo courtesy of Woods Hole Oceanographic Institution.)

used to determine gaseous and organic contents. An amino acid analyzer is used to detect protein (Figure 10-6).

10-3 THE CHEMISTRY OF SEAWATER

Seawater may be chemically defined as a solution of two components: *solvent* and *solute.* The water itself is solvent, and the dissolved salts in it are the solutes. Solutes include dissolved solids, gases, and organic and particulate matter (Figure 10-7).

Dissolved Solids

Seawater is a 96.6 percent pure compound composed of hydrogen and oxygen; the remaining 3.4 percent contains dissolved salts (Table 10-1). The bulk of dissolved salts is comprised of six elements: chlorine, sodium, magnesium, sulfur, calcium, and potassium. These elements are concentrated in more than 100 parts per million and are generally referred to as *major elements.* Seawater also contains several other ele-

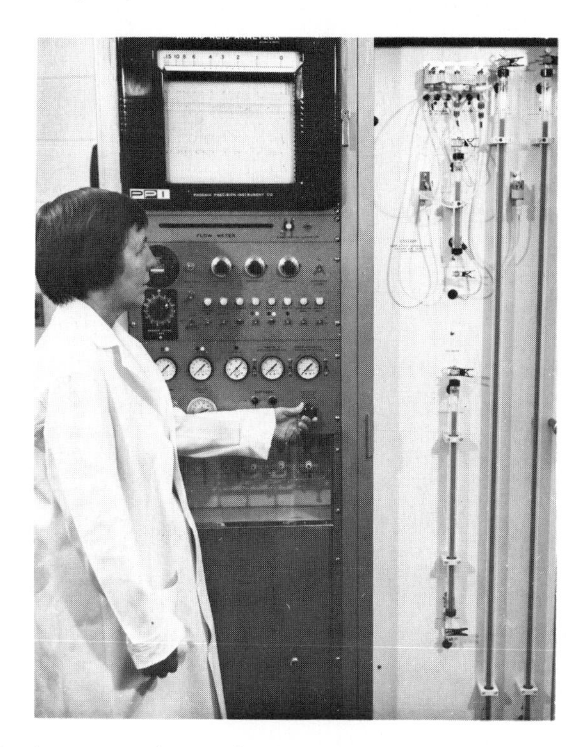

Figure 10-6. Instrumental setup for the determination of the organic content of seawater. (Photo courtesy of NOAA.)

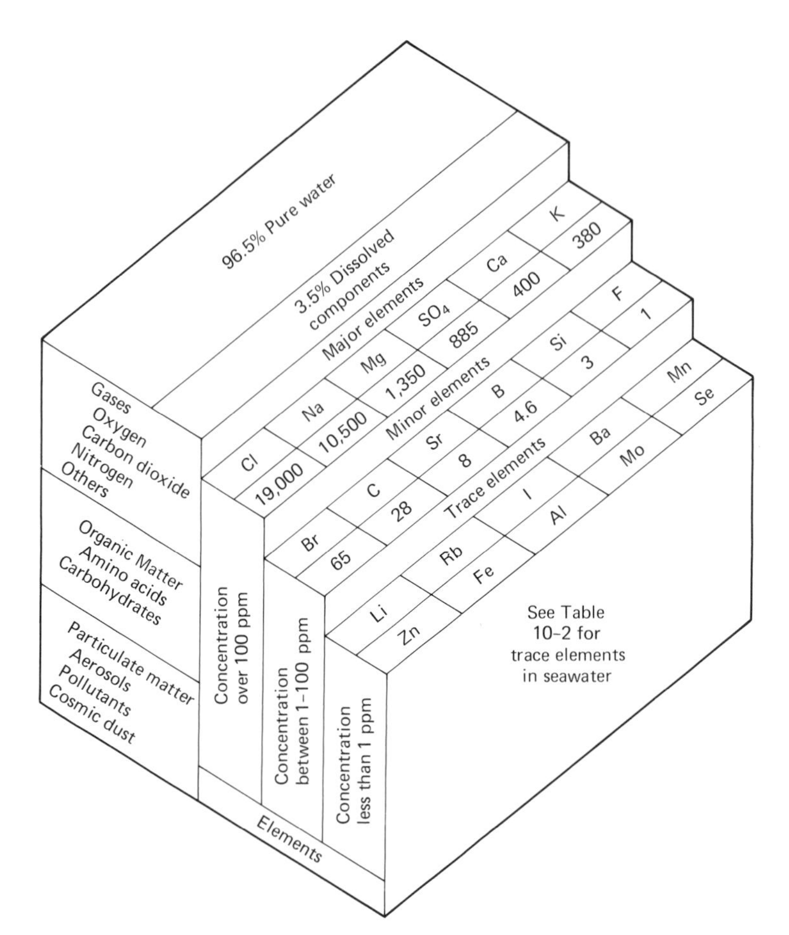

Figure 10-7. Chemical composition of seawater.

TABLE 10-1 Major Dissolved Constituents of Typical Seawater[a]

Constituent	*Content (in grams per kilogram)*
Chloride	19.353
Sodium	10.760
Sulfate	2.712
Magnesium	1.294
Calcium	0.413
Potassium	0.387
Bicarbonate	0.142
Bromide	0.067
Strontium	0.008

SOURCE: R. A. Horne, 1969, *Marine Chemistry,* New York, Wiley.

[a] 35‰ seawater.

ments concentrations of 1 to 100 parts per million; these are *minor elements* and include strontium, bromine, boron, iron, and silicon. *Trace elements* occur in concentrations of less than 1 part per million (Table 10-2). Their concentration in seawater is governed significantly by biological

TABLE 10-2 Concentration of Trace Elements in Seawater [a] (parts per billion)

Element	*Symbol*	*Concentration*
Lithium	Li	170
Rubidium	Rb	120
Iodine	I	60
Barium	Ba	30
Indium	In	20
Zinc	Zn	10
Iron	Fe	10
Aluminum	Al	10
Molybdenum	Mo	10
Selenium	Se	0.4
Tin	Sn	0.8
Copper	Cu	3
Arsenic	As	3
Uranium	U	3
Nickel	Ni	2
Vanadium	V	2
Manganese	Mn	2
Titanium	Ti	1
Antimony	Sb	0.5
Cobalt	Co	0.1
Cesium	Cs	0.5
Cerium	Ce	0.005
Yttrium	Y	0.3
Silver	Ag	0.04
Lanthanum	La	0.01
Cadmium	Cd	0.1
Tungsten	W	0.1
Germanium	Ge	0.06
Chromium	Cr	0.05
Thorium	Th	0.05
Scandium	Sc	0.04
Lead	Pb	0.03
Mercury	Hg	0.03
Gallium	Ga	0.03
Bismuth	Bi	0.02
Niobium	Nb	0.01
Thallium	Tl	0.01
Gold	Au	0.004
Protactinimu	Pa	2×10^{-6}
Radium	Ra	1×10^{-7}
Rare Earths		0.003–0.0005

SOURCE: E. D. Goldberg, 1965, in *Chemical Oceanography,* volume 1 (J. P. Riley and G. Skirrow, editors), New York, Academic Press.

[a]Excluding nutrients and dissolved gases.

activity. For example, plankton absorb many trace elements from seawater; nickel is likewise concentrated by sponges and strontium by radiolarians.

Particulate Matter

Particulate matter in seawater includes inorganic, clay-sized material and is derived from different sources. It may be windborne material from a desert or may be volcanic dust or ash supplied by a submarine eruption. Some particulate material is carried by waves and currents, which transport silt and clay from deltas or beach regions. In recent years, particulate matter resulting from aerosols has been brought to the sea.

Nutrients or Organic Matter

Nutrient elements are closely associated with living organisms and include carbon, oxygen, nitrogen, and phosphorus (see Chapter 12). These are found in solution as dissolved bicarbonates, phosphates, and nitrates. Nutrient elements are the most vital compounds in plants and animals and are usually found in the form of amino acids—the raw materials of protein, fats, starches, sugar, and phosphorus-bearing compounds of ATP (adenosine triphosphate). Carbon, nitrogen, and phosphorus are removed from oceans at shallow depths where light is available for *photosynthesis*.* In deeper waters, bacterial activity consumes and often destroys fine organic matter. Nutrient elements from organic matter are recycled by bacterial action into ionic forms in seawater. Through the mixing of waters, nutrient concentration in seawater is maintained and the survival of life thereby assured. Dissolved organic content in seawater off California, for example, increases with depth and ranges from 5 to 350 milligrams per gram (Figure 10-8).

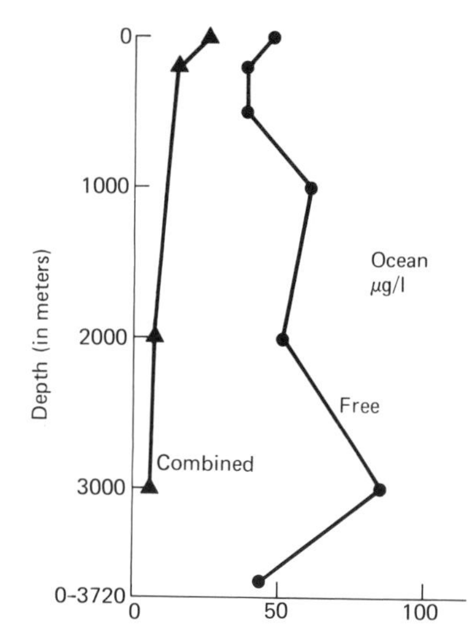

Figure 10-8. Depth distribution of amino acids in ocean water (µg/l). The term "Free" means that all constitutes pass an 0.4 micron filter. (*Source:* S. C. Rittenberg et al., 1963, *J. Sediment. Petrology*, *33*, 140–172.)

Dissolved Gases

Atmospheric gases are continuously in contact with the surface waters of oceans. Seawater contains several dissolved gases including oxygen, carbon dioxide, and nitrogen, and noble gases such as argon, helium, neon, and radon. Under certain conditions hydrogen sulfide gas (which has the smell of a rotten egg) is present, especially near a vol-

*Photosynthesis is a process in which energies of light and chlorophyll are used by plants to synthesize food from carbon dioxide and water.

canic source or in stagnant anaerobic (oxygenless) portions of the sea. There is, for example, a considerable amount of dissolved hydrogen sulfide gas in the Dead Sea. However, oxygen and carbon dioxide are the two most active gases dissolved in seawater.

Oxygen

Oxygen is dissolved directly from the atmosphere at the air–sea interface or is made available chemically by photosynthesis of plants in water. Photosynthesis takes place at or near surface layers. In oceanic conditions, oxygen is consumed at all depths. Also, oxygen is recombined through the decomposition of organic waste products, dead organisms, and and vegetation that sinks to the seafloor.

Nitrogen

Nitrogen occurs in the sea in combination with other elements such as ammonia, and in oxides of nitrogen as nitrite ion and nitrate ion. Nitrogen is vital for all organisms, because it helps to synthesize complex protein molecules that affect growth and reproduction. While they are alive, these organisms regularly remove complex nitrogen compounds as excreta. On the death of these organisms, complex nitrogenous compounds so formed are decomposed into simpler chemical compounds by bacteria.

The major obstacle to man's free descent into ocean abysses is the necessity of breathing nitrogen-laden air at high pressure. Nitrogen diffuses through the lungs and is retained for a long time. During ascent this high pressure is reduced and the nitrogen is released as bubbles into the bloodstream, causing a severe decompression sickness called the *bends*. The bubbles interfere with blood circulation and block the supply of oxygen to nervous tissue. This blockage can cause severe pain, paralysis, or death. In order to avoid the bends, divers are trained to ascend slowly to the surface, because a slow ascent allows a diffusion of nitrogen and provides the safe release of bubbles.

Carbon Dioxide

The amount of carbon dioxide in the earth's atmosphere is approximately 2.3×10^{12} (2,300 trillion) tons, which is only 2 percent of the total atmospheric mass. The oceans exchange with the atmosphere about 200 billion tons of carbon dioxide annually. The oceans contain approximately 1.3×10^{14} tons of carbon dioxide, about 50 times as much as the air. Some of this gas is dissolved in seawater, but most of it is locked in the form of calcium carbonate, as in limestone and lime sediments.

Both the atmosphere and the oceans continously exchange carbon dioxide in rocks and living organisms. Carbon dioxide in seawater may be increased by gaseous release from volcanoes or by the respiration and decay of organisms; it may be decreased by photosynthesis of plants.

Plants consume 60 billion tons of carbon dioxide each year for photosynthesis. The formation of new fossil fuel deposits withholds about 100 million tons of carbon dioxide, but this is less than 0.2 percent of the yearly generation of carbon through photosynthesis. About 350 million years ago, during the Mississippian and Pennsylvanian periods, when large deposits of coal and oil were formed, approximately 10^{12} (1000 trillion tons) tons of carbon dioxide were withdrawn from the atmosphere–ocean system.

10-4 RESIDENCE TIME

Residence time *The ratio of the total amount of an element in the ocean at a given time to its rate of replacement.*

The *residence time* of an element (T) is defined as the amount of time that an element remains in seawater before it is removed by organic or inorganic processes. The residence time of an element may be expressed as follows:

$$T = \frac{A}{(dA/dt)}$$

where A is the total weight of element suspended or dissolved in the oceans, and dA/dt is the annual rate of introduction of the element in the ocean. The importance of the residence time of an element is based on two assumptions: (1) that the amount of an element introduced into the ocean per unit time equals the amount deposited as sediment and (2) that elements are uniformly and rapidly mixed in the ocean and that this mixing time is small in relation to residence time. These assumptions stand up fairly well for the purpose of comparing relative residence times of various elements (Table 10-3). Elements with short residence times are highly reactive, such as silicon, iron, manganese, and aluminum. Elements of long residence times are less reactive, such as sodium, potassium, and lithium.

The short residence time of such elements as silicon and iron may be attributed to biological activities, for example, organisms concentrating different elements in their bodies, and chemical scavenging action of hydroxides, particularly in the case of iron and manganese, which absorb ions from solution during the formation of manganese nodules (see Chapter 12). The long residence time of such elements as sodium and potassium is attributed to their having relatively lower chemical reactivities.

10-5 ISOTOPE OCEANOGRAPHY

Isotope oceanography is a relatively new branch of chemical oceanography, which developed with the invention of such precision instruments as mass spectrometers. A standard mass spectrometer can identify two or more species of the same element (*isotopes*), which are insignificantly different in chemical behavior. An isotope of an element has an equal number of protons in its nucleus but a different number of neutrons. For example, regular hydrogen has only one proton in its nucleus, but its isotope, heavy hydrogen (deuterium), has one proton and one neutron. In other words, deuterium has the atomic number of 1 but an atomic mass of 2. Hydrogen and its isotope have a trivial difference in their masses; consequently, they behave nearly but not exactly identically in their physical and chemical properties.

Isotopes *Atoms possessing nuclei with the same number of protons but a different number of neutrons.*

The slight variations in chemical behavior between isotopes arise because atoms are being transferred from one molecule to another.

TABLE 10-3 Concentrations and Residence Times of Elements in Seawater

Category	*Element*	*Concentration (in parts per million)*	*Residence Time (in millions of years)*
Long residence time	Sodium	10,500	260
	Magnesium	1,350	
	Calcium	400	8
	Potassium	380	11
	Strontium	8.0	19
	Lithium	0.17	20
Intermediate residence time	Barium	0.03	0.084
	Zinc	0.01	0.18
	Copper	0.003	0.050
	Manganese	0.002	0.0014
	Cobalt	0.0001	0.018
Short residence time	Aluminum	0.01	0.00015
	Beryllium	6×10^{-7}	0.0001
	Iron	0.01	0.00014
	Chromium	0.5×10^{-5}	0.00035

SOURCE: E. D. Goldberg, 1965, in *Chemical Oceanography*, J. P. Riley and G. Skirrow (eds.), New York, Academic Press.

By employing carbon-14 (radiocarbon), an isotope of regular carbon-12, as a radioactive tracer, an oceanographer can determine the rates of deep-ocean circulation. The rates of ocean circulation (mixing of currents) are determined by following a particular designated water mass from its source. If the amount of time the water mass takes to cover a given distance is known, we can ascertain the rate of water movement. In this way, the dissolved radioactive tracer (^{14}C) signals the general speed of the water mass. Such radioactive tracer studies show that Antarctic bottom waters flow very slowly and have taken 600 years to complete their journey from 60°S to 30°N. Similar studies show that the average speed of Pacific deep waters, for example, is 0.05 centimenter per second.

Isotopic distribution varies in seawater according to depth and latitude. For example, surface waters are richer in deuterium than are bottom waters, and waters in temperate regions are poorer in heavy isotopes than are those at the equator. This uneven distribution of hydrogen isotopes occurs because of the greater evaporation on surface waters in equatorial regions. Evaporation transports lighter isotopes such as regular hydrogen and oxygen-16, thus leaving behind a higher content of residual deuterium and oxygen-18.

At different temperatures, an isotope of oxygen (^{18}O) is observed in calcium carbonate in relation to seawater. In this way, it is possible to apply the oxygen isotope ratio ($^{18}O/^{16}O$) of ancient carbonates to determine prehistoric seawater temperatures. In 1958 Caesar Emiliani deciphered a Pleistocene paleotemperature record (Ice Age) from $^{18}O/^{16}O$ data of carbonates in *Globigerina* ooze in deep-sea cores obtained from the Atlantic, the Caribbean, and the Mediterranean. Emiliani constructed a systematic climatic history of seawater that showed various fluctuations of temperatures during the last 300,000 years (Figure 10-9).

Radioactive Isotopes

Radioactive isotopes are employed in the age determination of deep-sea sediments, of rates of sedimentation, and of correlation of important

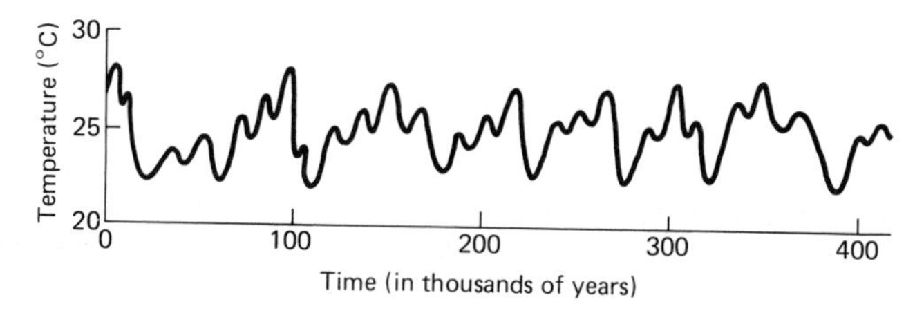

Figure 10-9. Generalized paleotemperature curve for Caribbean surface waters based on oxygen isotope variations in planktonic foraminifera for past 400,000 years. (*Source:* After C. Emiliani, 1972, "Quaternary Paleotemperatures and the Duration of the High-Temperature Intervals," *Science, 178,* 398–401.)

events in oceans and on land. The radioactive isotopes proactinium (^{231}Pa) and ionium (^{230}Th) are most commonly used in the dating of deep-sea sediments. ^{231}Pa has a half-life of 34,300 years, and ^{230}Th has a half-life of 75,000 years.

Half-life *The amount of time required for half the atoms of a radioactive isotope to decay to an atom of another element.*

In principle, ^{231}Pa and ^{230}Th behave alike chemically. Both are precipitated in seawater by the radioactive decay of uranium, and both are removed rapidly and deposited on the seafloor. The amounts of ^{231}Pa and ^{230}Th are greater on the seafloor then on land. Assuming that excess ^{231}Pa or ^{230}Th is deposited on the sea bottom at a constant rate, along with the sediment which too accumulates at a constant rate, deep-sea sediment can be dated. The ^{231}Pa method can date sediments up to 150,000 years old and ^{230}Th method up to 300,000 years old.

Radiocarbon (^{14}C)

Radiocarbon (^{14}C) is formed in the upper atmosphere when cosmic rays interact with nitrogen (Figure 10-10). Radiocarbon (^{14}C) disintegrates spontaneously, reverting to ^{14}N. The half-life of ^{14}C is 5565 years. ^{14}C dating has proved useful in establishing geologic records of the late Pleistocene and in archaeological dating of artifacts and related objects.

Carbon-14 *A radioactive isotope of carbon with atomic weight 14, produced by collision between neutrons and atmospheric nitrogen. Useful in determining the age of objects younger than 30,000 years. Half-life is 5,565 years.*

The ^{14}C dating technique is based on the fact that the ratio of the carbon isotopes (^{12}C and ^{14}C) in living plants and animals remains nearly identical with their corresponding ratios of carbon dioxide in the atmosphere. When a living organism, for example, an animal, ceases to live, the supply of radiocarbon from the air is stopped. As a result, the isotopic ratios between the radioisotope (^{14}C) and the stable isotope (^{12}C) begin to decline. The ratio of radiocarbon to the stable isotope of carbon in a dead organism compared with the corresponding ratio in the atmosphere provides a means of determining the time elapsed since the death of the organism.

^{252}Californium

Californium (^{252}Ca), an isotope of uranium, is used in nuclear prospecting for minerals. This isotope has a half-life of 2.5 years. Neutrons emitted during its decay are absorbed by minerals, which in turn emit characteristic gamma rays, and soon the entire ore body emits these rays. The germanium-lithium detector sealed on the probe head of the sensing device can quantitatively reveal the concentration of seafloor mineral deposits. This method is used for locating ores of gold, silver, uranium, vanadium, copper, aluminum, flourine, tin, silicon, sodium, and iron.

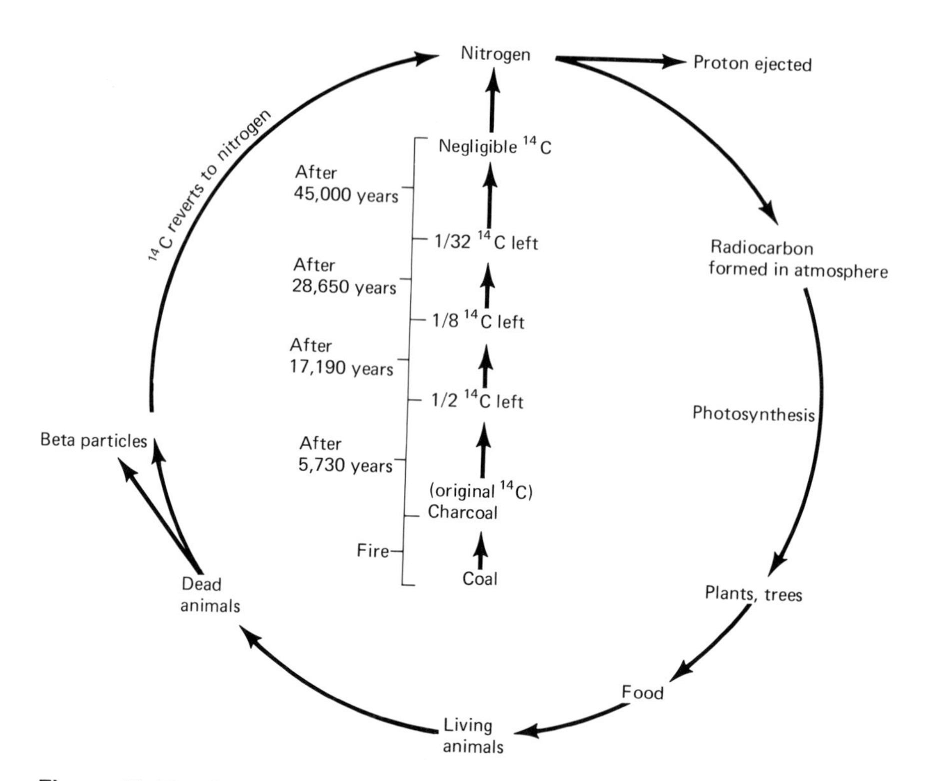

Figure 10-10. Radiocarbon (^{14}C) is formed in the upper atmosphere by the bombardment of cosmic rays on nitrogen. Subsequently ^{14}C, along with regular carbon (^{12}C), is passed on to plants via photosynthesis and then on to living animals as food. Up to this stage, ^{14}C:^{12}C are equal, but once the animal is dead, ^{14}C reverts to nitrogen. As a result ^{14}C:^{12}C gradually change. If plants are buried and turn into charcoal again, ^{14}C will revert to nitrogen at a constant rate. The amount of ^{14}C residue either in a dead animal or charcoal is the index of age.

SUMMARY

1. Chemical oceanography is the exploration of the chemical nature of oceans, the composition of seawater, and the distribution of elements and their isotopes.
2. A variety of methods is used to determine the composition of seawater.
3. Seawater is a solution composed of solvent and solute. Water is the solvent, and dissolved matter is its solute.
4. Seawater is a 96.6 percent pure compound of hydrogen and oxygen; the remaining 3.4 percent contains dissolved salts of chlorine, sodium, magnesium, sulfur, calcium, and potassium.
5. Seawater contains minor elements, trace elements, particulate matter, and organic matter.

6. Seawater contains several dissolved gases; oxygen and carbon dioxide are the most important.
7. The amount of time that an element spends in seawater before its removal by organic or inorganic activity is its elemental residence time.
8. Isotopic studies of various elements found in seawater or in sediments provide useful clues to the ages of sediment, rates of sedimentation, rates of mixing, and temperatures of the geologic past.

Suggestions for Further Reading

Broecker, W. S. 1974. *Chemical Oceanography.* New York: Harcourt Brace Jovanovich.

Martin, D. F. 1970. *Marine Chemistry,* volume 2: *Theory and Application.* New York: Marcel Dekker.

McIntyre, F. 1970. "Why the Sea is Salt." *Scientific American, 893.*

National Academy of Sciences. 1971. *Marine Chemistry: A Report of the Marine Chemistry Panel of the Committee on Oceanography.* Washington, D.C.: U.S. Government Printing Office.

Riley, J. P., and R. Chester. 1971. *Introduction to Marine Chemistry.* New York: Academic Press.

Wilson, T. R. S. 1975. "Salinity and the Major Elements of Sea Water." In *Chemical Oceanography,* volume 1, second edition (J. P. Riley and G. Skirrow, editors). New York: Academic Press.

11 Life in the Ocean

11-1 INTRODUCTION

DURING THE *Challenger* EXPEDITION BETWEEN 1872 AND 1876, scientists collected thousands of marine organisms. This collection was the basis for the systematic description and classification of many marine plants and animals. In the twentieth century, however, marine biologists shifted their emphasis from a purely taxonomic one to an internal study of organisms themselves; to the chemistry of seawater; to extracting vital vitamins and drugs from marine life; to training dolphins and porpoises for undersea missions; and to developing new techniques in fisheries and mariculture.

11-2 MAJOR ENVIRONMENTAL DOMINIONS

The marine environment is characterized by complex and intricately interdependent organic systems and is divided into two principal realms: pelagic and benthic (Figure 11-1).

The pelagic realm is divided into the *neritic province* and the *oceanic province*. The neritic province is the entire watermass above the continental shelf. It is a near-shore environment extending to a depth of 200 meters. The oceanic province consists of all the water seaward from the continental shelf. These two provinces meet at the edge of the continental shelf. The neritic province is characterized by the following: (i) it is the most populated area in the ocean because of its shallow depth, resulting in its receiving the most sunlight; (ii) it is much more influenced by continental processes such as river inflows and seasonal variations affecting salinity and temperature distribution; (iii) the upwelling nutrient-rich currents have made it the most fertile area in the

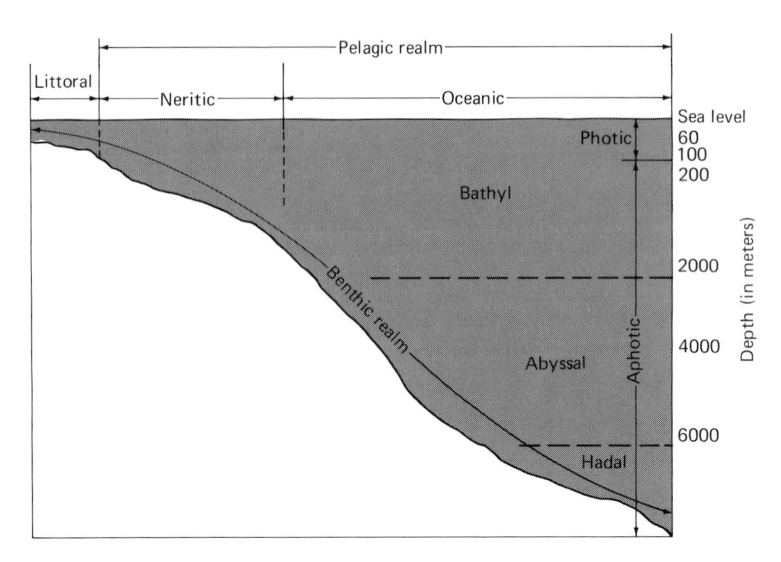

Figure 11-1. Major biozones of the ocean.

ocean, as witnessed by the well-developed fisheries on the continental shelf (see Chapter 13). The oceanic province is characterized by the following: (i) it has great geographic distribution and unmatched ranges of depth; (ii) it has relatively uniform temperature and salinity distribution; and (iii) it receives greater sunlight both horizontally (in terms of geographic distribution) and vertically (in terms of depth). The greater penetration of light with depth is largely due to the relatively clear water of the open ocean.

Benthic *Referring to that part of the ocean bottom populated by organisms.*

Littoral *The nearshore environment.*

The benthic realm is divided into the *littoral, bathyal, abyssal,* and *hadal* zones.

The littoral zone is divided into two subdivisions: the *eulittoral* and the *sublittoral*. The eulittoral zone extends from the high-tide level to a water depth of 60 meters. Most holdfast plants grow in this zone. The sublittoral zone marks the marine environment between 60 to 200 meters deep.

Bathyal *An oceanic zone between depths of 200 to 2,000 meters.*

Abyssal *Referring to the ocean zone of depths 2000 to 6000 meters.*

Hadal *Refers to the deepest marine environment, over 6000 meters deep.*

The deep-sea benthic zones include the bathyal, abyssal, and hadal. The bathyal zone extends from the continental shelf from depths of 200 meters to 2000 meters. The bottom of this zone is usually either rocky or muddy. There are a number of bottom-dwelling organisms living in this zone. The abyssal zone extends from depths of 2000 to 6000 meters. The general population is much lower than in the upper zones. The hadal zone extends from 6000 meters to the deepest trenches. In 1960 the bathyscaphe *Trieste* found the presence of life in the *Challenger Deep* near the Marianas Trench at a depth of 11,000 meters. On the whole, the population of organisms decreases with depth. Food gathering techniques also change. Most organisms in the deep-sea zones (abyssal or hadal) are carnivorous, living on the organic debris of dead organisms of the upper zones. Moreover, deep-ocean currents play an important role in supplying the necessary oxygen and essential nutrients to these organisms.

The marine environment is also divided into three zones on the basis of penetration of light: (1) the euphotic (well-lighted) zone, (2) the disphotic (poorly lighted) zone, and (3) the aphotic (lightless) zone. The *euphotic zone* extends from the surface layers to a depth of about 80 meters. In this zone, photosynthesis is carried out effectively (Table 11-1).

11-3 MODES OF MARINE LIFE

Marine plants and animals are categorized on the basis of their modes of locomotion and habitats: (1) plankton, (2) benthos, and (3) nekton.

Plankton

Plankton are floaters or wanderers, which, since they have no means of self-propulsion, are carried passively by currents. Most plankton live in shallow water where they absorb sunlight and mineral nutrients (Table 11-2). Plankton are usually microscopic organisms, but there are notable exceptions such as jellyfish and a brown alga, *Sargassum*.

Phytoplankton *Collective term for planktonic plants.*

Zooplankton *Animal plankton.*

Plankton are either phytoplankton (plant plankton) or zooplankton (animal plankton). *Phytoplankton* include microscopic *diatoms,* the bodies of which consist of silicon dioxide and which prefer cooler waters (Figure 11-2). *Dinoflagellates* also precipitate silicon dioxide into their bodies. They have tiny whiplike flagellae that aid in locomotion (Figure 11-3). Dinoflagellates are commonly found in warmer waters and often cause "red tide." *Coccolithophores* are microscopic organisms that have

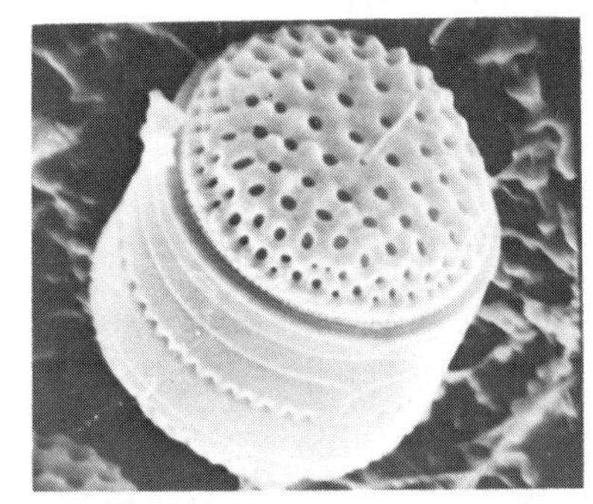

Figure 11-2. A scanning-electron micrograph of a marine diatom (Thallocrassia). Magnification is 5000×. (Photo courtesy of Paul E. Hargraves, Graduate School of Oceanography, University of Rhode Island.)

TABLE 11-1 Basic Data on the Principal Biozones

Name	*Depth (meters)*	*Average Temperature Range (°C)*	*Area Covered (percent of total ocean floor)*
Littoral	up to 200	20–25	8
Bathyal	200–2000	5–15	16
Abyssal	2000–6000	less than 4	75
Hadal	more than 6000	3.5–1.2	1

TABLE 11-2 Vertical Distribution of Plankton in the South Atlantic Ocean near Ascension Island

Depth (in meters)	*Cells per liter*
0	10,147
50	9,443
100	2,749
200	726
400	216
700	114
1,000	87
2,000	57
3,000	18
4,000	17
5,000	15

SOURCE: E. Hentschel in G. Dietrich, 1963, *General Oceanography,* New York, Wiley-Interscience.

protective calcium carbonate plates called coccoliths (Figure 11-4). Coccolithophores are responsible for depositing white flourlike material as ooze (see Chapter 9). Often a dense distribution of phytoplankton in the ocean produces a slimy appearance. The Sargasso Sea is characterized by a heavy population of floating seaweed, particularly by eight com-

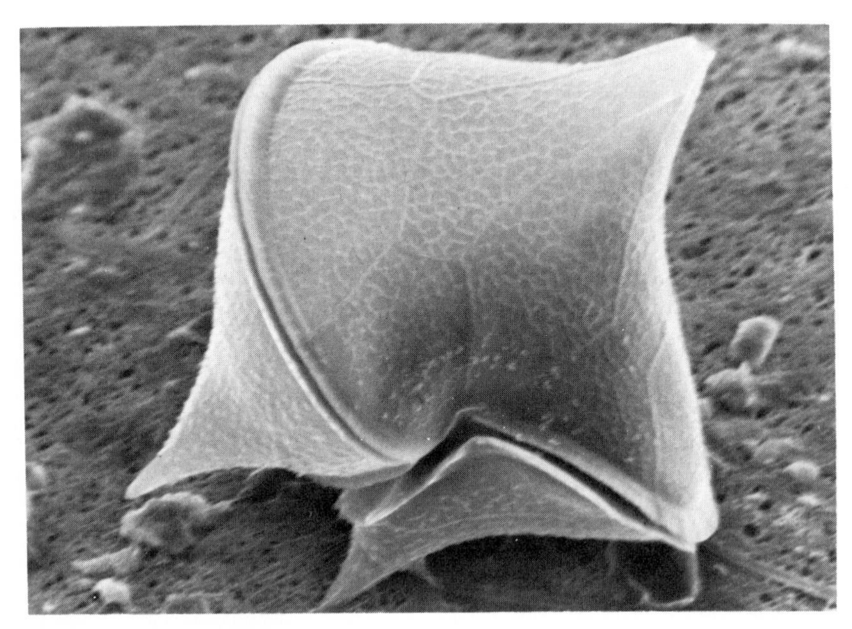

Figure 11-3. A scanning electron micrograph of the marine dinoflagellate *Peridinium.* Magnification 1000×. (Photo courtesy of Paul E. Hargraves, Graduate School of Oceanography, University of Rhode Island.)

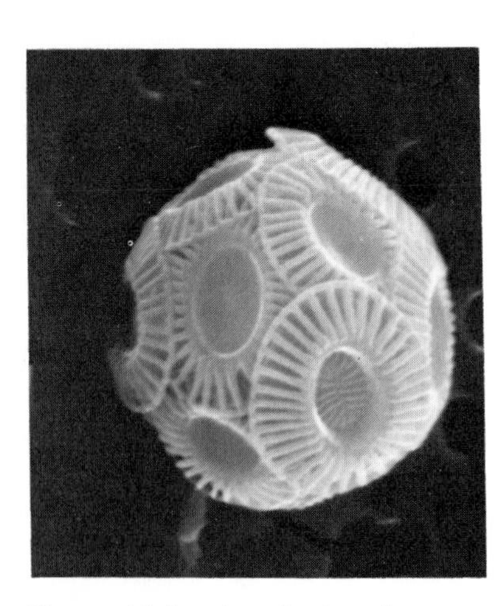

Figure 11-4. A typical marine coccolithophore, *Coccolithus* is a protozoan that provides raw material for lime deposits. Magnification 10,000×. (Photo courtesy of Paul E. Hargraves, Graduate School of Oceanography, University of Rhode Island.)

mon species of *Sargassum* (Figure 11-5), which is inhabited by microscopic plankton. Plankton production in the Sargasso Sea runs from 10 to 1000 million tons per day. Considering the size of the sea, however, this amount is negligible. It is believed that the characteristic seaweed was originally transported in the eddy of the waters to the surface from the seabed but that the weed has since adapted and can now sustain itself without roots.

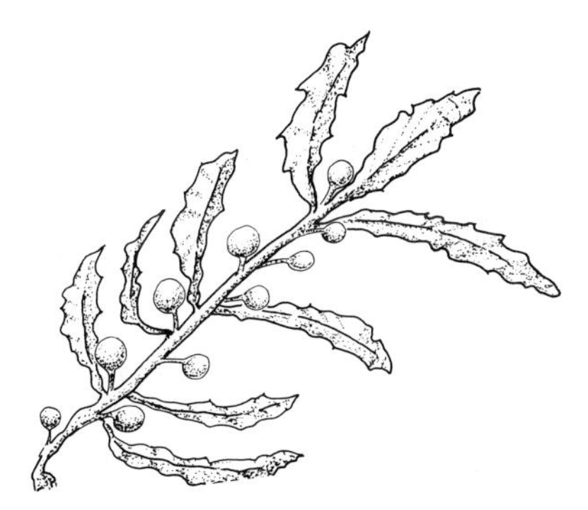

Figure 11-5. *Sargassum* species (in Portuguese, Sarga, meaning grape). This seaweed grows in great quantities in the Sargasso Sea.

Zooplankton are larger and more complex in character than phytoplankton. They include tiny jellyfish, arrowworms, and small crustaceans (copepods). Zooplankton live under varying conditions of temperature, salinity, currents, light, and nutrient supply. Two types of zooplankton are *holoplankton,* which are permanent drifters, and *meroplankton,* which are temporary drifters, spending the earlier stages of their lives as larvae and later modifying into either nektonic (swimmers) or benthic (bottom-dwelling) forms.

Copepods are tiny crustacean organisms (0.3 to 10 millimeters). They can concentrate in their bodies large quantities of phytoplankton, which they then pass on to larger animals (see Chapter 12). Crustaceans larger than copepods are known as krill (Figure 11-6). Krill is the most common food of some whales.

Benthos

Benthos *Organisms that live on the ocean bottom.*

Sessile *Refers to an organism permanently stationary or fixed.*

Benthos are bottom-dwelling organisms which usually follow three modes of life: sessile, creeping, and burrowing. *Sessile* (Latin: *sedere,* low or dwarf) benthos are firmly attached to the seafloor. Sessile plants, largely confined to the euphotic zone, include virtually all seaweeds and eelgrasses. Sessile animals include corals, sponges, barnacles, and oysters. *Creeping* benthos are mainly animals and include lobsters, crabs, and snails, which crawl or bounce on the seafloor. *Burrowing* benthos, such as clams, worms, and some crustaceans, exist by digging into the sediment or rocks of the seafloor (Figure 11-7).

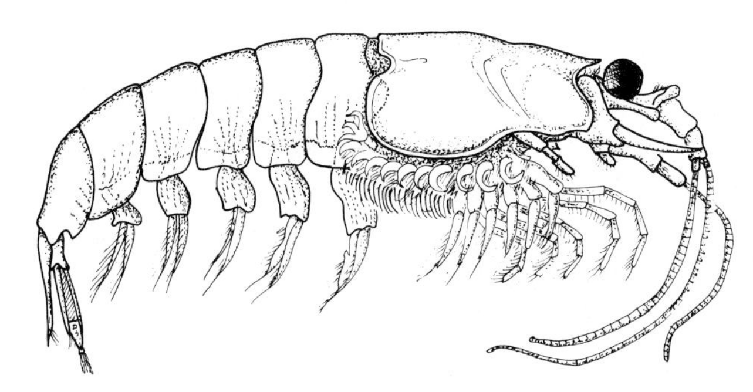

Figure 11-6. The crustacean *Euphausia superba* (5 cm), the krill of the Antarctic Ocean, is a popular food of whales.

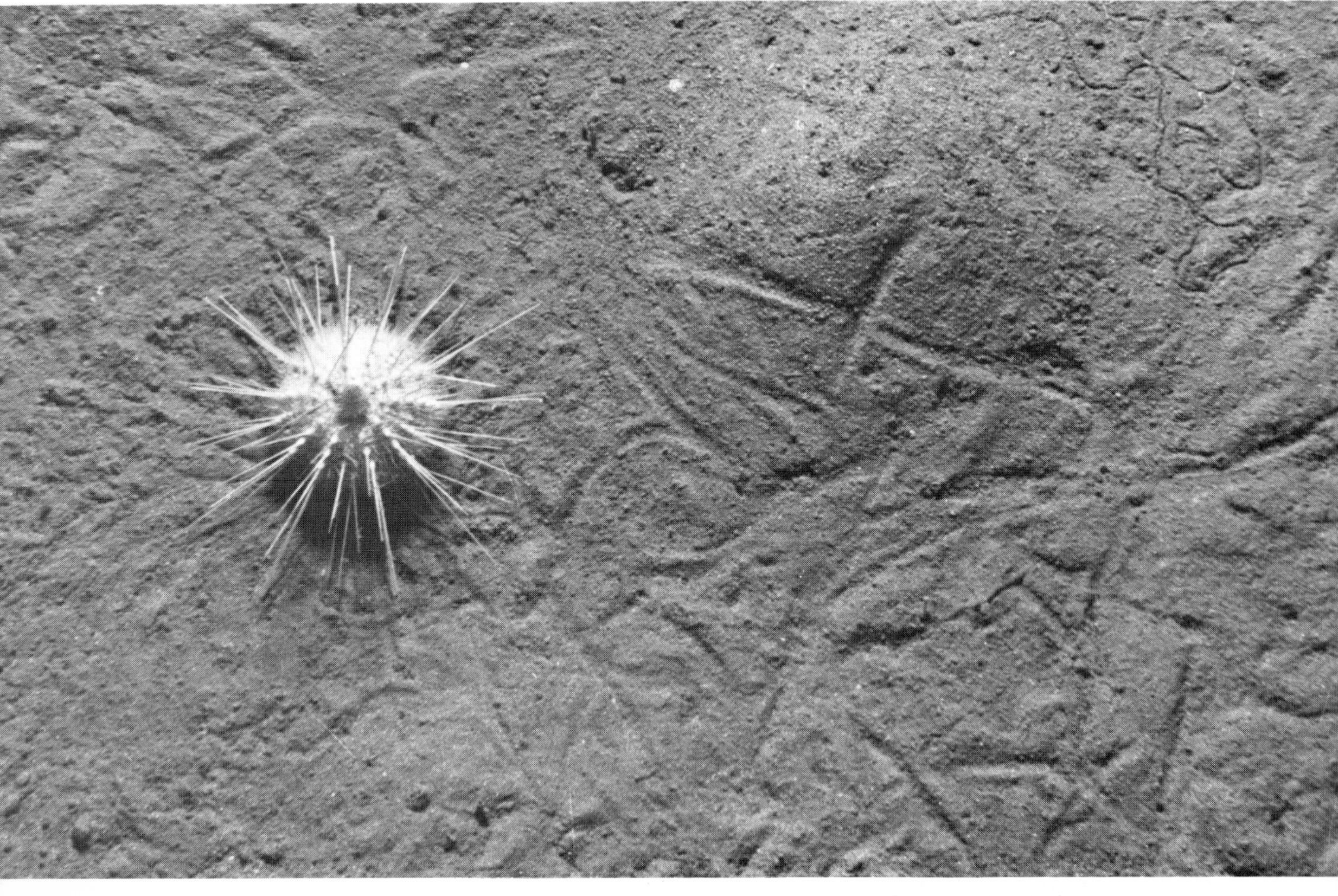

Figure 11-7. A typical benthos may be composed of crinoids (Class Echinoidea), important contributors of organic debris for the formation of marine limestones. Note trail marks left by the burrowing animals. (Photo courtesy of Woods Hole Oceanographic Institution.)

Nekton

Nekton *Pelagic animals that are able to swim independently.*

Nekton are free swimmers and include fishes, whales, dolphins, porpoises, and other mammals. Nekton feed primarily on zooplankton such as krill. In relation to benthos and to plankton, nekton are an advanced form of animal (Figure 11-8). Dolphins and porpoises are regarded among the most intelligent animals of the world. They are easily trained and are capable of carrying out responsible missions, as, for example, messengers, transporters of mail, and carriers of tools to aquanauts.

11-4 OCEAN LIFE

In the eighteenth century, Carolus Linnaeus, a Swedish naturalist, devised a remarkable scheme for organizing countless numbers of animals and plants in terms of their structural similarities. According to his classifications, animal and plant kingdoms are divided first into *phyla* (Greek: stock or race); each phylum includes organisms that share fundamental anatomical characteristics. For example, animals with jointed legs and bodies, such as insects, spiders, and crabs, form Phylum Arthropoda (Greek: *arthron*, joint; *poda*, foot).

Each phylum is divided into *classes*, in which similarities are greater. Classes of vertebrates (Latin: *vertebratus*, having a backbone) include Pisces (fish), Aves (birds), Mammalia (mammals), and Reptilia (reptiles).

Classes are divided into *orders*. Class Mammalia includes, for example, Orders Carnivora (flesh-eaters) and Rodentia (gnawers). Orders are divided into families, which are in turn divided into genera and further into species. The species is the smallest unit in the Linnaeus scheme (Table 11-3). Generic names are always capitalized and that of

Figure 11-8. Food often brings nekton (fish) and benthos (starfish) together. (Photo courtesy of National Marine Fisheries Service.)

TABLE 11-3 The Taxonomic Classification

Linnaeus Scheme	*Vertebrate*
Kingdom	Animalia
Phylum	Chordata
Class	Mammalia
Order	Primates
Family	Hominidae
Genus	*Homo*
Species	*Homo Sapiens*

the species is not; both are italicized. For example, *Homo sapiens* (Greek: *Homo,* man; *sapiens,* wise), designates first the generic, then the specific, name for man. The ocean population in this chapter is treated in a simplified fashion by considering only more common plants and animals, with a brief note on marine bacteria. The interested reader is advised to review one of the references on marine biology listed at the end of this chapter.

11-5 MARINE PLANTS

The ocean is overwhelmingly populated by primitive plant forms, which include various types of algae. Many of these forms live at or near the surface of the ocean and, unlike their terrestrial counterparts, are constantly bathed by water, nutrients, and many tiny organisms. Except for eelgrass (Figure 11-9) advanced forms of plants are virtually absent

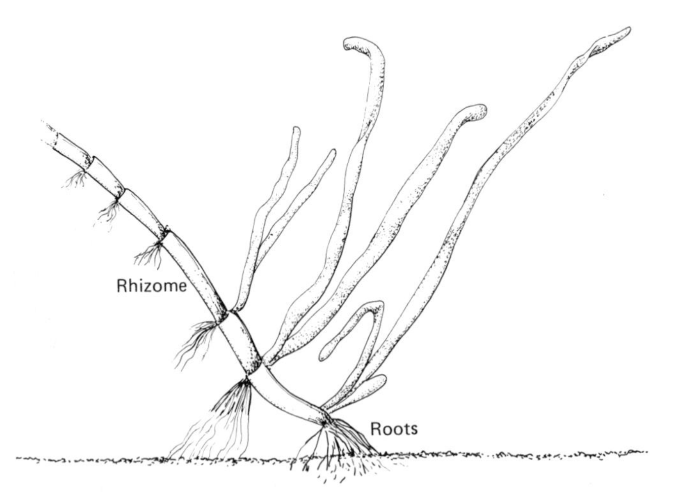

Figure 11-9. Eelgrass *(Zostrea)* is a rare advanced sea plant that grows in shallow water (less than 5 meters) in the nearshore zone. This plant has true roots attached to a stem or rhizome.

from the ocean. Most common marine algae include blue-green, green, red, and brown algae, diatoms, and dinoflagelletes.

Blue-Green Algae

The most familiar type of blue-green algae (Cyanophyta) is *Trichodesmium.* Its cells contain excessive pigment, *phycocrythrin,* which will stain reddish on drifting to the surface. *Trichodesmium* slicks are common in the Red Sea, the Indian Ocean, and the Java Sea, and have occasionally been observed along coasts for many miles. The filamentous blue green algae off Andros Island in the Bahamas form an algal mat in the intertidal zone, trapping and binding fine sediment.

Green Algae

Green algae (Chlorophyta) are confined to shallow waters, usually to depths of less than 10 meters, because of their dependence on sunlight for photosynthesis. These algae contain chlorophyll and are not influenced by other pigments. Common examples of green algae are sea lettuce (*Ulva)* and *(Halimeda)* (Figure 11-10). *Halimeda* is very common in tropical waters. It secretes calcium carbonate and on its death supplies a considerable amount of sand-sized carbonate, raw material for future

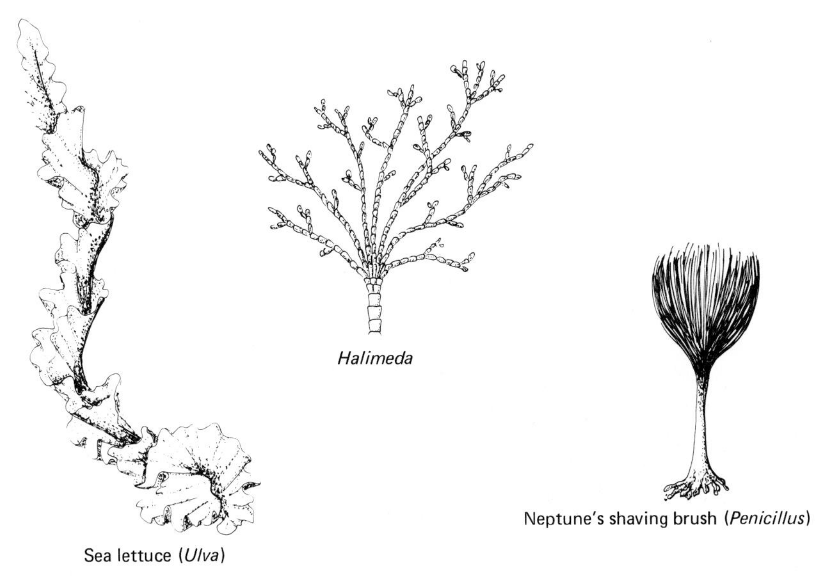

Figure 11-10. Common forms of marine green algae.

Irish moss
(*Chondrus crispus*)

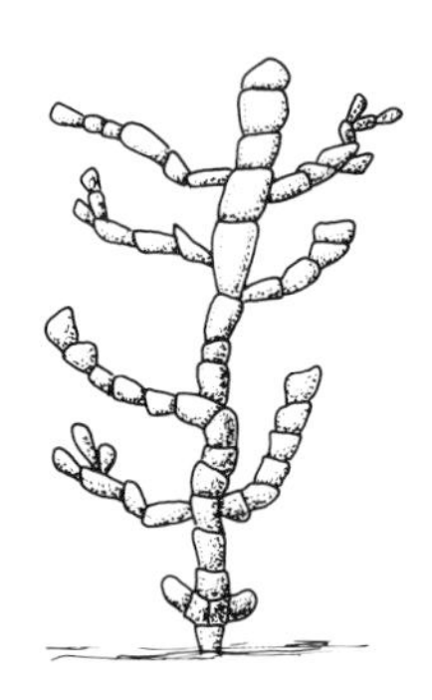

Corallina

Figure 11-11. Common forms of marine red algae.

limestone. Another green algae, Neptune's shaving brush (*penicillus*), is characterized by tiny needle-shaped crystals of calcium carbonate, which provide raw material for the mud-sized carbonate sediment (Figure 11-10).

Red Algae

Red algae (Rhodophyta) are one of the most beautiful of sea creatures, particularly Irish moss (*Chondrus crispus,* Figure 11-11). Red algae usually grow in different shades of red, purple, brown, and green (Figure 11-12). Red algae çontain chlorophyll, which enables them to carry out photosynthesis. Many red algae, notably *Lithothamnion,* contain calcium carbonate in their bodies, which is often rich in magnesium. *Lithothamnion* is one of those important groups of algae that participate in the growth of coral reefs (see Chapter 12). Large quantities of *agar-agar,* a colorless, odorless, tasteless, jellylike substance, are extracted from certain red algae. Agar-agar is used in oriental countries as a gelling or stabilizing agent in foods. It is also used in the manufacture of ice creams and glues.

Brown Algae

Brown algae (Phaeophyta) are the most advanced type of algae. They are exclusively marine and include familiar forms such as kelp and *Sargassum.* Some brown algae such as *Macrocystis and Nereocystis* grow over 50 meters long (Figure 11-13). They are set firmly into the seafloor by a *holdfast.* The holdfast runs into a cylindrical hollow stipe that

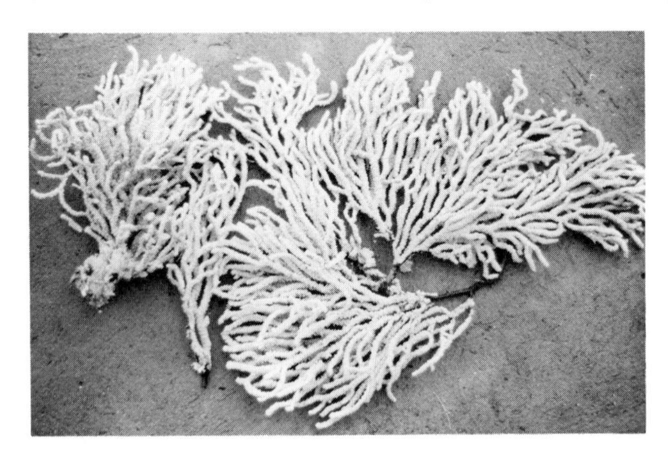

Figure 11-12. A typical Atlantic Ocean red alga, *Primnoa resedaeformis.* (Photo courtesy of R. B. Theroux, National Marine Fisheries Service.)

extends to or near the surface. At the end of the stipe is a gas-filled bulb which gives buoyancy to the plant. Several fronds jetting from the bulb are photosynthetic (Figure 11-13). Brown algae are a vital ocean resource for iodine and potash.

Diatoms

Diatoms are unicellular algae characterized by a bivalved shell known as the *frustule*. The two valves, the *epitheca* and the *hypotheca*, fit together like the halves of a petri dish. These valves are composed of silica, which gives them a glasslike appearance. The glasslike structure, often described as a silica box, is characterized by numerous perforations that help the cell to absorb dissolved gases and nutrients from the sea. The silica box also contains *chloroplasts*, pigments that absorb sunlight to convert carbon dioxide and water to organic matter.

Diatom *Group of algae. Microscopic, one-celled plants usually covered by siliceous matter.*

Diatoms are asexual; they divide, they vary in size, but one cell will match the size of the original cell, leaving the others to decrease until they can get no smaller and still survive (Figure 11-14). The smallest cells, or "spores," begin to grow from their diminutive size into a large cell. The cells of diatoms will grow continually if they are not subjected to severe environmental hazards.

When diatoms die, their silica boxes sink to the seafloor, and extensive deposits of silica-rich *diatom ooze* accumulate. With the passage of geologic time, diatom ooze is solidified into *diatomaceous ooze* or *earth*, a valuable filtering material.

Dinoflagellates

Dinoflagellates are next to diatoms in abundance. Most dinoflagellates are *autotrophic*; that is, they contain chlorophyll and are

Dinoflagellates *Group of algae. Microscopic, one-celled organisms with pairs of hair-like flagella used for locomotion.*

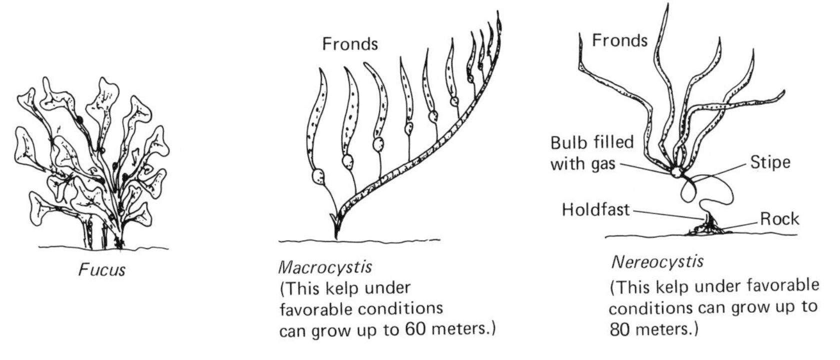

Figure 11-13. Common forms of marine brown algae.

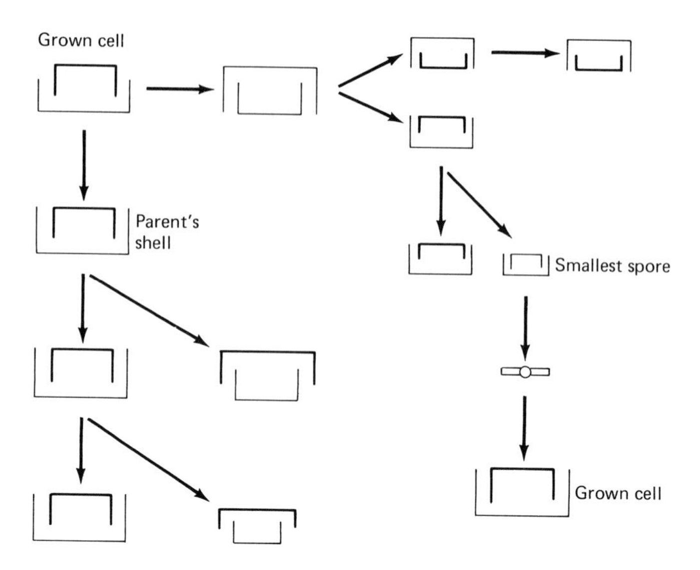

Figure 11-14. Cell division in a diatom.

therefore regarded as photosynthetic, like diatoms. Many are also capable of ingesting particulate matter. A few dinoflagellates are *heterotrophic,* lacking pigments; they obtain their energy from dissolved organic matter in seawater. Dinoflagellates often manufacture significant amounts of extracellular matter, and many are trapped in large globs of gelatinous material.

Coccolithophores are microscopic phytoplankton (5 to 10 microns in size) with intricate structures. They absorb calcium carbonate from seawater and concentrate it in their bodies. Thus these organisms contribute to the formation of lime *ooze* (see Chapter 9).

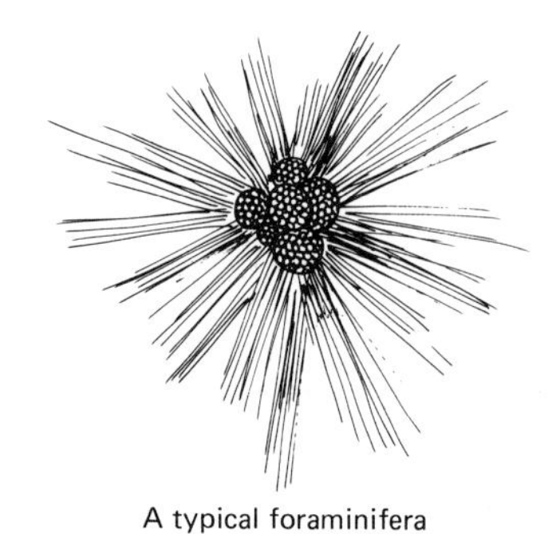

A typical foraminifera

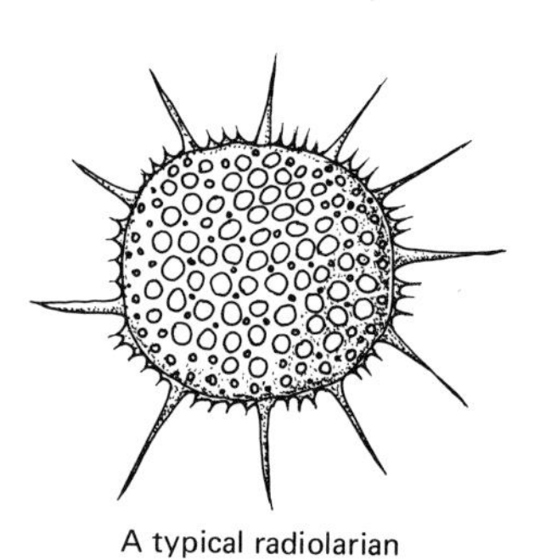

A typical radiolarian

Figure 11-15. Common protozoans.

11-6 MARINE ANIMALS

Animals in the ocean are more diversified than plants in terms of shape, size, and form. They are also more widespread in terms of depth. Marine animals are able to live at great depths where there is permanent darkness owing to their lack of dependency upon sunlight. Plants must depend upon sunlight to synthesize their own food; marine animals do not.

Sarcodina

These primitive microscopic animals lack visceral organs for reproduction, digestion, or blood circulation, and feed on microscopic plants. Two major examples include foraminifers and radiolarians, which are important marine rock builders (Figure 11-15).

Foraminifers are exclusively marine organisms that build tiny-chambered shells of calcium carbonate. Most live either on the seafloor or in seaweeds at surface waters. Among the many types of foraminifers, the plankton foraminifers, notably *Globigerina,* are the most common and extensively distributed (Figure 11-16). Because of their calcium carbonate shells, *Globerigina* form the soft and fine-grained limy deposits of *Globigerina ooze.* It is estimated that about 50 million square miles of the ocean floor are covered to unknown depths by *Globigerina ooze* deposits. *Radiolarians* are characterized by their siliceous shells and form deposits of *radiolarian ooze* in deeper sections of the ocean floor.

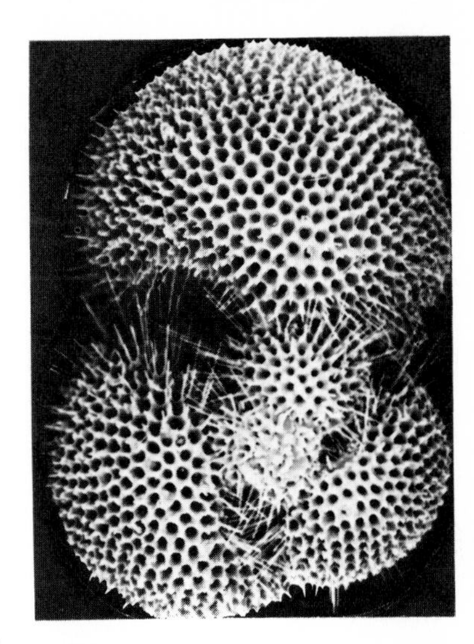

Figure 11-16. The globigerinan *Globigerinoides sacculifer* Brady (190×) is the most extensively distributed species in warm current waters. This specimen was taken from Core V12-126, at 26°06.9′N and 78°12.4′W in waters 902 meters deep. (Photo courtesy of Tsunemasa Saito, Lamont-Doherty Geological Observatory of Columbia University.)

Poriferans

Poriferans include modern sea sponges (Figure 11-17). These are the simplest of multicellular animals and have little specialization of tissues. Because Porifera are primitively developed in relation to other phyla, scientists believe that modern sponges evolved from them. Sponges are self-generating; if they are destroyed or crushed, they can regenerate into a well-developed animal. The wall of sponges are perforated to facilitate the flow of water and nutrients into the body. Once utilized, ingested material is discarded through one or more of the openings called *oscula* at the top of the organism (Figure 11-17).

Coelenterates

Coelenterata *Animals characterized by radial symmetry and a single body cavity performing all the vital functions. Examples include all corals, jellyfish, and sea anemones.*

Coelenterates (Greek: *koinus,* hollow; *enteron,* cavity) are exclusively marine organisms that adapt either to solitary or colonial life styles. Each has a saclike body cavity with a ring of tentacles surrounding the mouth. These animals are like sponges in that they consist of a body wall and have no internal organs. Coelenterates include sea anemones (Figure 11-18), corals, Portuguese man-of-war, sea pens, and jellyfish (Figures 11-19, 11-20). Many of them lead active predatory lives, feeding on plankton, molluscs, crustaceans, and fish. They attack or capture these organisms by paralyzing them with stinging cells (nematocysts).

Coelenterates are divided into three classes: (1) hydrozoans, (2) scyphozoans, and (3) anthozoans. *Hydrozoans* include colonial-living forms, such as the Portuguese man-of-war and jack-sail-by-the-wind. *Scyphozoans* include floating animals such as the jellyfish, which propels itself through the water by oscillating its large bell. Jellyfish may be as large as 2 meters in diameter. *Anthozoans* include primarily benthic and colonial animals, such as sea anemones and corals. Although corals and

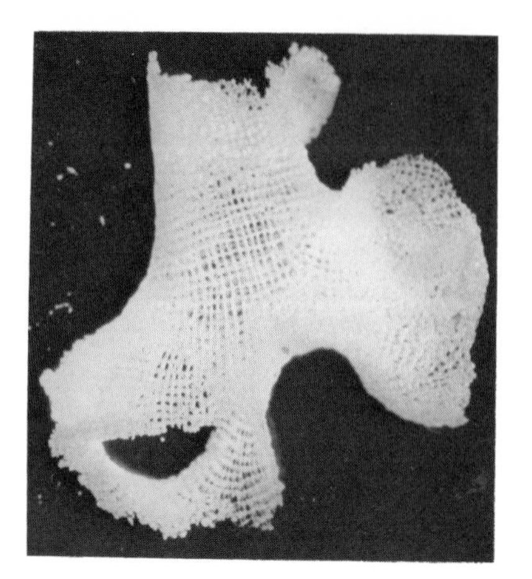

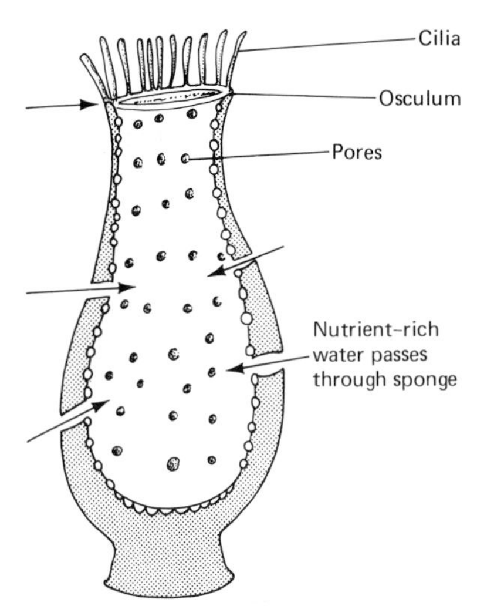

Figure 11-17. *Top:* Siliceous sponge. (Photo courtesy of R. B. Theroux, National Marine Fisheries Service.) *Bottom:* Details of a typical sponge.

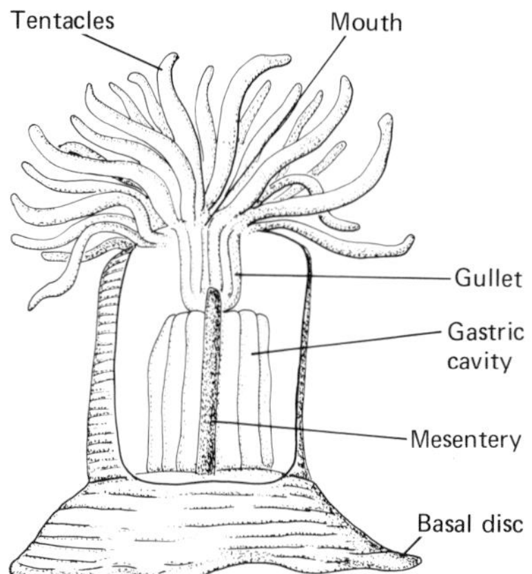

Figure 11-18. A sea anemone maintains a typically benthic life style. Note organic debris on the sea floor. (Photo courtesy of R. B. Theroux, National Marine Fisheries Service.)

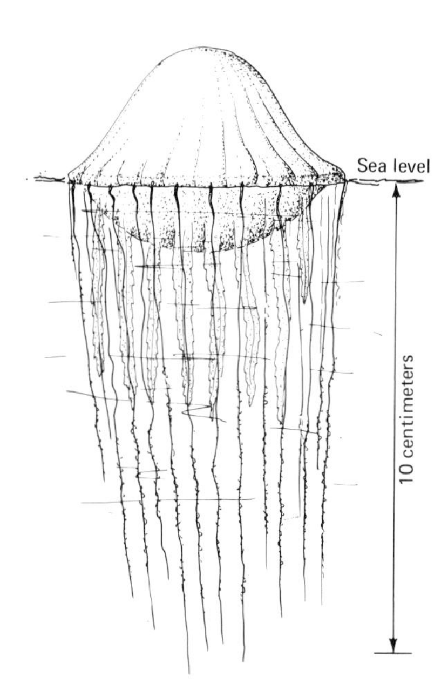

Figure 11-19. A jellyfish.

sea anemones belong to the same class, they differ in their structures; corals form protective calcareous exoskeletons around their polyps.

Echinoderms

Echinoderms (Greek: *echinos,* spine; *dermata,* skin) are exclusively marine animals characterized by spiny skins. They include starfish, sea urchins, sea cucumbers, and sea lilies. The five classes of echinoderms are: (1) asteroideans, (2) ophiuroids, (3) holothurians, (4) crinoids, and (5) echinoids.

Asteroids, including sea stars and starfish, are characterized by five arms that radiate from a central plate, which contains the mouth (Figure 11-21). *Ophiuroids,* in contrast to asteroids, are characterized by elongated, brittle arms that are longer than the central plate; they live in deeper waters. The brittle star is a good example (Figure 11-22). *Holothurians,* or sea cucumbers, are characteristically long and slender, preferring a benthic life (Figure 11-23). The *Crinoids,* such as sea lilies, are solitary animals that live attached to the seafloor. *Echinoids* include sea urchins and sand dollars (Figures 11-24, 11-25).

Figure 11-20. A coral (Phylum Coelenterata). The body of this organism is composed of calcium carbonate, a raw material in marine limestones often rich in petroleum. (Photo courtesy of R. B. Theroux, National Marine Fisheries Service.)

Molluscs

Molluscs include several organisms that share a fundamentally similar body plan, characterized by a soft, unsegmented body and a well-developed foot. Five molluscs include: amphineurans, scaphopods, gastropods, pelecypods, and cephalopods.

Amphineurans (chitons) are bottom-dwelling animals with shells usually ornamented with eight plates. Scaphopods (tusk shells) have long and slightly curved shells (Figure 11-26)

Gastropods include snails and slugs (Figure 11-27). Over 20,000 species have been identified. Gastropods are found in sea- and freshwater and on land in arid, wet, and cold environments. Gastro-

Figure 11-21. The starfish, an exclusively marine invertebrate, has five radiating symmetrical arms and a calcareous skeleton. It usually lives in shore waters. (Photo courtesy of R. B. Theroux, National Marine Fisheries Service.)

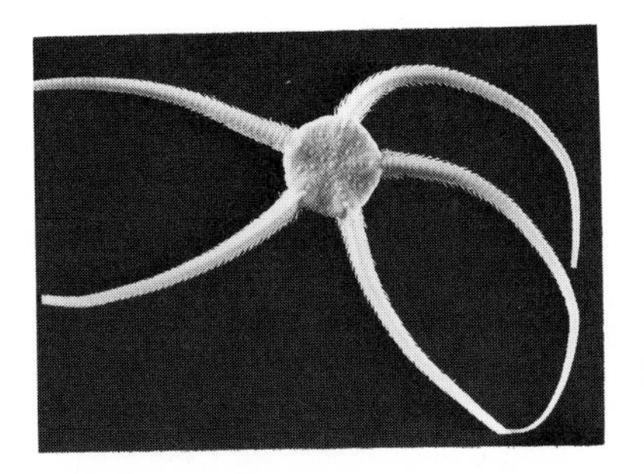

Figure 11-22. The brittle star, an exclusively marine invertebrate, has five radiating symmetrical arms and a calcareous skeleton. It usually lives in deep waters. (Photo courtesy of R. B. Theroux, National Marine Fisheries Service.)

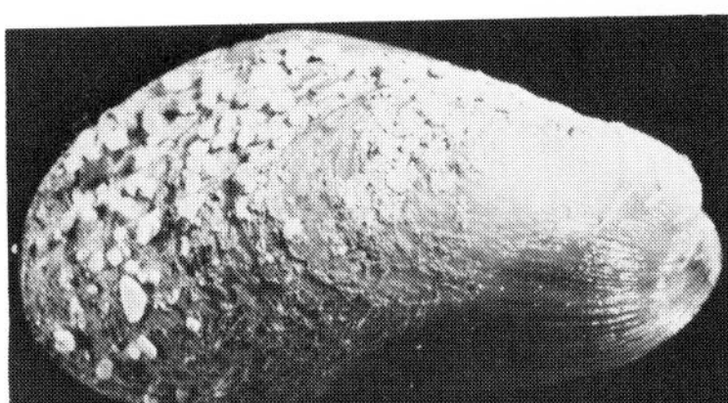

Figure 11-23. The sea cucumber is a free-living marine invertebrate. (Photo courtesy of R. B. Theroux, National Marine Fisheries Service.)

Figure 11-24. Sea urchins are exclusively marine invertebrates. (Photo courtesy of National Marine Fisheries Service.)

(a)

(b)

Figure 11-25. Top and bottom views of the sand dollar, a typical benthonic invertebrate characterized by calcareous plates and five radiating arms at the center. (Photo courtesy of R. B. Theroux, National Marine Fisheries Service.)

pods have well-developed heads equipped with eyes and, in some cases, with sensory organs. They also have well-developed feet, which are usually withdrawn inside the shell when the animal is at rest. Gastropods have a highly specialized mouth equipped with *radulae*. A radula has thousands of tiny, sharp teeth that constitute a rasp for tearing food. The radula is the main defensive mechanism of gastropods. Most gastropods are herbivores feeding on the plants that grow on the seafloor. Some, however, are predatory carnivores.

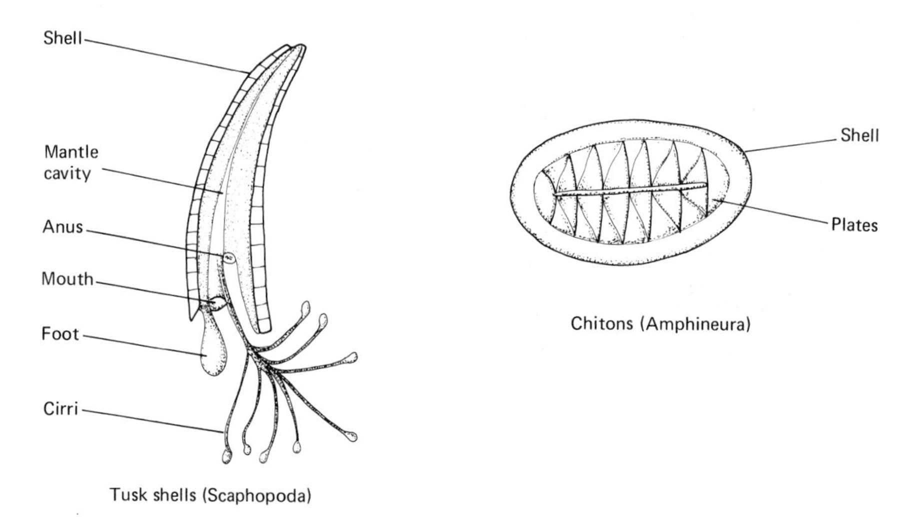

Figure 11-26. Examples of scaphopods and amphineurans.

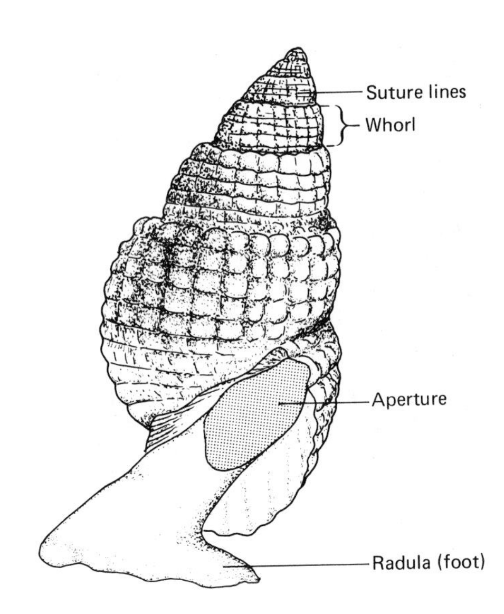

Figure 11-27. A typical gastropod and details. (Photo courtesy of R. B. Theroux, National Marine Fisheries Service.)

Pelecypods are marine bivalves represented by such well-known seafoods as oysters, clams, scallops, and mussels. The shells are similar in shape and articulate along a hinge line attached by teeth and muscles. Bivalves have right and left shells. These organisms usually live buried in bottom mud and feed by way of long siphons that draw nutrient-rich water to their mouths (Figure 11-28).

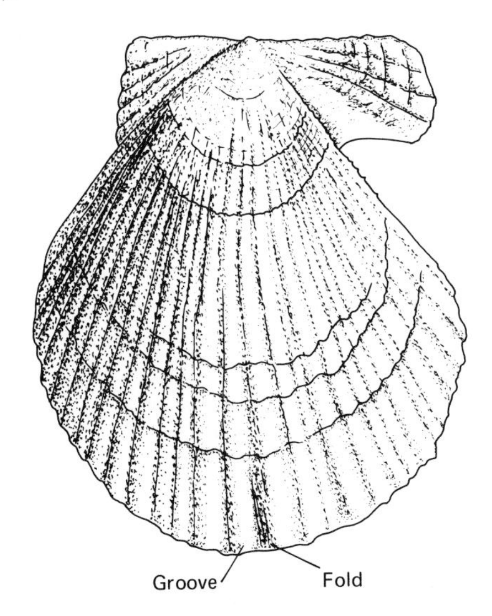

Figure 11-28. The typical clam *Pecten* is a mollusc. (Photo courtesy of R. B. Theroux, National Marine Fisheries Service.)

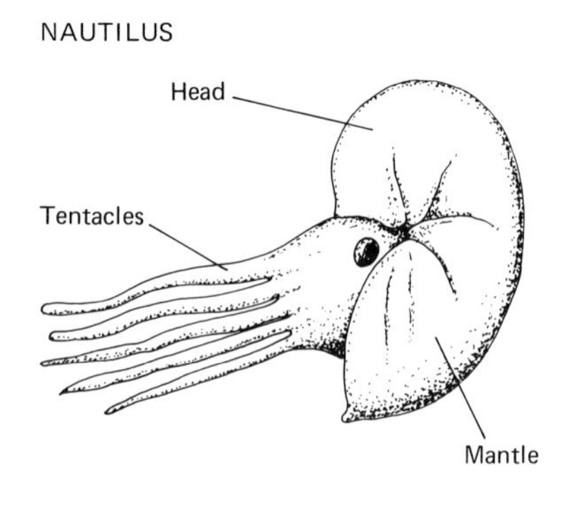

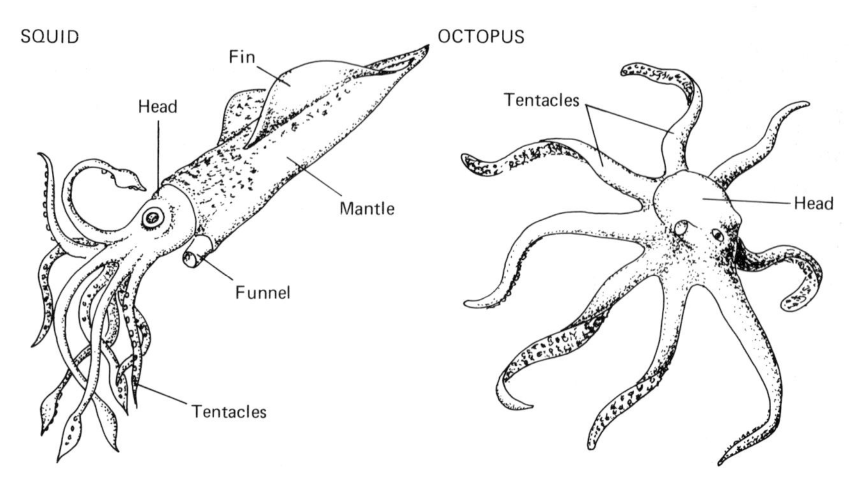

Figure 11-29. Common cephalopods.

Cephalopods are the most advanced molluscs and include squids, nautili, and octupi (Figure 11-29). Cephalopods are exclusively marine and lead active predatory lives. They have well-developed heads engulfed by tentacles. Some squids attain a length of 16 meters on spreading their 10 tentacles. An octopus has 8 tentacles.

Brachiopods

Brachiopods, commonly known as lampshells, resemble common bivalves such as pelecypods. However, brachiopods are different from clams; they have top and bottom shells. These organisms are attached to rocks by tubes (Figure 11-30).

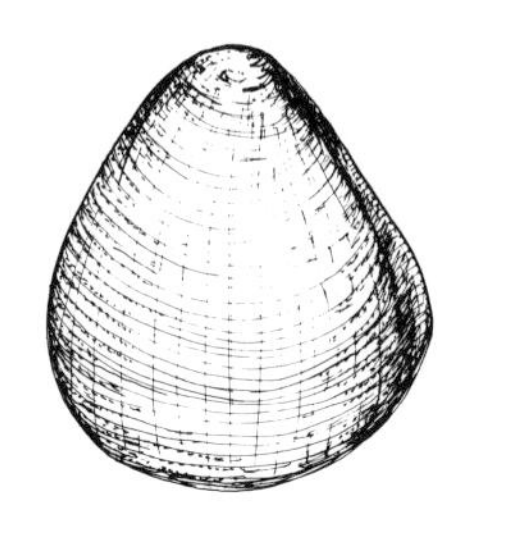

Bryozoans

Bryozoans (Greek: *bryon,* moss) or moss animals are colonial marine organisms that secrete chitinous or calcareous matlike skeletons. They are commonly attached to rocks, boats, and seaweed in coastal areas. Bryozoans show a superficial resemblance to corals because of their tentacles. But structurally they are quite advanced; each has a complete digestive tract.

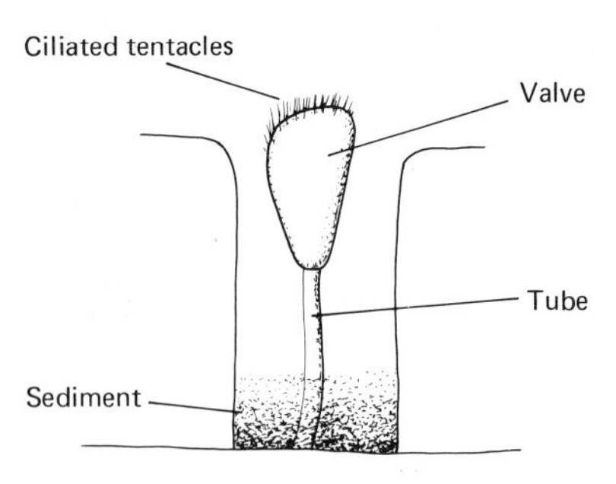

Figure 11-30. Lampshell *(top)* and *(bottom)* burrowing brachiopods.

Ctenophores

Ctenophores are small, fragile jellylike plankton. The most common example of ctenophores are sea gooseberries and comb jellies. These are exclusively marine animals and feed primarily on smaller organisms.

Marine Worms

Marine worms are usually long and slender and are either benthic or planktonic in their life styles. They are grouped into three phyla: (1) flatworms, (2) arrowworms, and (3) segmented worms.

Flatworms are primarily benthic and commonly live on or under rocks and sediment. Many flatworms are carnivorous and supplement their food requirements by consuming dead fish, clams, and other soft-bodied animals. Flatworms often live in these animals as parasites.

Arrowworms are exclusively marine planktonic organisms. They are widely distributed, both geographically and through sea depth. They eat fish, fish larvae and, other small soft-bodied animals. The most common arrowworm is *Sagitta,* which has grasping spines near its head (Figure 11-31). These spines are active in the capture of prey.

Segmented worms are called annelids and are mainly planktonic.

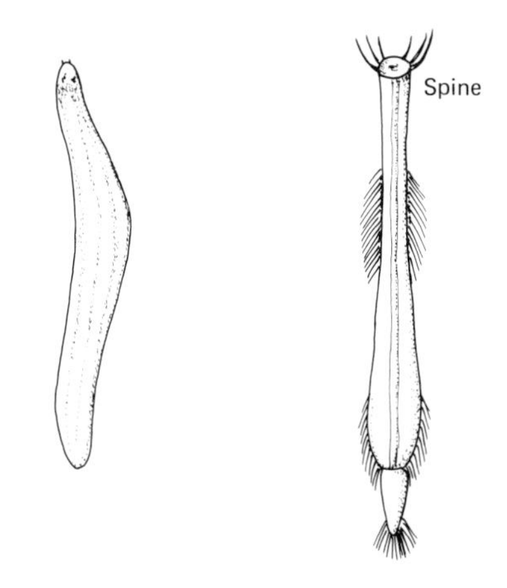

Figure 11-31. The flatworm *(left)* and the arrowworm *(right)* are examples of marine worms.

Arthropods

Arthropoda *A phylum of segmented animals with jointed appendages on each segment. Examples are insects, crabs, and lobsters.*

Arthropods are the largest group in the animal kingdom and comprise over 90 percent of all known living species. Arthropods have segmented bodies; paired, jointed limbs; and hard body shells with movable joints. These rather advanced animals have well-developed reproductive, circulatory, digestive, and nervous systems. Arthropods are divided into various classes of which three are most important: insects, arachnids, and crustaceans.

Although *insects* dominate the animal population on earth, they are primarily terrestrial and are not within the scope of this book. Marine *arachnids* include sea spiders. But the most dominating arthropods are *crustaceans*. Marine crustaceans include crabs, lobsters, shrimps, and barnacles. Crustaceans have two pairs of antennae, and most breathe through gills (Figure 11-32, 11-33, 11-34).

Copepods are important crustaceans in terms of numbers, life style, and overall economy of the seas. More than 5000 species of copepod exist, about 70 percent as free-living organisms and the remainder parasites. Free-living copepods are generally small, between 0.3 and 8 millimeters (Figure 11-35). *Calanus finmarchicus* is a copepod about the size of a wheat grain (Figure 11-36). This genus is abundantly distributed in the North Atlantic and often imparts a red color to water. It is also found in the Arctic and in the North Pacific. The copepod *Calanus finmarchicus* is a principal food of herring, mackerel, and baleen whales. The *Calanus* species migrates vertically off the Scandinavian coast. They migrate to deeper water during winter and move to the surface in spring. Most of their breeding activity takes place during summer. As winter approaches, they again move down into water of about 300 meters (see Chapter 12).

Figure 11-32. The crab is a thick-shelled arthropod characterized by a jointed or segmented body, ten appendages, and compound eyes. (Photo courtesy of R. B. Theroux, National Marine Fisheries Services.)

Figure 11-33. The lobster is a thick-shelled arthropod characterized by two large pincer claws, eight additional appendages, and a jointed tail. (Photo courtesy of National Marine Fisheries Service.)

Marine Fishes

Over 25,000 species of fish are found in nearly all water bodies, including oceans, lakes, and rivers. However, marine fish dominate the overall population in terms of numbers, diversity, and size.

Fish belong to Phylum Chordata and all possess gill slits. Fish fall into three classes: (1) agnathans, (jawless fish), (2) cartilaginous fish (sharks and rays), and (3) bony fishes. *Agnathans* once were common in the oceans in the geologic past. Lampreys and hagfish are today the most common species of agnathans (Figure 11-37). *Cartilaginous* fish include such predators as sharks and rays, which are characterized by skeletons composed of cartilage rather than bone, open gill slits and well-developed teeth (Figures 11-38, 11-39). *Bony fish* are the most

Figure 11-34. A relative of the krill found in the Atlantic Ocean. (Photo courtesy of R. B. Theroux, National Marine Fisheries Service.)

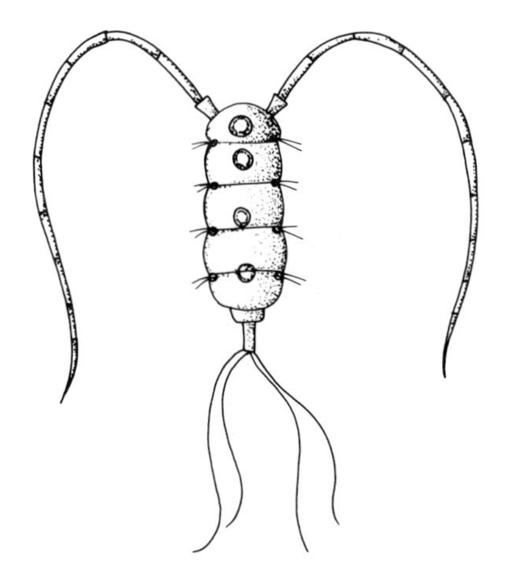

Figure 11-35. A typical copepod.

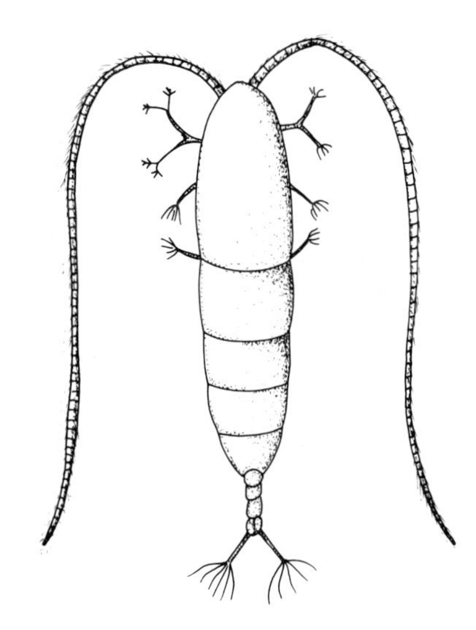

Figure 11-36. The copepod *Calanus finmarchicus.*

abundant and widely distributed group of fish (Figure 11-40). They are characterized by bony skeletons and slimy, scale-covered bodies, large mouths, and wide eyes. Lungfish are a special type of bony fish that has a modified air bladder for breathing air.

Fish are also classified as demersal or pelagic (Figure 11-41). All fish that reside near the bottom of the sea are termed *demersal.* They include roundfish (such as cod, haddock, and hake) and flatfish (such as flounder, sole, and halibut). Demersal fish lay floating eggs in large numbers. For example, a well-developed female European flounder may lay a half million eggs, but only a few of the eggs manage to survive and grow. *Pelagic* fish are found in the waters above the seafloor. Part of their lives are spent near the surface. They include important commercial fish, such as herring, pilchard, sardine, and mackerel (see Chapter 13).

Demersal fishes *Those species living near the sea bottom.*

Reptiles, Birds, and Mammals (Phylum Chordata)

Marine reptiles are relatively scarce. Approximately 50 species of sea snake are commonly found in shallow waters (less than 200 meters). Most sea snakes are found in the Pacific and Indian oceans.

Birds associate closely with the sea for food, but they prefer land for nesting. Penguins are skilled swimmers. They are highly concentrated in the Antarctic, except for a single species found in the Galapagos Islands, off Ecuador (Figure 11-42). Albatrosses are found in all oceans of the world, and skuas can be seen in the Northern Atlantic. Another common sea bird is the herring seagull.

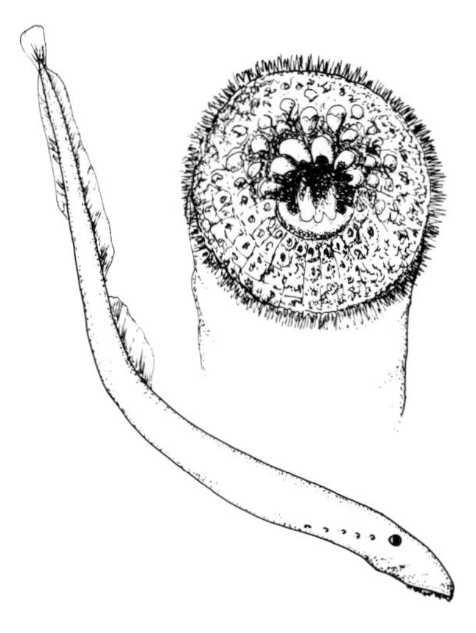

Figure 11-37. *Left:* The lamprey is an agnathan (jawless fish). *Right:* Detail of sucking mouth.

Figure 11-38. Cartilaginous fish. *Top left:* Ray *(Raja clarata)*. *Top right:* Whale shark. *Middle:* Basking shark. *Bottom:* Great white shark.

Figure 11-39. A zebra-striped fish does not seem to fear a blue shark. (Photo courtesy of Harold Pratt, National Marine Fisheries Service.)

Marine mammals include whales, porpoises, dolphins, walruses, seals, manatees, and dugongs (Figure 11-43). They are usually excellent swimmers and divers and spend a considerable amount of time in the open sea. They often migrate hundreds of kilometers in groups.

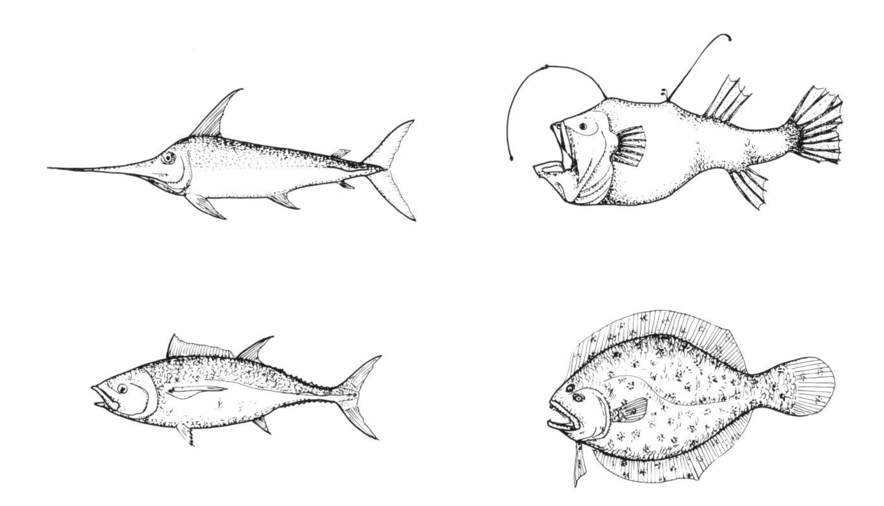

Figure 11-40. Bony fish. *Top left:* Swordfish. *Top right:* Angler fish. *Bottom left:* Tuna. *Bottom right:* Flounder.

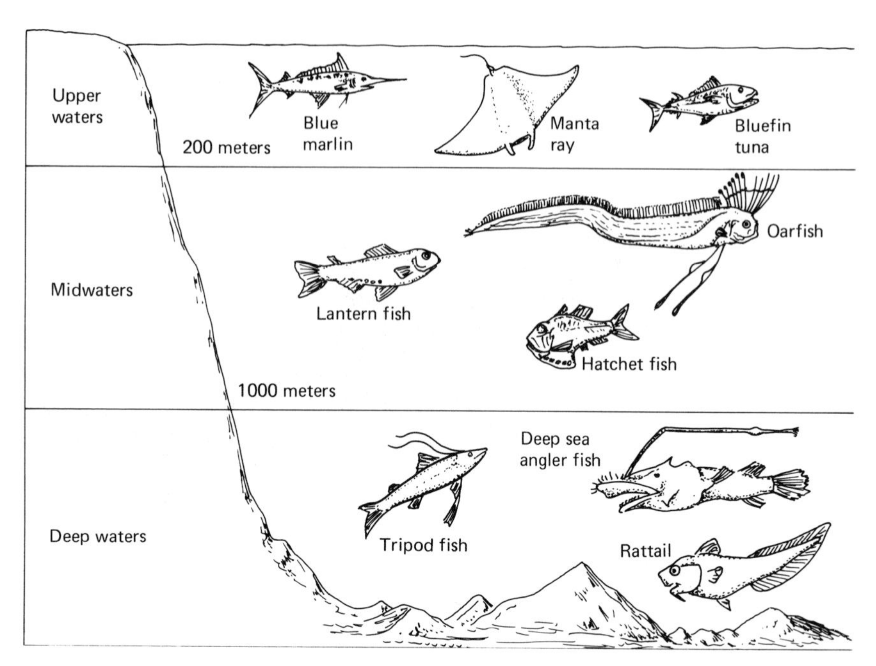

Figure 11-41. Demersal fish live at or near the sea floor; pelagic fish live at or near surface waters.

Figure 11-42. Penguins in the Antarctic. (Photo courtesy of Woods Hole Oceanographic Institution.)

Figure 11-43. Examples of sea mammals. *Top left:* Killer whale. *Top right:* Porpoise. *Bottom left:* California sea lion. *Bottom right:* Sperm whale.

Marine Bacteria

Bacteria have survived not only since the dawn of life but have played vital roles in marine ecology because of their numbers and their ability to conduct chemical and physical transformations in oceans and in deep-sea sediments (Table 11-4). Bacteria are the oldest organisms, the most extensively distributed, and the most effective mobilizers (decomposers) in the marine environment. In 1952, Claudius Zobell, a contemporary leader in marine microbiology, showed that bacteria are capable of living and reproducing in the deepest ocean sediments, withstanding pressure up to 1000 atmospheres on the seafloor. Zobell also demonstrated that bacteria play an important role in the formation of petroleum. Detailed studies of marine sediments in the California Basin further showed that bacteria restore the water bulk of nutrients that sink

TABLE 11-4 Bacteria in Sediment Samples from the Philippine Trench

	Incubation Pressure	
Depth (in meters)	*MPN[a] (at 1 atmosphere)*	*MPN[a] (at 1000 atmospheres)*
10,060	2,300	760,000
10,190	930	3,500,000
10,210	680	210,000
10,160	8,400	920,000
1,000	540,000	0
1,960	2,300,000	0
10,120	5,900	2,800,000

SOURCE: C. E. Zobell and R. Y. Morita, 1957, *Journal of Bacteriology*, 73, 563–568.
[a]MPN is most probable number per gram wet weight.

to the bottom. Bacteria also play an important part in the formation of limestone, silicate rocks such as chert, and sedimentary iron deposits.

Collection of Organisms

Plankton net *A plankton trapping device that is towed through the water.*

Several types of dredges are used to collect benthic organisms. Nektons are collected by trawls and coarse nets. Traps are sometimes employed to catch organisms that are difficult to collect by trawl or by net. For collecting small plankton, specially designed "bongo" nets are used (Figure 11-44). Essentially, the plankton-sampling technique involves filtering and centrifuging prior to microscopic study. For the collection of benthos, a benthic sled is employed (Figure 11-45), and for quahogs (New England clams) a rock-type quahog net is commonly used (Figure 11-46).

SUMMARY

1. Life is organized into two kingdoms: plants and animals. In addition, marine life is divided into pelagic and benthic realms.
2. Marine life is also grouped by locomotion and habitat: plankton, benthos, nektons.
3. Plants and animals are classifed into phyla, classes, orders, families, genera, and species.
4. Common marine plants include numbers of blue-green, green, red, brown, and yellow-green algae.

Figure 11-44. Plankton nets, popularly known as "bongo" nets, are used to sample plankton in deep water. (Photo courtesy of National Marine Fisheries Service.)

5. Marine invertebrates include protozoans, poriferans, coelenterates, echinoderms, molluscs, brachiopods, bryozoans, worms, and arthropods.
6. Marine vertebrates include fish, reptiles, birds, and mammals.
7. Bacteria are widespread in marine sediments. They decompose organic matter and sediments.
8. Oceanographers use trawls and coarse nets to catch nektons, "bongo" nets to collect plankton, and benthic sleds and quahog dredges for benthos.

Figure 11-45. A benthic sled is commonly used to collect bottom dwelling organisms. (Photo courtesy of Woods Hole Oceanographic Institution.)

Suggestions for Further Reading

Fell, H. B. 1975. *Introduction to Marine Biology*. New York: Harper & Row.
Greenwood, P. H. 1975. *A History of Marine Fishes,* third edition. New York: Wiley.

Figure 11-46. A scallop dredge. (Photo courtesy of NOAA.)

Heezan, B. C., and C. D. Hollister. 1971. *The Face of the Deep*. London: Oxford University Press.

McConnaughey, B. N. 1974. *Introduction to Marine Biology,* second edition. St. Louis: C. V. Mosby.

Menzies, R. J., R. Y. George, and G. T. Rowe. 1973. *Abyssal Environment and Ecology of the World Oceans*. New York: Wiley.

Parsons, T. R., and M. Takahasi. 1973. *Biological Oceanographic Processes*. Oxford: Pergamon.

Thorson, Gunner. 1971. *Life in the Sea*. New York: McGraw-Hill.

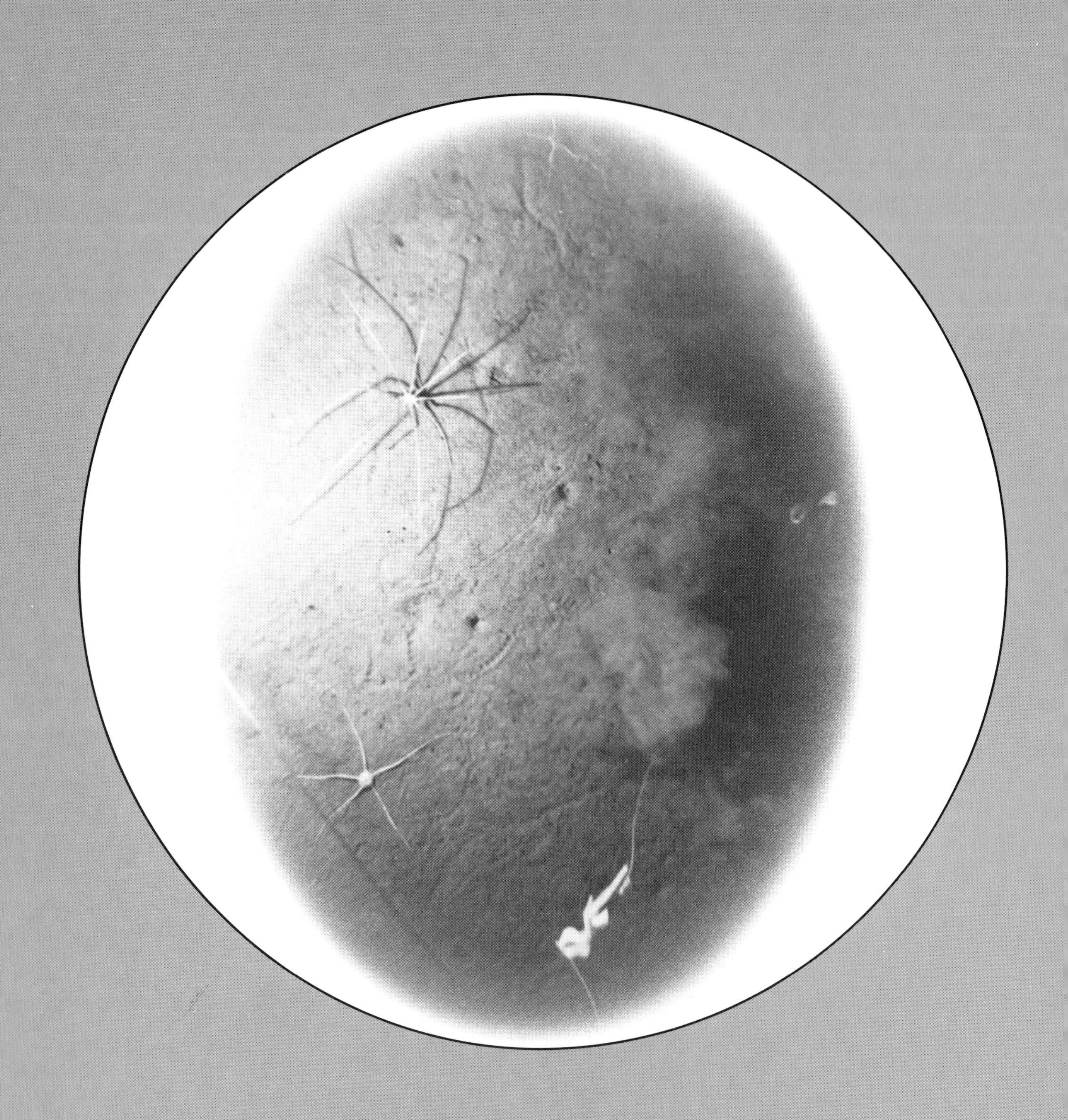

12 Marine Ecology

12-1 INTRODUCTION

Ecology IS THE STUDY OF RELATIONSHIPS BETWEEN ORGANISMS or groups of organisms and their environments (Figure 12-1). In its broadest sense, ecology may be considered as the science of life and environment. Marine ecology is concerned with the ecological aspects of oceans.

Ecology *The study of the relationship between organisms and their environment.*

12-2 THE MARINE ECOSYSTEM

The entire system of organisms and the nonliving environment with which they interact constitutes the *ecosystem*. An ecosystem is commonly composed of two primary specialized functional organisms, autotrophs and heterotrophs. *Autotrophs,* which are self-nourishing, are capable of fixing solar energy. They utilize simple inorganic compounds, including most green plants and some purple bacteria, to build complex substances. *Heterotrophs,* which are other-nourishing, are dependent on autotrophs for their food (or energy). They include most herbivore zooplankton, which obtain their food from phytoplankton.

Autotrophs *Organisms that use only inorganic materials as a source of nutrients.*

Heterotrophs *Animals and bacteria that depend on the organic compounds produced by other animals and plants for food. Such organisms are not capable of producing their own food by photosynthesis.*

An ecosystem is characterized by four constituents: (1) abiotic substances, (2) producers, (3) consumers, and (4) decomposers (Figure 12-2). *Abiotic substances* include organic or inorganic compounds of the environment such as air, water, carbon, phosphorous, nitrogen, and oxygen. *Producers* are organisms, notably green plants including phytoplankton, that are capable of producing food through photosynthesis. Most animals are *consumers;* that is, they consume other organisms. Zooplankton, for example, eat phytoplankton for energy. *Decomposers* are chiefly bacteria with the ability to break down complex organic compounds into simpler ones and to make them available as nutrients for

Decomposers. *Organisms such as bacteria that have the ability to break down nonliving organic matter.*

Figure 12-1. Marine ecology is concerned with the study of the relations of organisms or groups of organisms to each other and to their environment. (Photo courtesy of National Marine Fisheries Service.)

the plants. Bacteria can also fix inorganic substances, especially nitrogen, from the environment, thus making them available to organisms.

Habitat *A geographic area occupied by a particular plant or animal.*

Often the role of an organism is described as its habitat or niche. A *habitat* is the residence of an entire community or group of communities of an organism, such as a pond, a lake, or an estuary. On the other hand, *niche* refers to what organisms do in order to live. For example, the green algae *Zooxanthelle,* because it has chloroplasts, is able to

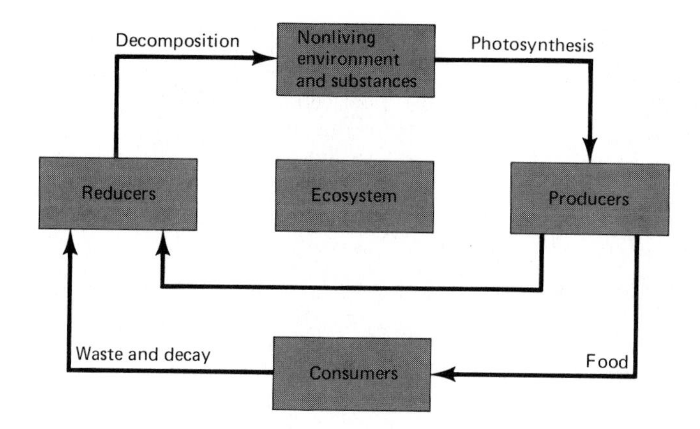

Figure 12-2. Basic elements of an ecosystem.

synthesize food. The ecological niche of *Zooxanthelle* is thus its producer role in a reef community.

12-3 THE NATURAL ASSOCIATIONS OF ORGANISMS

The relationships between organisms are commonly categorized as symbiotic, commensalistic, or parasitic. In *symbiosis,* the relation between interacting organisms is mutually beneficial. An example is the close association between one-celled *zooxanthelles* and marine nudibranchs in the Sargasso Sea (Figure 12-3). Zooxanthelles live within nudibranchs, and through their photosynthic mechanism provide food to their hosts. The host in turn benefits the zooxanthelles by supplying waste carbon dioxide and nitrogenous matter. Similarly, algae and fungi have adopted a symbiotic relation to form lichens; the former synthesizes food, and the latter provides protection. Without this symbiosis, lichens cannot grow. Symbiosis is also common among relatively advanced animals, between, for example, hermit crabs and sea anemones. Sea anemones provides protection and camouflage, and crabs provides transportation.

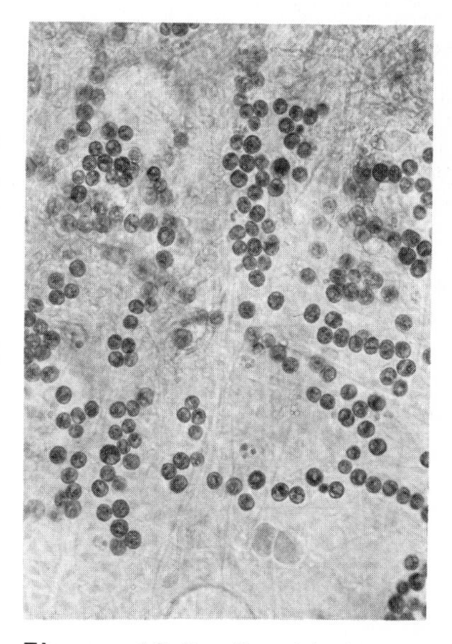

Figure 12-3. Symbiotic dinoflagellates *(zooxanthelle)* in the tissue of a marine nudibranch from the Sargasso Sea. Magnification is 500×. (Photo courtesy of Paul E. Hargraves, Graduate School of Oceanography, University of Rhode Island.)

In *commensalism,* the association between interacting organisms is such that the parasitic partner benefits without causing harm to the host. The benefitting organism is termed commensal. Commensalism is widespread; nearly every host, such as a shellfish, a worm, or a sponge, contains various harmless guests. Most oysters contain small delicate crabs in their bodies, and these commensal crabs usually leave their hosts healthy.

Symbiosis *A mutually beneficial relationship between two or more organisms.*

Commensalism *An association between two organisms in which one is benefited by the other without any hurt or benefit to the other organism.*

Parasitism *An involuntary association between two or more organisms in which the host is harmed and the parasite is benefited.*

In *parasitism,* the predator benefits at the expense of its victim or prey. In certain lichens, fungi penetrate algal cells, consequently adopting a parasitic role.

12-4 ECOLOGICAL FACTORS

The relations between organisms and their marine environments must consider physical, chemical, and biological factors.

Light

The limitation of light penetration to less than 100 meters of water directly restricts photosynthesis and accordingly the overall size of an organic community.

Temperature

Temperature effects on an organic community are vividly observed by the restricted growth of reef-forming corals. These do not grow in water of less than 21°C.

Salinity

Isotonic *The property of two fluids having equal osmotic pressure. If two such fluids are separated by a semipermeable membrane that will allow osmosis to occur, there will be no net transfer of water molecules across the membrane.*

Hypotonic *The property of an aqueous solution having a lower osmotic pressure (salinity) than another aqueous solution from which it is separated by a semipermeable membrane that will allow osmosis to occur. The hypotonic fluid will lose water molecules to the other fluid through the membrane.*

Salinity on the whole is relatively uniform in the ocean (35‰), except where freshwater inflow is high or where the evaporation rate is high. The body chemistry of a fish in seawater is more constant than that of a fish in freshwater. The chemistry of a seawater fish is an *isotonic* condition; the chemistry of a freshwater fish is a *hypotonic* condition. Saltwater fish use less energy to fight isotonic solutions than do freshwater fish. As a result, possibly some saltwater fish tend to be larger than those in freshwater.

Pressure

Hydrostatic pressure in the ocean is one of the controlling factors of life. In water, pressure increases 1 atmosphere for every 10 meters of depth. In the deepest parts of the ocean, the pressure can be as high as 1000 atmospheres. Many organisms can withstand wide ranges of pressure changes, particularly when their bodies do not contain free air or gas.

Carbon Dioxide

The amount of carbon dioxide in seawater is variable. Carbon dioxide enters into chemical combination with water to form carbonic acid solution, which reacts with available limestones to form carbonate and bicarbonate. These compounds provide nutrients and, more important, act as buffers against drastic changes in *pH*. The availability of carbon dioxide determines to a large degree the rate of photosynthesis. If the supply of carbon dioxide increases significantly, it could be fatal to animal organisms.

Oxygen

The supply of oxygen in seawater is extremely important. Without it life cannot exist, except for some anaerobes (bacteria that can live in oxygenless environments).

Nutrients

Nutrients such as phosphates and nitrates are important for the growth and reproduction of nearly all organisms. For example, phosphate nutrients form an energy-yielding compound called *adenosine triphosphate* (ATP). Energy-rich phosphate bonds contain between 800 and 10,000 calories per gram mole. Large quantities of calcium are required by molluscs and vertebrates. Magnesium is an essential constituent of chlorophyll. Cobalt is a vital ingredient in vitamin B_{12}, and molybdenum is essential for nitrogen-fixing microorganisms.

pH

Hydrogen ion concentration (*pH)* is one factor that controls the growth of organisms. The *pH* of a solution is the negative logarithm of the hydrogen ion concentration. *pH* determines a solution neutral if its hydrogen ion concentration is 10^{-7}. That is, $pH = -\log 10^{7} = 7$. If *pH* is between 7 and 14, the solution is alkaline; if it is between 1 and 7, the solution is acidic. *pH* influences organic activity, particularly enzyme activity. Under critical *pH,* enzyme activity is greatest.

Depth

The bulk of basic organic material that pumps energy into an infinite number of marine animals is synthesized by one-celled microscopic plants within an extremely thin layer of surface water (up to depths of 100 meters).

12-5 PHOTOSYNTHESIS

Photosynthesis *The process in which the energies of light and chlorophyll are used by plants to manufacture food from carbon dioxide and water.*

Chlorophyll *A green pigment found in plants that absorbs energy from sunlight, making it possible to synthesize organic matter from atmospheric carbon dioxide and water through photosynthesis.*

Respiration *Oxidation- and energy-providing process of living organisms in which oxygen is consumed and carbon dioxide is removed from their body system.*

Photosynthesis is the single most vital biological process that has made organic matter available on a massive scale, especially in the upper surface of oceans. It is defined as the process in which plants, water, carbon dioxide, nutrients, and sunlight interact to produce food. Certain bacteria can photosynthesize in the absence of light, but photosynthesis is mainly carried on by plants. The most important factor in photosynthesis is the green pigment *chlorophyll.* Chlorophyll is composed of minute bodies known as *chloroplasts,* which are generally found in leaves, and in certain stems and roots. They are also present in green, red, and brown algae and diatoms (see Chapter 11). Chlorophyll absorbs the radiant energy of the sun, initiating biochemical reactions involving water and carbon dioxide. Chemically, the process may be formulated as follows:

$$\underset{\text{carbon dioxide}}{CO_2} + \underset{\text{water}}{H_2O} \xrightarrow{\text{sunlight}} \underset{\text{organic matter}}{CH_2O} + \underset{\text{oxygen}}{O_2} \uparrow$$

Photosynthesis is confined to shallow water and takes place primarily during daylight. Phytoplankton, such as certain types of algae, are the chief beneficiaries of photosynthesis, because they float on surface waters when sunlight is plentiful. A great quantity of phytoplankton, seawater, dissolved carbon dioxide, air, and sunlight interact to produce mass organic matter on which countless organisms feed. At night, photosynthesis is reversed, resulting in *respiration*. In respiration, organic matter and oxygen interact to produce carbon dioxide, water, and energy:

$$\underset{\text{organic matter}}{CH_2O} + \underset{\text{oxygen}}{O_2} \longrightarrow \underset{\text{carbon dioxide}}{CO_2} + \underset{\text{water}}{H_2O} + \text{energy (heat)}$$

12-6 NUTRIENT CYCLES

Nearly all chemical elements, particularly those essential to life, such as carbon, phosphorus, nitrogen and sulfur, flow in the biosphere from the environment to the organism and back to the environment. This circulatory behavior of these elements constitutes a *biogeochemical cycle*. This term implies the occurrance of a series of organic and inorganic chemical reactions during the back-and-forth flow of elements.

The Nitrogen Cycle

The nitrogen cycle is a complex but fairly well-regulated flow of nitrogen from the environment to the organisms and back to the environment (Figure 12-4). In this cycle, nitrogen present as protein molecules in organisms is broken down into inorganic forms by a series of decomposers, mainly denitrifying bacteria. The result is turned nitrate, the form most readily consumed by green plants. Further bacterial breakdown of nitrate leads ultimately to the formation of gaseous ammonia. Thus nitrogen is released to the atmosphere as ammonia. Nitrogen-fixing bacteria utilize nitrogen from the environment as it becomes available to plants as nutrients.

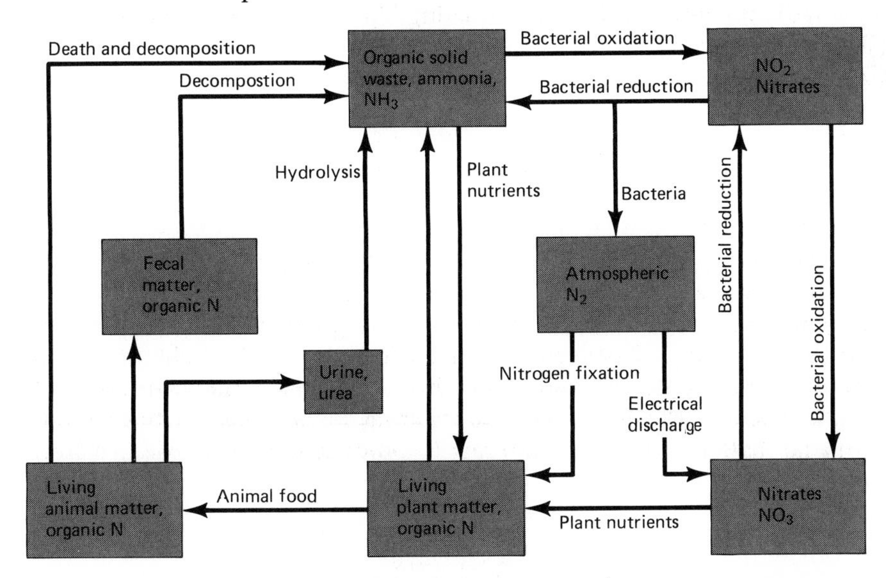

Figure 12-4. The nitrogen cycle.

The amount of nitrogen fixed from air (noncyclic nitrogen) is estimated at between 140 and 700 milligrams per square meter. Some nitrogen is also fixed by lightning.

A number of microorganisms are able to extract molecular nitrogen from the atmosphere as their source of nitrogen. The conversion of molecular nitrogen into nitrogenous compounds is called *nitrogen fixation*. The process of nitrogen fixation is accelerated by phosphates and trace elements dissolved in sea water. Among marine algae, nitrogen fixation is largely confined to blue-green algae. These algae can produce excess nitrogenous compounds which in turn can be used by a number of other algae that cannot otherwise fix nitrogen. This ability has two important implications: it means that the ocean is a relatively independent system as far as its sources of nitrogen are concerned and this provides a unique means of increasing the food production from the ocean.

The Phosphorus Cycle

The phosphorus cycle is relatively simpler but less perfect than the nitrogen cycle. Phosphorus is a vital element of adenosine triphosphate, required for the growth and reproduction of all organisms. When phosphorus is broken down by bacteria, it results in phosphates, important nutrients for the growth of plants. It is reutilized and thus returned to organic life (Figure 12-5).

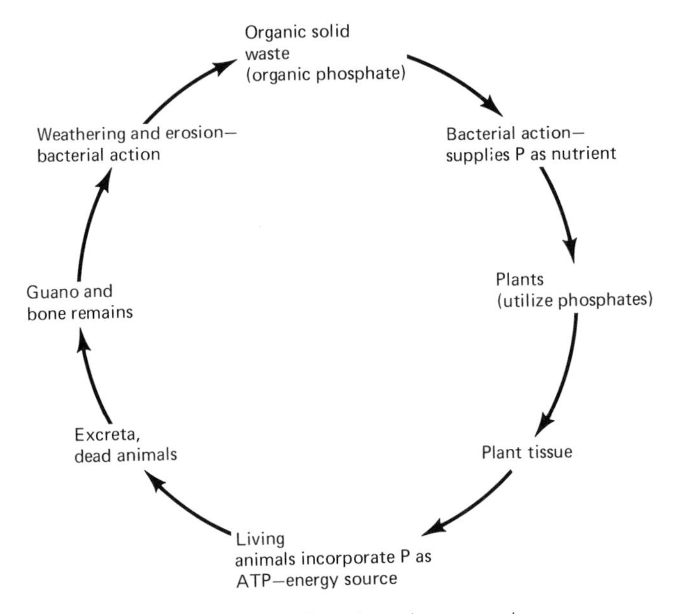

Figure 12-5. The phosphorus cycle.

The Sulfur Cycle

Sulfur is present as sulfate in water. Taken by plants in reduced form, sulfur is utilized by certain amino acids, which are converted into protein. Protein is consumed by plants and animals and is passed on as excreta. When the organic waste is decomposed by bacteria, hydrogen sulfide is released. Some hydrogen sulfide is oxidized to sulfate by purple and green sulfur bacteria; the remainder is oxidized by aerobic sulfide oxidizers. The sulfate in the water is once again utilized by plants, and thus the sulfur cycle is completed (Figure 12-6).

The Carbon Cycle

The carbon cycle begins with the removal of carbon from the atmosphere primarily by photosynthesis (Figure 12-7). In photosynthesis, carbon dioxide is absorbed; and oxygen is liberated. Withdrawn carbon from the atmosphere is furthur passed on as reduced organic matter for use by animals. Subsequently, bacteria and fungi decompose the bulk of organic matter from dead animals and plants, as well as the various products excreted by animals. Carbon dioxide is reutilized and is finally returned to the atmosphere.

Carbon as carbon dioxide is also withdrawn from the atmosphere and permitted to dissolve as carbonate and bicarbonate in water, especially in oceans. Carbonate in seawater combines with dissolved calcium ions and, under slightly alkaline conditions (that is, $pH = 7.8$),

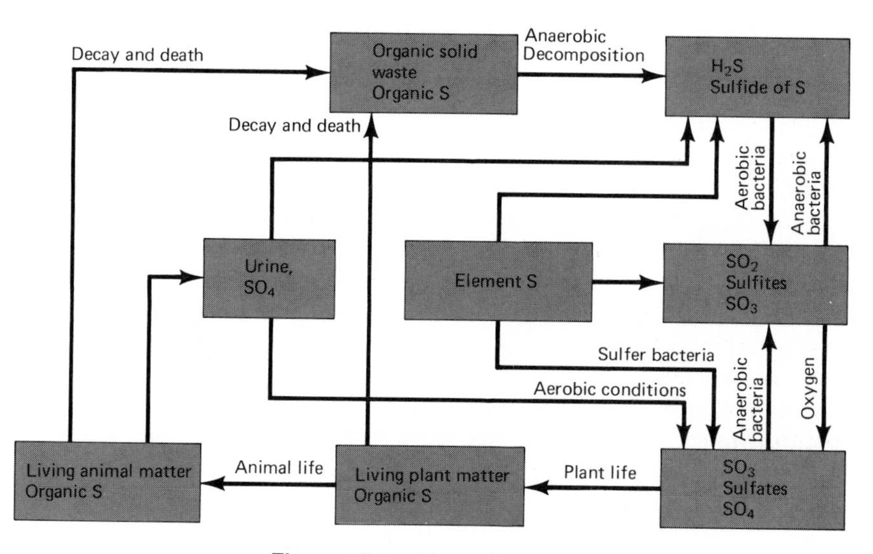

Figure 12-6. The sulfur cycle.

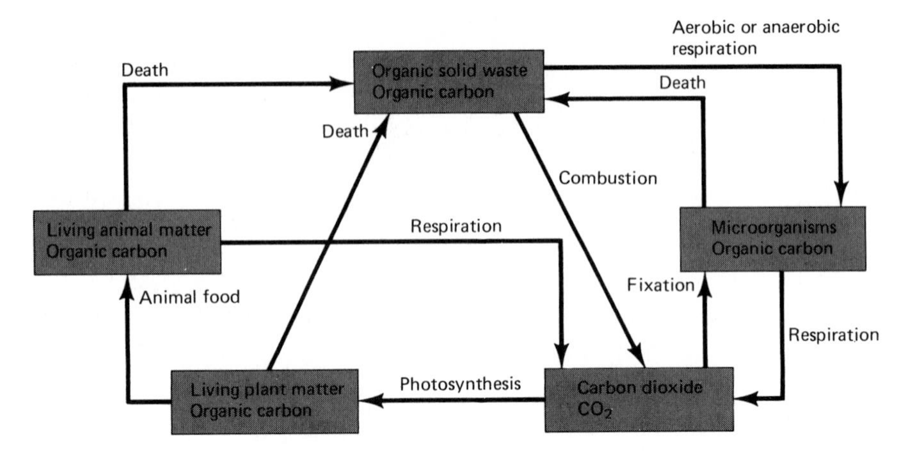

Figure 12-7. The carbon cycle.

Calcium carbonate *A white insoluble solid occurring naturally as chalk, limestone, marble, and calcite. Used in the manufacture of lime and cement.*

precipitates as calcium carbonate. The precipitation of calcium carbonate depends largely on *pH* conditions and on biological activity. It is deposited biologically in the shells of protozoans, corals, and molluscs. *Globigerina* ooze exemplifies the carbon cycle well (see Chapters 5 and 11). When limestones are subjected to weathering, carbon as carbon dioxide is released to the atmosphere.

12-7 THE DEEP-SCATTERING LAYER

Deep-scattering layer *Sound (or sonic) reflecting layer above the sea floor formed by a dense concentration of zooplankton and fish.*

Many marine organisms follow daily vertical migration patterns. This type of migration was proved by echo-sounding techniques. Echo-sounding reflects a "false bottom" as well as true seafloor. These false-bottom reflections are collectively known as the *deep-scattering layer* (Figure 12-8). The scattering layer migrates toward the surface at dusk and descends from the surface at dawn, because certain animals tend to avoid light. Some animals find it advantageous to gather food in darkness in order to escape from predators; others just cannot tolerate light. Various species of copepods, for instance, migrate vertically in the Antarctic Ocean. One copepod remains near 100 meters below the surface during the day and ascends to the surface at sunset, staying there until before sunrise. These animals feeds on surface phytoplankton in the absence of light.

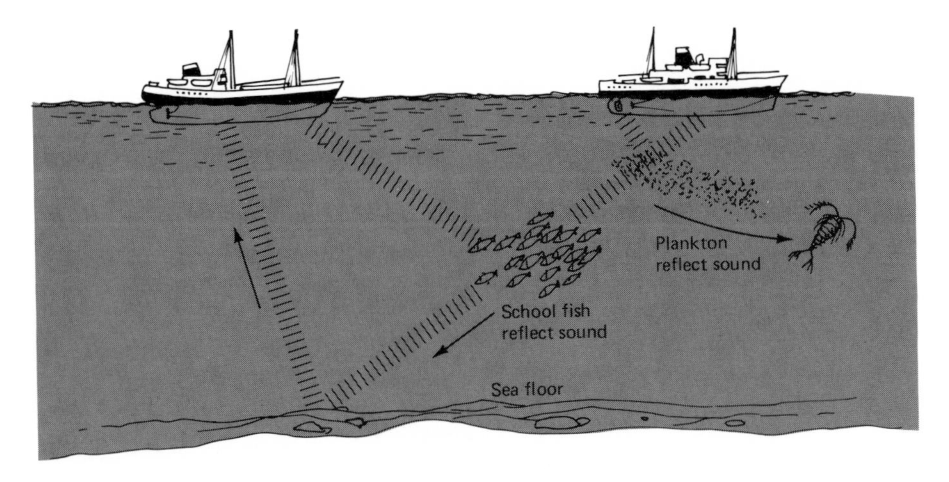

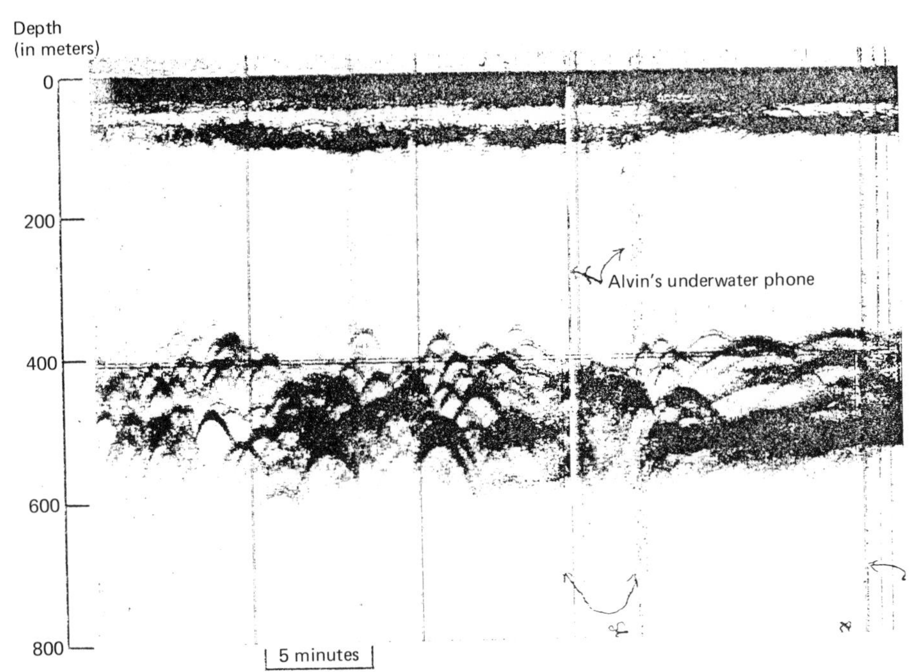

Figure 12-8. Deep-scattering layer by echo-sounding method. *Top:* The principle of the method. *Bottom:* The result. (Photo courtesy of Woods Hole Oceanographic Institution.)

12-8 THE SEAFOOD WEB

The overall transfer by repeated ingestion of food from minute single-celled organisms, such as phytoplankton, through to successively larger animals constitutes the *food chain* (Figure 12-9). Three specialized

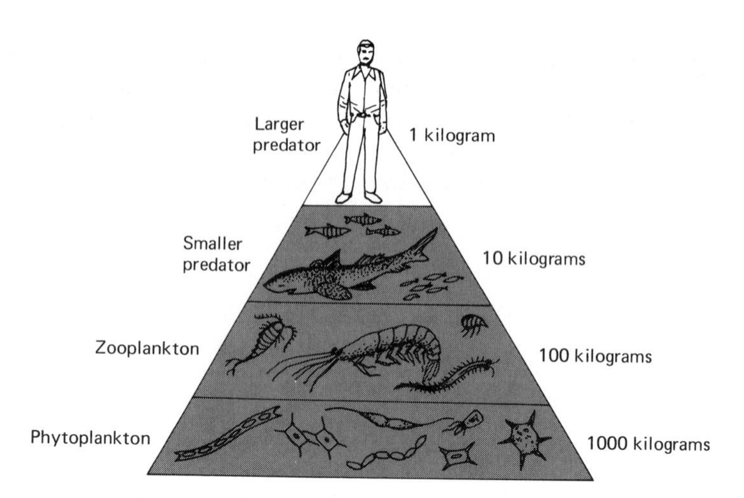

Figure 12-9. The seafood pyramid.

cycles of the food chain include: (1) the *predator chain,* which begins from a plant base and moves upward to small and then to larger animals; (2) the *parasitic chain,* which goes from larger to smaller organisms; and (3) the *saprophytic chain,* which moves from dead matter to microorganisms. These food chains together form an inextricable and complex chain system called the *seafood web.*

The underlying principle of the seafood web is that most basic organic material that supplies energy to marine life is synthesized within the thin layer of surface water (where much light is available) by ordinary one-celled microscopic plants, particularly tiny phytoplankton. In fact, photosynthesis by phytoplankton is the hub of the seafood web. Phytoplankton such as diatoms are consumed by herbivores or by zooplankton (foraminifers, pteropods, and krill), which in turn are eaten by larger animals and ultimately by man. It is estimated that 1000 kilograms of phytoplanktonic food will produce 100 kilograms of herbivores, which will further produce 10 kilograms of medium-sized nekton and in turn 1 kilogram of larger nekton.

The actual seafood cycle is not ideal. A number of small seafood chains may operate independently outside the theoretical cycle. Sinking organic debris, for example, supports animals on the seafloor and is outside the seafood web.

12-9 OCEAN PRODUCTIVITY

Photosynthesis by phytoplankton in the upper surface of ocean water is the basic mechanism of organic production. Estimates of organic productivity in oceans vary geographically because of variations in intensity of light, depth, currents, nutrients, and dissolved gases. According to Steeman-Nielsen,* the total annual net organic production averages 55 grams of assimilated carbon per square meter for oceans. He estimated the net production by considering 1.5 x 10^{10} tons of fixed carbon per year in oceans and assumed a 25-percent loss of carbon through respiration. Marine organic productivity in different bodies of water is given in Table 12-1.

Phytoplankton vary regionally, seasonally, and habitationally. The seasonal phytoplankton bloom in upper latitudes is the most notable example (Figure 12-10). During winter, turbulence caused by storms provides a significant quantity of nutrients in the surface layers to support flagellates. As a result, the species grows rapidly and soon dominates the region. In early spring, the increase in sunlight favors the growth of diatoms, which then predominate over flagellates. Later in spring, the diatom population declines because of consumption by zooplankton and because of a decrease in the supply of nutrients, caused by relatively quiet weather conditions. The lowest level of the diatom population is reached in midsummer, and the flagellates take over as winter approaches.

Plankton bloom *An enormous growth of plankton in a given specific area due to the abundant supply of nutrients by upwelling currents.*

12-10 CORAL REEFS: A CASE STUDY IN MARINE ECOLOGY

A reef is a biological community built on a wave-resistant framework of calcium carbonate; it is generally dominated by corals and algae. The growth of a reef depends on shallow, warm, and saline waters generally less than 50 meters deep and temperatures of about 21°C.

Reef *A predominantly organic deposit made by living or dead organisms that forms a mound or ridge-like elevation.*

TABLE 12-1 Primary Productivity of Marine Ecosystems as Determined by Gas Exchange Measurements

Ecosystem	*Rate of Production (grams per square meter per day)*
Averages for long periods—six months to one year	
Infertile open ocean (Sargasso Sea)	0.5
Shallow, inshore waters (Long Island Sound) year average	3.2
Texas estuaries (Laguna Madre)	4.4
Coral reefs (average of three Pacific reefs)	18.2

SOURCE: E. P. Odum, 1959, *Fundamentals of Ecology*, second edition, Philadelphia, Saunders.

*Steeman-Nielsen, E., 1954, "On Organic Production in the Ocean," *J. Cons. Int. Explor. Mer.*, *49*, 309-328.

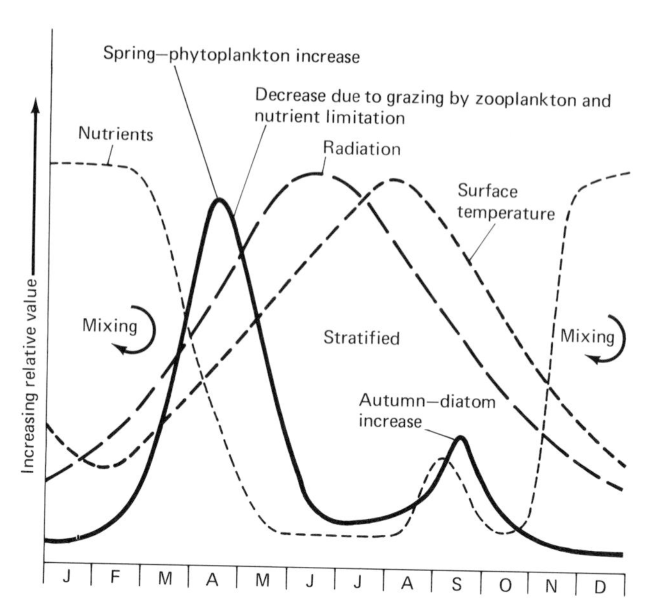

Figure 12-10. Diagrammatic presentation of seasonal variations in phytoplankton cycle, nutrients and light. (*Source:* T. F. Smayda, 1971, in *Deep Oceans*, P. J. Herring and M. R. Clark, editors, New York, Praeger.)

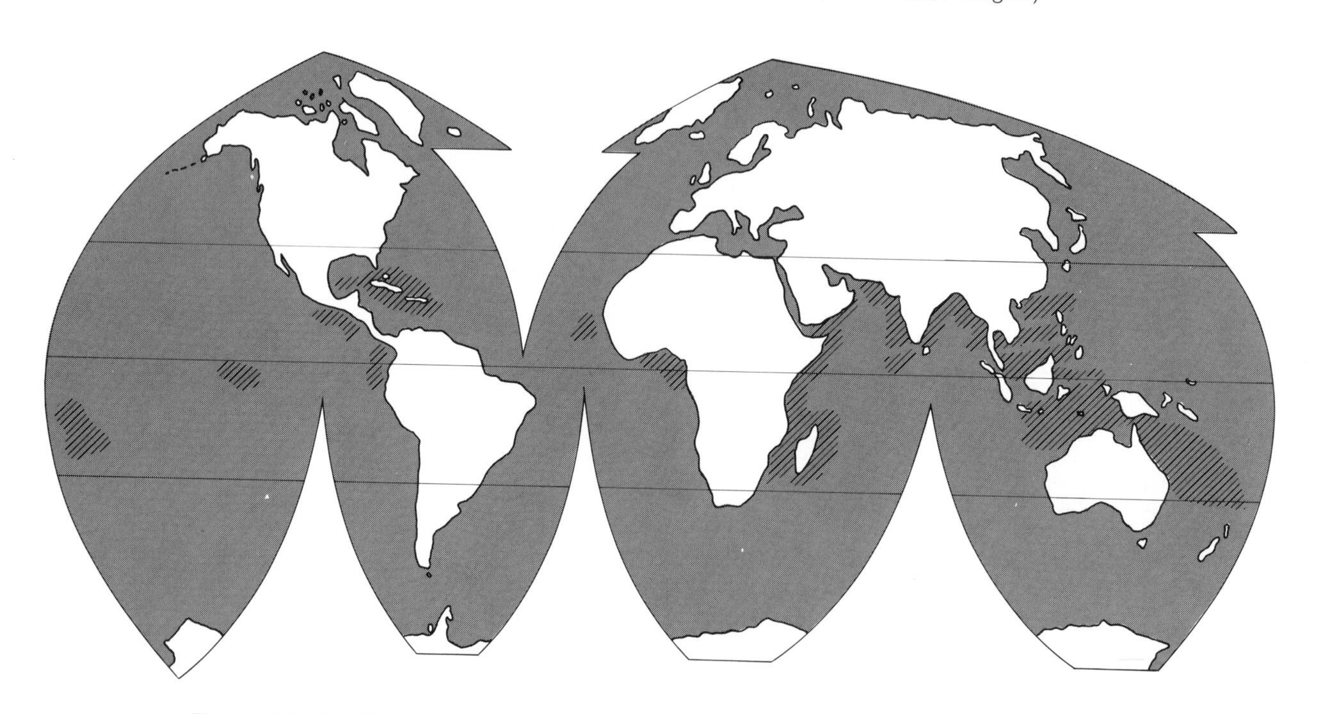

Figure 12-11. Worldwide distribution of coral reefs. Note their concentration in tropical and subtropical zones.

Most reefs develop in tropical and subtropical waters (between 30°N and 30(°S) (Figure 12-11). Fossil coral reefs were more extensively developed in the geologic past. At present, coral reefs occupy about 200 million square kilometers in oceans, especially in the tropics (Figure 12-12).

The three principal types of reefs are: fringing, barriers, and atolls (Figure 12-13). *Fringing reefs* grow in shallow water along coasts. The reef along Florida Keys is a typical fringing reef. A *barrier reef* is separated from a continent by a lagoon. The Great Barrier Reef of Australia extends over 2000 kilometers along its eastern margin. An *atoll* is a circular or an oval reef surrounding a lagoon. Rongelop and Kwajalein atolls in the Marshall Islands (Pacific Ocean) are good examples.

Coral *A group of benthic animals belonging to the phylum Coelenterata. These animals live as individuals or in colonies and secrete calcium carbonate skeletons. In collaboration with certain algae, and under favorable conditions, corals can build reefs.*

Fringing reef *A reef attached directly to the shore.*

Barrier reef *A coral reef that runs parallel to the land but is separated from it by a lagoon.*

Atoll *A ring-like "coral" island or group of islands of reef origin encircling a lagoon in which there are no islands of noncoral origin.*

The Formation of Coral Reefs

Charles Darwin explained the formation of coral reefs about 150 years ago. During the voyage of the *Beagle* (1831–1836), he made a detailed study of coral reefs and atolls in the Pacific. Darwin proposed that coral reefs evolved in stages: first, volcanic activity formed an island with a shallow water base on which a fringing reef grows. Second, when the volcanic activity stopped, the island subsided, but continued reef growth and the deposition of large amounts of *coralline* and algal debris

Figure 12-12. View to north over Bimini Lagoon. Small mangrove islands are at center and right. Dark areas at lower right are grass bottoms. (Photo courtesy of John Imbrie, Department of Geology, Brown University.)

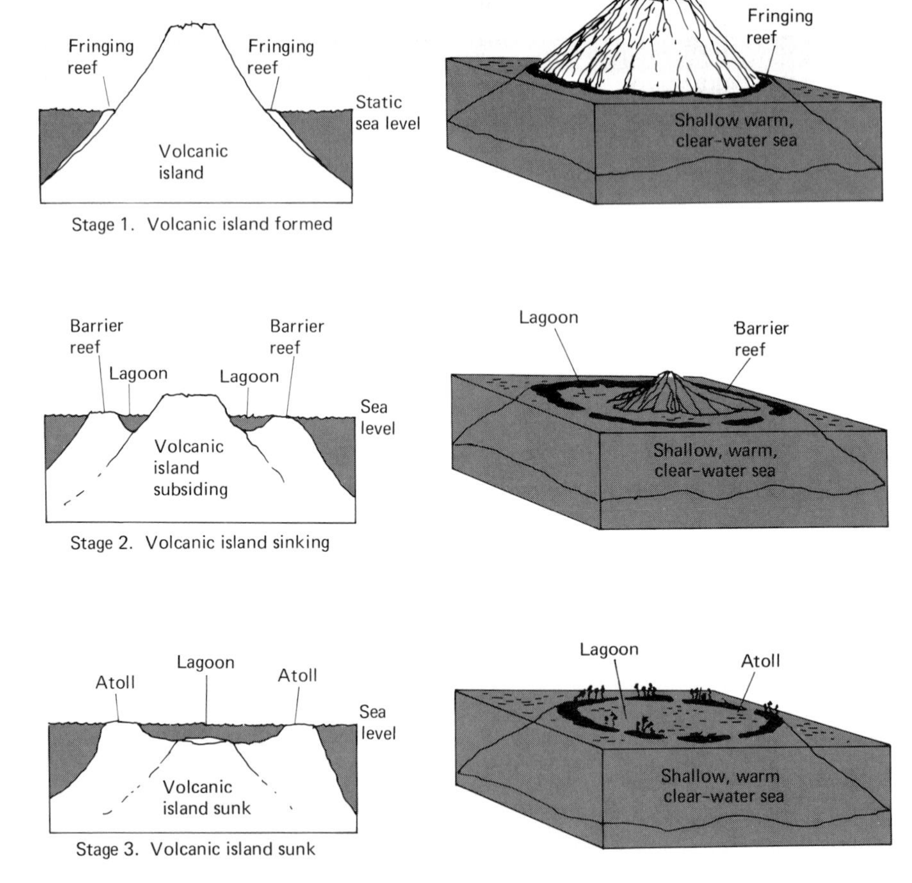

Figure 12-13. Origin and evolution of reefs according to Darwin's sinking-island hypothesis.

permitted the reef to survive in modified form. As a result, the modified reef was separated from the island by a lagoon and thus became a barrier reef. Continued subsidence permitted the island to submerge, resulting in an atoll. Darwin's view was verified by the presence of volcanic material in core samples from the bottoms of Eniwetok and Bikini atolls in the Marshall Islands.

Zonation of Coral Reefs

Coral reefs are complex but well-organized communities of organisms. A characteristic change is observed in many reefs as one moves from the shore seaward (Figure 12-14). Most of the change is attributable to varying depth, light, and temperature. As a result, reefs are

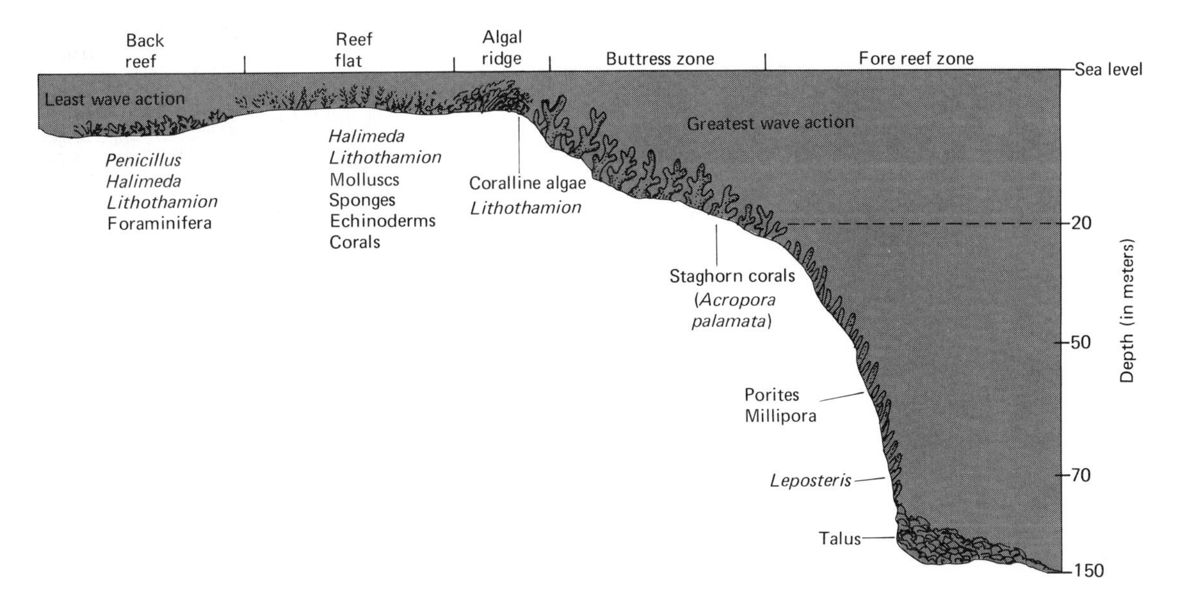

Figure 12-14. Major zonation of a reef. (*Source:* Modified after J. J. Bhatt, 1976, *Geochemistry and Geology of South Wales Main Limestones (Mississippian),* Cranston, R.I., Modern Press.)

divided into four zones: (1) the back reef, (2) the reef flat, (3) the buttress zone, and (4) the fore reef. In the *back reef,* wave action is weak and the seafloor is covered by sand-sized carbonate sediments of organic material. *Halimeda* possess broad thin plates of calcium carbonate and comprise the bulk of the calcareous sand (Figure 12-15). Calcareous sand

Figure 12-15. An oceanographer in Bimini Lagoon samples a muddy carbonate-sand bottom, which is partly stabilized by grasses and algae. (Photo courtesy of John Imbrie, Department of Geology, Brown University.)

also contains remains of associated organisms, such as foraminifers, corals, and red algae. The skeleton of *Penicillus* is composed of fine calcium carbonate needles. After the death of the organisms, these needles are broken up by waves and are dispersed into deeper waters of the reef, where they become lime mud.

The *reef flat* is located farther toward the sea. Here wave action is stronger than that at the back reef. Corals, sponges, sea urchins, and fish and algae proliferate (Figure 12-16). Calcareous red algae are important here, because they cement coral sediment into a wave-resistant solid framework (Figure 12-17). Photosynthesis is quite efficient in shallow reef flats. *Lithothamnion* and *coralline* algae are abundant in algal ridges.

The *buttress zone* and the *fore reef* are at the outermost end of the reef, where they are constantly under attack by strong waves and currents. The reef height is about 20 meters below the surface. The reef is covered by large treelike corals such as staghorn coral (Figure 12-18).

As depth increases to more than 20 meters, light diminishes, but wave action continues to be strong. Thus, no significant accumulation of organic debris is possible. At greater depth (from 30 to 70 meters), the slope of the seafloor becomes steep (usually between 45 and 70 degrees). At this depth, both light and wave action become weaker, consequently permitting deposition of calcareous sand- and silt-sized sediments.

(a)

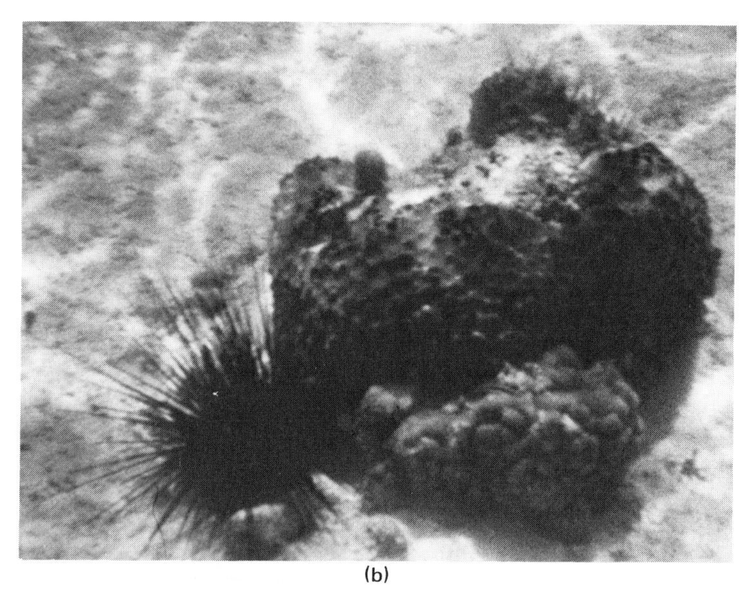

(b)

Figure 12-16. (a) General view of a reef flat community. (b) Close-up view of the community showing the echinoid *Diadema* and a sponge on a carbonate-sand bottom in Frazer Hog Cay, West Indies. (Photo courtesy of John Imbrie, Department of Geology, Brown University.)

The Coral Reef Crisis

The prolific growth of the Crown-of-Thorns starfish in coral reefs has provided a crisis of survival for reefs (Figure 12-19). This starfish has caused extensive damage in the Great Barrier Reef. As a result, conservation of this reef has become a national issue in Australia. A detailed study of the coral reefs in Hawaii showed that because of the delivery of raw sewage and of uncontrolled sedimentation, reefs have been choked to death. Coral reefs are becoming endangered species.

Figure 12-17. Dead man's fingers, an erect tabular sponge observed in the reef flat community (3 meters deep) near Bimini Island, West Indies. (Photo courtesy of John Imbrie, Department of Geology, Brown University.)

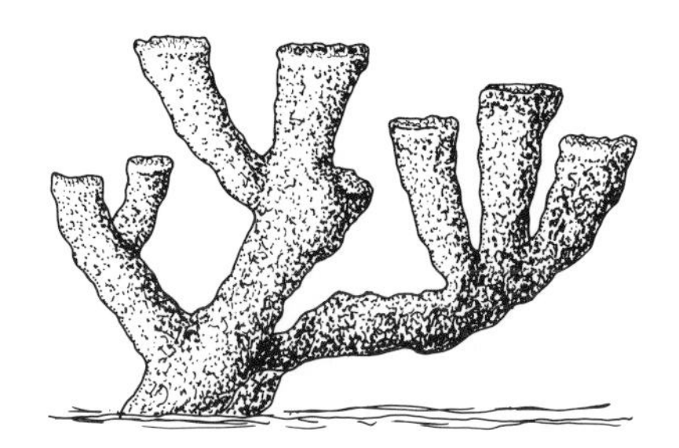

Figure 12-18. Staghorn coral thrives best in high-energy zones, especially in buttress and fore reefs.

Figure 12-19. Crown-of-thorns starfish causes great destruction of coral reefs.

SUMMARY

1. Marine ecology is concerned with the relations between organisms or groups of organisms to their environments.
2. An ecosystem may include autotrophic (self-feeding) and heterotrophic (other-feeding) organisms.
3. Abiotic substances, producers, consumers, and decomposers are ecosystem constituents.
4. A natural association of organisms may be symbiotic, commensalistic, or parasitic.
5. Principal ecological factors influencing the distribution of organisms in the ocean environment include light, temperature, salinity, pressure, carbon dioxide content, oxygen content, nutrient content, *pH*, and depth.
6. Organic productivity in oceans is determined partially by photosynthesis and nutrient cycles.
7. The deep-scattering layer of organisms migrates vertically every day.
8. The predator chain, the parasite chain, and the saprophytic chain contribute to the seafood web.
9. Darwin proposed the island-sinking hypothesis for the formation of reefs. Coral and algae play important roles in forming reefs on wave-resistant frameworks. Three types of reefs are fringing reefs, barrier reefs, and atolls. Many reefs are destroyed by the crown-of-thorns fish. This has alerted marine conservationists.

Suggestions for Further Reading

Baker, J. M. (editor). 1976. *Marine Ecology and Air Pollution*. New York: Wiley.

Cameron, A. M. *et al*. (editors). 1974. *Proceedings of the Second International Symposium on Coral Reefs*, 2 volumes. Brisbane, Australia: Great Barrier Reef Committee.

Cushing, D. H. 1975. *Marine Ecology and Fisheries*. London: Cambridge University Press.

Jones, O. A., and R. Endean (editors). 1973. *Biology and Geology of Coral Reefs*. New York: Academic Press.

Kinne, Otto. (editor). 1975. *Marine Ecology: A Comprehensive, Integrated Treatise on Life in Oceans and Coastal Waters*. New York: Wiley.

Menzies, R. J., R. Y. George, and T. R. Rowe. 1973. *Abyssal Environment and Ecology of the World Oceans*. New York: Wiley.

Odum, E. P. 1971. *Fundamentals of Ecology,* third edition. Philadelphia: Saunders.
Ryther, John. 1969. "Photosynthesis and Fish Production in the Sea." *Science, 166,* 72–76.
Tait, R. V., and R. S. Desanto. 1972. *Elements of Marine Ecology.* New York: Springer.

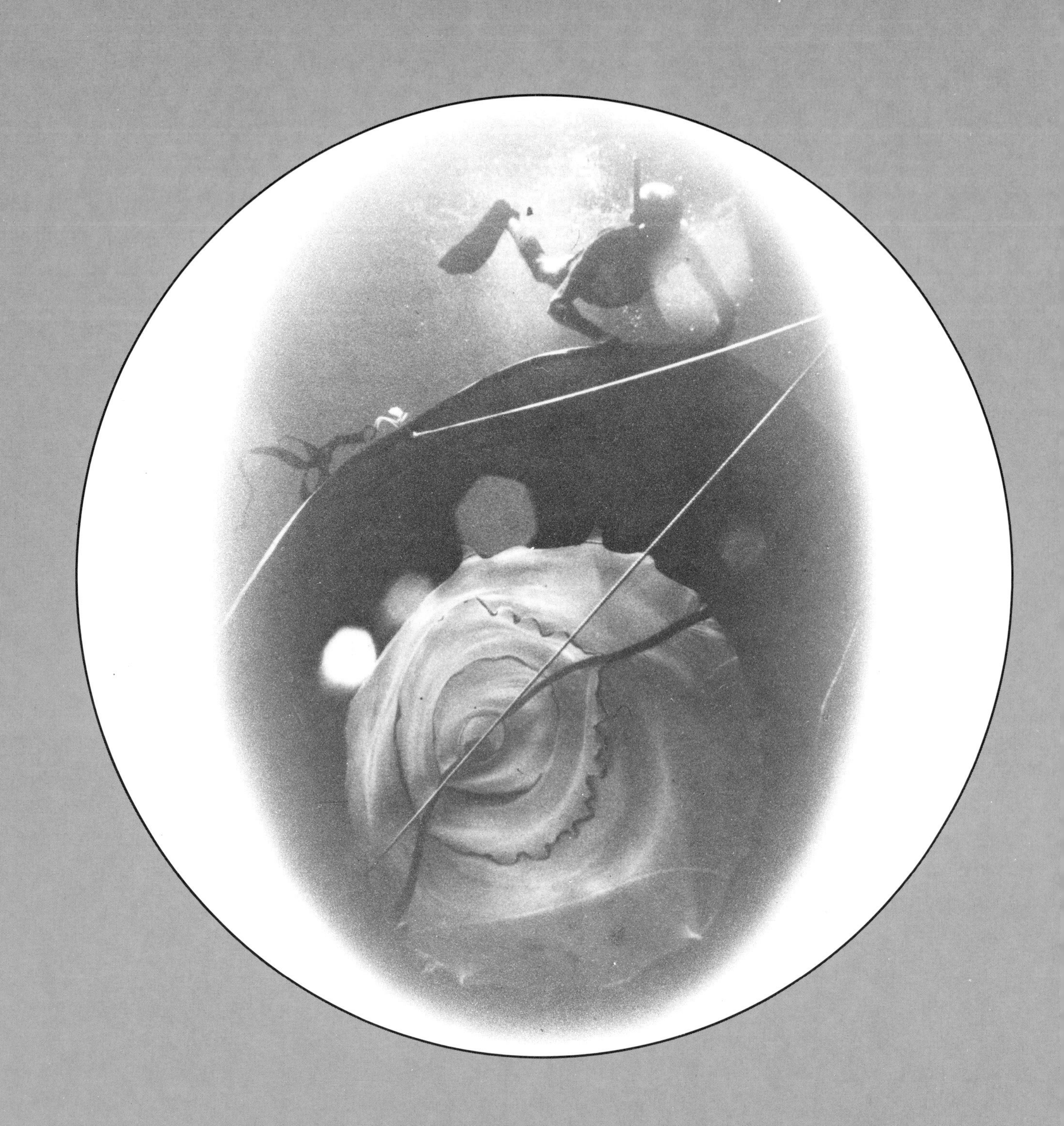

13 Biological Resources

13-1 INTRODUCTION

ALTHOUGH MAN HAS EXPLOITED MARINE BIOLOGICAL RESOURCES FOR centuries, he has done so only on a limited scale. Even so, he consumes fish, molluscs, crustaceans, and even whales. He uses trash fish in the manufacture of oil, of cholesterol free fish-protein concentrate, and of fertilizer. He utilizes fish skin to make glue, and kelp and other algae to make foods and curative medicines. Fishing and shipping are the oldest marine enterprises.

The biological wealth of the oceans is overwhelming (Figure 13-1). A drop of water may contain countless microscopic plants on which hundreds of millions of medium-sized large animals depend for their growth and existence. In addition, the oceans contain 40,000 species of molluscs and crustaceans and 25,000 species of fish. Vitamins and medicinal drugs are present in seawater. The single most important feature of marine biological resources is that they are *renewable.* Consequently they afford potential for vast food resources.

13-2 WORLD FISHERIES

Countries exploit marine foods for different reasons. The Scandinavian countries and Japan depend literally on marine foods for sustenance. Others, such as Peru and Chile, utilize their seafood resources not only for direct domestic consumption but also as export commodities. Most countries, however, depend little on marine foods. Because it is estimated that by the year 2000 there will 6 billion people on earth, we must look elsewhere than land for food resources. Many countries in the world are already protein poor because of general

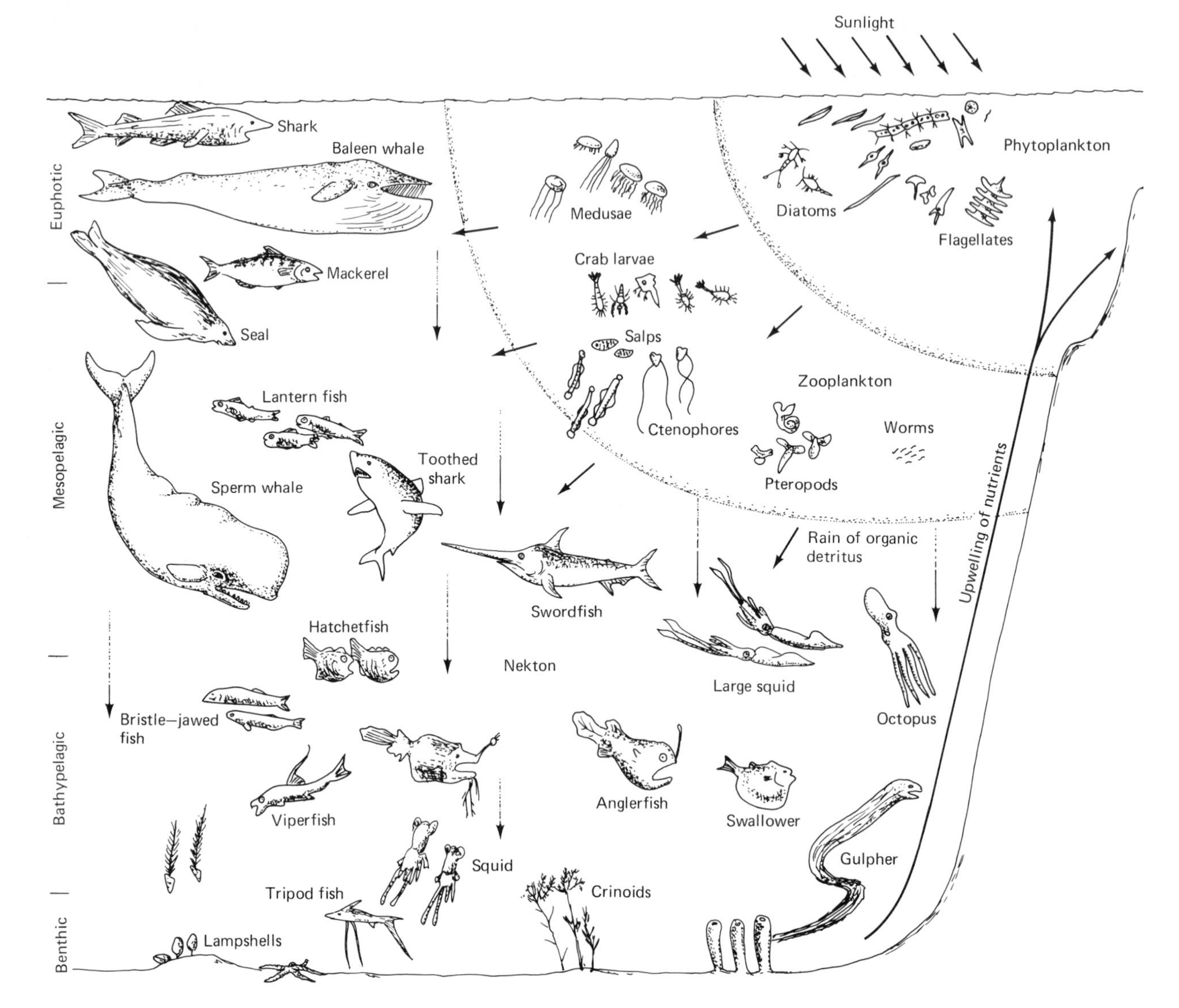

Figure 13-1. Biological wealth of the ocean.

shortages of meat. Accordingly, the world's oceans will become even more widely attractive places for food gathering.

In 1970, only four countries (Peru, Japan, Russia, and Norway) were responsible for 75 percent of the record world fish catch of 69.9 million metric tons (Figure 13-2). The bulk of the fish catch for that year was comprised of fin fish, but a fraction included whales, crustaceans, and invertebrates such as oysters, clams, and shellfish (Figures 13-3, 13-4, 13-5). For these countries fishing was an $8-billion business. This

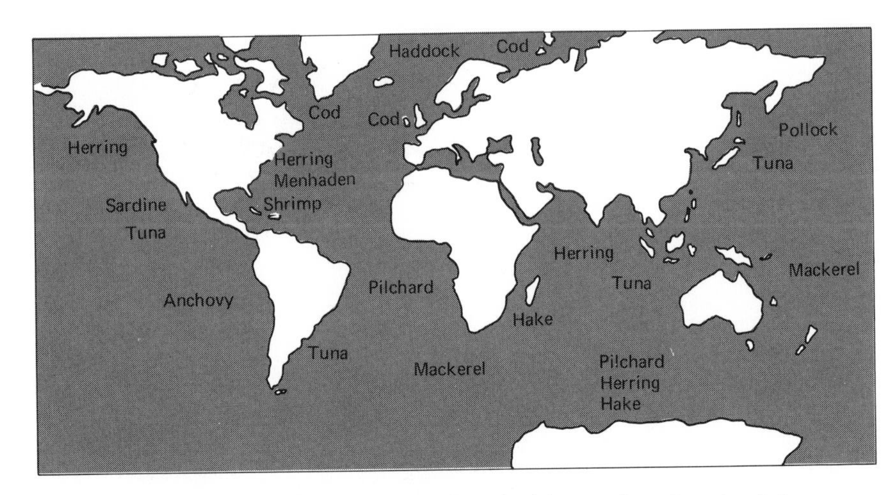

Figure 13-3. Some commercially valuable species of marine fish.

Figure 13-2. The world's fish catch in 1975 was 69.7 million metric tons. This catch was slightly less than the record high of 69.9 million metric tons of 1970. (Photo courtesy of National Marine Fisheries Service.)

was twice as much money gained from offshore production of oil and natural gas for the same year. For a number of years Peru has been the leading country in developing and catching single-species fish: it caught, for example, 10 million tons of anchovies annually, almost all of which was reduced to meal. U.S. fish consumption in 1970 was about 80 pounds per person, but the bulk of this went to the pet industry.

In 1975, world fish production showed a stationary trend of 69.7 million metric tons (compare Tables 13-1 and 13-2). One reason for the stationary fish production was the severe effects of *El niño* on the

Figure 13-4. Common commercial fish from North American and Northwestern European Atlantic waters. *Top left:* Herring. *Top right:* Cod. *Bottom left:* Menhaden. *Bottom right:* Haddock.

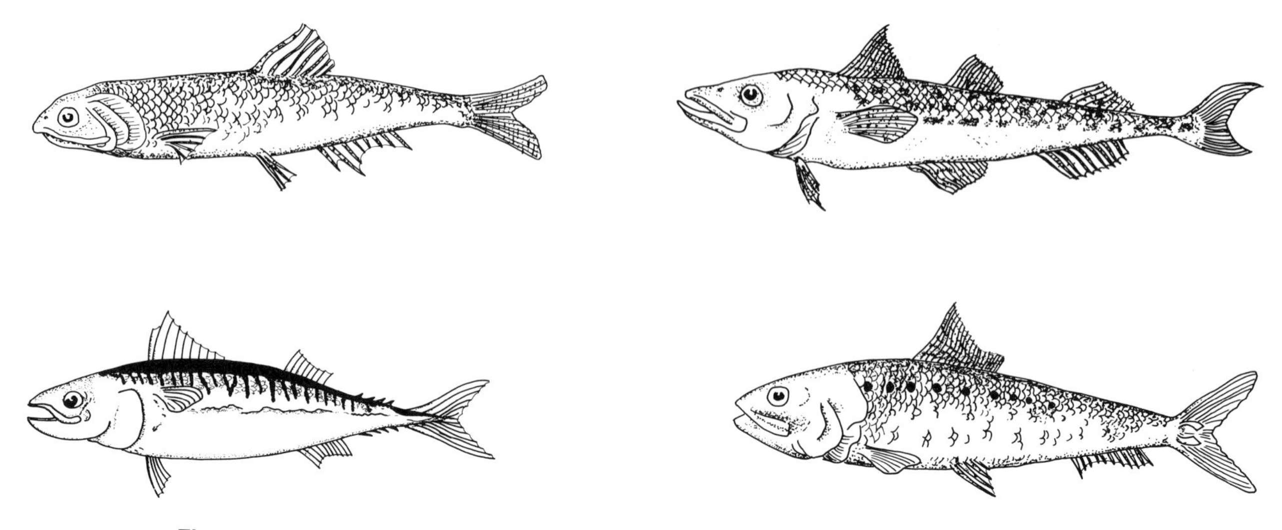

Figure 13-5. Common commercial fish. *Top left:* Anchovies are taken in water off Peru. *Top right:* Pollack, especially Alaska walley pollack, are concentrated in the northern Pacific. *Bottom:* Mackerel and pilchard are found in the southern Atlantic.

Peruvian anchovy in 1972 and 1973. The phenomenon *El niño* occurs as a result of the replacement of the upwelling, cold, nutrient-rich Peru Current by nutrient-poor tropical waters. This replacement takes place every seven years and is caused largely by the reduction in wind, which in turn reduces upwelling and thus nutrients for plankton and fish, often ending in their massive destruction. Popularly called the spoiler, *El niño* explains why Peru dropped from first to fourth in total fish catch (Tables 13-1 and 13-2). During 1975, Alaska pollack, anchovy, North Atlantic cod and haddock, and hake constituted the bulk of the world fish

TABLE 13-1 World Fish Catch, Selected Countries, 1970

Country	*Catch (in millions of metric tons)*
Peru	12.6
Japan	9.3
USSR	7.2
Norway	3.0
United States	2.7
India	1.7
Thailand	1.6
Spain	1.5
Indonesia	1.2
Chile	1.2
Denmark	1.2
United Kingdom	1.1
World total	69.9

SOURCE: United Nations Food and Agriculture Organization.

TABLE 13-2 World Fish Catch, Selected Countries, 1975, and Average Catch, 1971–1975

Country	*Catch (in millions of metric tons)*	*Average Catch (in millions of metric tons)*
Japan	10.5	10.4
USSR	9.9	8.6
China	6.9	6.9
Peru	3.4	5.0
United States	2.8	2.7
Norway	2.5	2.9
India	2.3	2.0
South Korea	2.1	1.6
Denmark	1.8	1.6
Spain	1.5	1.6
South and South West Africa	1.4	1.3
Indonesia	1.4	1.3
Thailand	1.4	1.6
Philippines	1.3	1.2
Chile	1.1	1.3
Canada	1.0	1.3
Vietnam	1.0	0.9
World total	69.7	68.8

SOURCE: United Nations Food and Agriculture Organization

catch. In 1976, world fish catch estimates ran between 72 and 73 million metric tons, indicating possible rejuvenation of the world's fisheries industry. Basic information on United States fisheries, in terms of most common species caught and their dollar value, is summarized in Table 13-3.

Despite growth in the industry, fish contribute only one-tenth of the animal protein to our diet. Fish is a valuable protein source, because it contains a favorable balance of amino acids that are required in human diets but are not commonly found in plants. But in 1969, for example, fishery products supplied only 3 percent of the world's food supply, although for Japan and Norway it made up 10 percent of their food supply.

Important fishing areas of the world include: (1) North American waters, (2) Northwestern European waters, (3) South American waters, (4) east Indian waters, and (5) east Asian waters. North American waters are located particularly in the northwest Atlantic Ocean off the coasts of New England and Nova Scotia. Northwestern European waters are located primarily in the North Sea between Great Britain and the European mainland. The North Sea is one of the richest fishing areas in the world. Iceland and Scandinavia are heavily dependent on seafood from these waters. South American waters, particularly those off the coasts of Peru and Chile, are potentially rich fishing grounds. East Indian waters include mainly the Indian Ocean and, to a lesser extent, parts of the

TABLE 13-3 U.S. Fisheries: Catch and Value of Principal Species, 1975

Species	*Preliminary Catch (millions of pounds)*	*Value (millions of dollars)*
Fish		
Menhaden	1803	49
Tuna	391	108
Pacific Salmon	202	116
Flounder	156	43
Sea Herring	119	5
Atlantic Cod	56	13
Whiting	42	4
Atlantic Ocean Perch	32	3
Mackerel (Pacific and Jack)	30	1
Pacific Halibut	16	5
Shellfish		
Shrimp	344	226
Crabs	301	84
Clams (meat)	120	41
Oysters (meat)	53	43
Lobsters (northern)	28	49

SOURCE: U.S. Bureau of the Census, 1976.

Pacific Ocean. These waters are extremely rich, and the fishing industry of India is developing rapidly. East Asian waters are the most intensively fished in the world. Russia, Japan, China, and smaller nations take large quantities of tuna, sardines, herring, mackerel, and shellfish. In the Pacific Ocean, the waters off Peru and Chile and the Gulf of Thailand are primary catch areas.

The proportionate fish catch from the world's oceans in 1969 was as follows: the Pacific yield was 53 percent; the Atlantic, 40 percent; the Indian, 5 percent; and the Mediterranean Sea, 2 percent (Figure 13-6). The increasing world fish catch is aided by large factory ships that employ modern fishing methods (Figure 13-7). Estimated targets for the world's future annual fish harvest vary from 120 million metric tons to over 2 billion metric tons. According to the United Nations' Indicative World Plan (an agency of the Food and Agriculture Organization), by 1985 the annual world's fish catch will be 100 to 200 million metric tons.

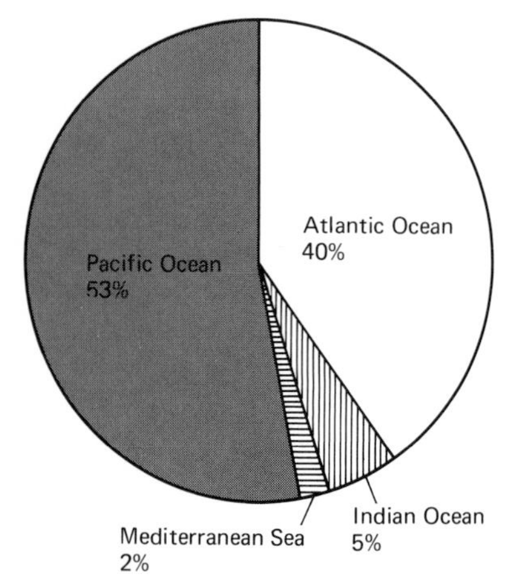

Figure 13-6. Percent of fish caught in world's major oceans and in the Mediterranean Sea in 1970.

13-3 MARICULTURE

Mariculture is generally defined as an operation in which fish or shellfish are temporarily kept captive until they can be sold com-

Figure 13-7. Modern Russian fishing ships are 40,000-ton factories, usually carrying their own fleet of catcher boats such as the boats on either side of the ship. The fishing ship is fitted with a computer-controlled sensor for locating potential fishing grounds. Helicopters fitted with fish-finding gear are often employed for expanding fish-hunting operations. In addition, the ship has a huge fish processing factory, living accommodations for factory personnel, swimming pool, library, cinema, dining room, and entertainment center. (Photo courtesy of National Marine Fisheries Service.)

mercially (Figure 13-8). Mariculture is not new; the Japanese and the Chinese were familiar with oyster culturing many centuries ago. The predominant biological resources commonly utilized in mariculture are

Figure 13-8. A marine aquaculture experiment is conducted on the Woods Hole pier during a 1972 visit of the Alcoa Seaprobe to the institution. (Photo courtesy of Woods Hole Oceanographic Institution.)

oysters, mussels, scallops, shrimp, and a small number of fish, including carp, salmon, trout, catfish, Asian milkfish, mullet, and yellowtail tuna (Figure 13-9). In recent years, total worldwide mariculture production has been only 2 million metric tons annually, which amounts to about 4 percent of the total world catch; this is carried out in less than 1 percent of the world's water.

In 1972, John Bardach and his associates concluded from their worldwide survey of marine mariculture that many countries employ different techniques. In Japan, yellowtail tuna juveniles are caught at sea and transported to floating cages where they are reared until they attain commercial size. In this way, Japan produces over 20,000 tons of yellowtail. In the United States and Canada, fish are caught in large pots and traps and are kept captive for a temporary time in pans near the sea until they reach marketable size. In the United States, commercial farming of pompano, Pacific sardines, and mackerel has been under study for some time. Lobster rearing in New England is a well-known example of sea farming. In Figure 13-10, the life history of shrimp in the Gulf of Mexico

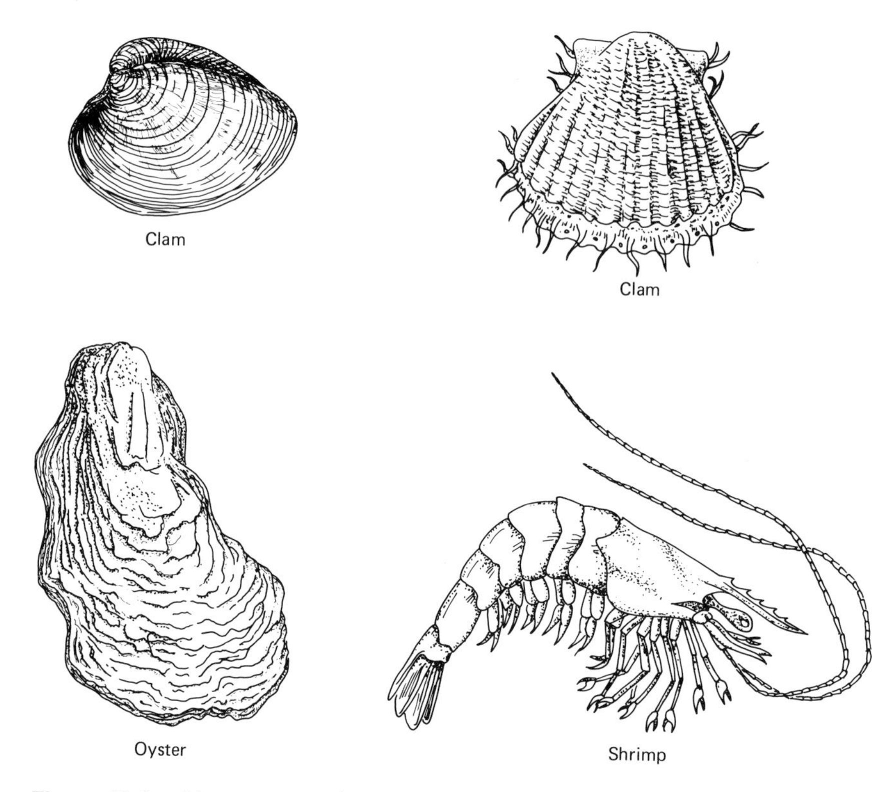

Figure 13-9. Many common invertebrates and crustaceans are grown and subsequently sold through the mariculture technique. *Top:* Two clams. *Bottom left:* Oyster. *Bottom right:* Shrimp.

is outlined as a case study to demonstrate how shrimp can be raised under controlled conditions.

In recent years, rearing prawns has been quite successful in India, Thailand, and the Philippines. The prawns are first put in saltwater tanks of 100-to-200-ton capacity. Each captive prawn lays over a half million eggs. After a week, the larval stage begins. At this stage, the larvae are fed well with microscopic plants and small brine shrimp. In the postlarval stage, the prawns are released into ponds and are fed on

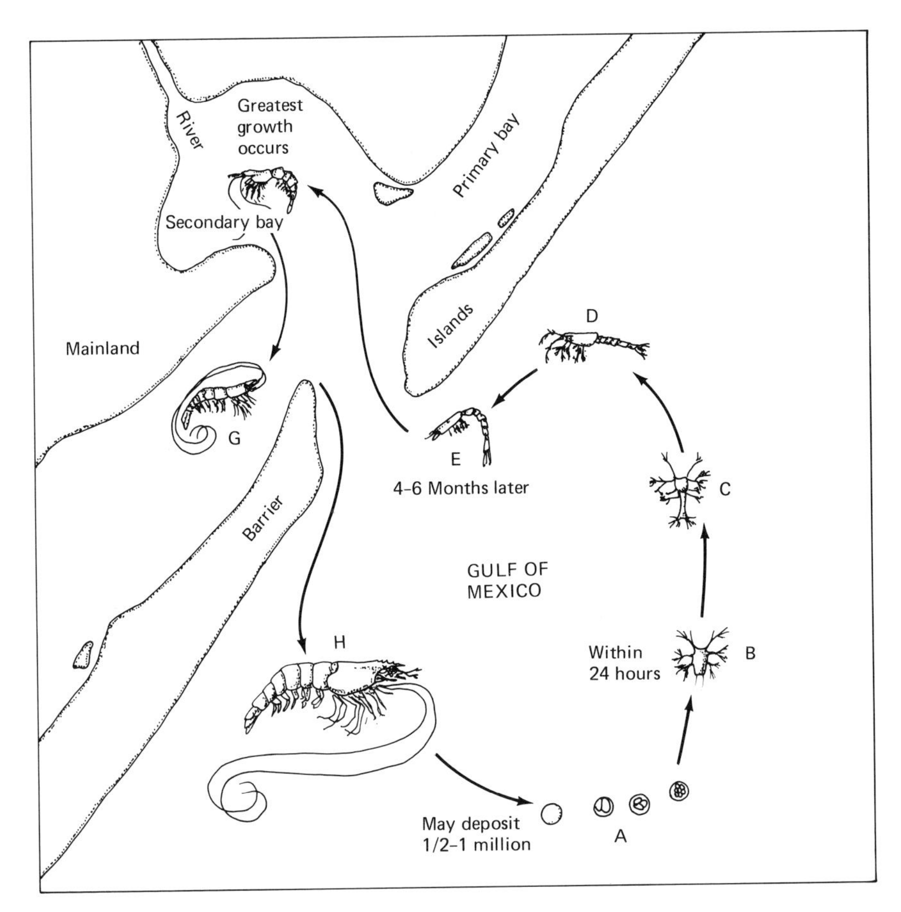

Figure 13-10. Life history of white, pink, brown, royal red, and seabob shrimp caught in waters off Gulf of Mexico. (A) Young are hatched. A mature female deposits between ½ and 1 million eggs, but only a few young survive and enter the larval stage. (B) Larvae are carried by onshore water movements to shallow estuaries and coastal lagoons, where they feed, grow rapidly, and attain adolescent stage. (C) Young adults continue their journey toward the bay for spawning. (D, E, F) Final phase of journey to the spawning grounds and the spawning period. (G and H) Mature adults return to the Gult of Mexico to lay eggs once again. (*Source:* Redrawn from A. W. Moffett, 1967, "The Shrimp Fisheries in Texas," *Texas Parks and Wildlife Bulletin, 50,* page 36.)

nutritious fish and clams. Prawns are kept in the ponds for about nine months, until they attain commercial size.

In Australia, oysters are grown on wooden sticks. In Spain, mussels are hooked to ropes and lowered into the water where they consume phytoplankton and zooplankton.

Limitations of Mariculture

Although mariculture seems to be a promising technique for increasing seafood production, certain problems are inherent: (1) distance from shore, wave and current conditions, depth, and light penetration restrict mariculture along the fringes of the sea; (2) fishermen prefer to be hunters rather than herders, because hunting entails no maintenance costs for equipment, storing, or herding facilities (also, fishermen are not prepared to give up their traditional freedom of the seas); (3) most areas considered suitable for mariculture are rapidly polluted; (4) because mariculture is still experimental, in that it is limited by a lack of scientific data.

In 1965, John Bardach suggested ways to improve the mariculture of fish and shellfish, proposing that a satisfactory candidate species have these qualities: (1) the species must be able to reproduce in captivity or in semiconfinement; (2) eggs must be capable of being hatched in a controlled environment; (3) larvae must be able to tolerate rapidly increasing food dosages; (4) adults must be able to gain weight rapidly from cheaply supplied food. Organisms that meet these criteria become economically significant.

Red tide *The red coloration of seawater caused by an overgrowth of certain organisms (e.g., dinoflagellates). It is poisonous to many organisms, particularly those that feed upon dinoflagellates.*

Mariculture of algae from the ocean on a massive scale is one of the most potentially effective means of meeting food crises, particularly in areas of high population and protein deficiency. The "red tide" organism, *Gonyaulax,* a dinoflagellete, has a protein content of 27.5 percent and an amino acid distribution similar to that of casein (the principal protein of milk). Chicks fed on a *Gonyaulax* diet have growth rates comparable with those living on casein diets. However, caution must be exerted in selecting the species. One species of *Gonyaulax* causes paralytic shellfish poisoning. Farming "red tide," then, would be one manner in which mariculture could significantly help to increase future food production.

13-4 OCEAN RANCHING

Although ocean research in the training of sea mammals such as porpoises, dolphins, and whales is still embryonic, it has long been realized that these mammals can communicate and can follow instructions. Scientists such as John Lilly at the Communication Research Institute in Miami, Florida, have shown through various experimental communication studies that porpoises can mimic sounds produced by humans. Scientists at the University of California at Los Angeles have demonstrated that porpoises can be trained to obey by remote control. Porpoises released into the ocean will return at the call of a "porpoise whistle."

Dolphins are being trained to transport maintenance and repair tools between anchored ships and divers at work undersea. Dolphins may be trained for rescue missions, including those entailed in naval warfare. In the future, porpoises may be trained as "aqua-cowboys" to herd schools of fish to fertile grazing regions. "Aqua-cowboys" would be guided by remote control. Upon a command issued to porpoises via a SONAR device attached to their bodies, they would return the school of fish to the ranch. A school of fish could be raised like livestock on a land-based ranch. When the fish reached a commercially profitable size, they would be sold or processed for food at a factory.

Ocean ranching of fish has a potential future growth for the following reasons: (1) Of 25,000 species of fish, only 200 to 300 species are presently caught for food. (2) Ocean ranching will be economically rewarding, because most of the required raw materials would be free; the major cost would involve the initial investment for leasing sea space and training of porpoises. (3) The demand for protein will continue to increase as long as population continues to rise.

13-5 WHALES AND WHALING

Whales

Whales are the largest known animals and represent the acme of adaptation to sea life. *Cetaceans,* which include whales, dolphins, and porpoises, are among original land animals that have returned to the sea.

Cetacea *An order of sea mammals that includes whales and dolphins.*

Approximately 90 species of whales live in oceans and adjoining seas. Whales are the most diversified of sea mammals in terms of their feeding habits, sizes, shapes, and weights. They vary in length from 2 to 35 meters and weigh from 1 to more than 120 tons. In general, they feed on plankton, krill, fish, squid, and sometimes shellfish. Whales are grouped into toothed whales and toothless whales.

Toothed whales comprise about 75 species, including intelligent creatures such as dolphins and porpoises. Sperm whales are the largest species in this group. Toothed whales feed on fish, squids, and crustaceans. Most toothed whales are marine animals, but occasionally toothed whales are found in estuaries and rivers.

Killer whales are the most feared of these marine animals (Figure 13-11). They feed preferably on seals, porpoises, penguins, and other whales. Killer whales weigh approximately 1 ton and are usually less than 7 meters long. They normally hunt in groups, and they launch sophisticated, well-planned precision attacks on their victims. Sperm whales are characterized by a large blunt heads with narrow underslung jaws containing many teeth. Sperm whales dive to 1300 meters below the surface to feed on giant squids. They weigh from 30 to 60 tons, and they may be 20 meters long. Sperm whales have been hunted extensively for their oil.

Baleen (Whalebone) *The filtering organ, consisting of numerous plates with fringed ends, of plankton-feeding whales.*

Toothless whales are popularly known as *baleen* or whalebone whales. Instead of teeth, these whales use a sieve of whalebone plates (baleen) in their mouths to filter food such as plankton and krill from seawater. Most of the 12 species of baleen whales are exclusively marine. Notable examples of the baleen whales include blue, fin, humpback, sei, and gray whales (Figure 13-12).

Figure 13-11. Common toothed whales. *Top:* Killer whale. *Bottom:* Sperm whale.

Baleen whales follow a yearly alternating migration pattern of breeding and feeding, in two distinct geographic regions of the ocean. They breed in the warm waters of subtropical zones in winter, and they feed in the cold waters of polar zones in summer. During their feeding period baleen whales are most vulnerable to their main predator— man.

Blue whales are the largest baleen whales, weighing as much as 112 tons and ranging from 26 to 33 meters. Blue whales are on the verge of extinction; in the 1930s there were an estimated 40,000; there are now approximately 2,000. They feed on krill, herring, and various small fish. Many blue whales live near the Antarctic because of the availability of plankton in its nutrient-rich waters.

Unlike other whales, fin whales are exceedingly vulnerable to predators because of their slow movements and their lack of fear of man. Fin whales are also hunted extensively for their oil.

The Pacific gray whale is one of the most widely known whales in North America, because it migrates along the West Coast of the United States from the Arctic to Lower California. Most Pacific gray whales are 10 to 13 meters long and weigh 25 to 40 tons.

Figure 13-12. Common baleen (toothless) whales. *Top:* Blue whale. *Bottom:* Humpback whale.

Whaling

Whaling has been a lucrative commercial business, because whales provide huge quantities of meat and blubber for food, oil for lighting, and ambergris. Ambergris, used in perfume as a scent fixative, is a waxy substance regurgitated by sperm whales. At present, baleen whale oil, is used as a raw material for manufacturing margarine and cooking fats. Sperm whale oil, although not edible, is used extensively as an industrial lubricant, in textiles dressing, and in soap and cosmetics manufacture. Whale meat is used to feed livestock, and whalebone meal is used in fertilizers. The best quality of whale meat is used for human consumption and in pet food industries.

Techniques of whaling have improved with time. Until the mid-nineteenth century, open-boat whaling with hand harpoons was common, particularly in catching slow-swimming whales (Figures 13-13, 13-14). Somewhat later the steam whaler, equipped with a harpoon cannon, was introduced to worldwide whaling. Once the whale was killed, compressed air was forced into its body in order to keep it afloat. Most whaling was done from shore-based stations and was carried out within a range of 100 miles. Since the beginning of the twentieth century, pelagic whaling has become more common. An average whaling expedition is equipped with a factory ship, catching boats, and a couple of refrigerator ships for the storage of meat. More whales have been killed since World War II than ever before. Unfortunately, modern harpoons and the use of aircraft and remote-sensing techniques have made whal-

Figure 13-13. Bequians are professional West Indian whale hunters who still use nineteenth-century hunting gear, a hand harpoon with a dart gun. (Photo courtesy of J. E. Adams, University of Minnesota, Duluth.)

ing so effective that some species of whales are nearly extinct. Whale catches in terms of countries and by species are presented in Tables 13-4 and 13-5.

Conservation of Whales

Concern is growing over the rapid depletion of certain species of whales (Table 13-6). Eight species of whales have been declared to be on the verge of extinction. Large baleen whales have suffered severely from overwhaling in recent years. More recently, blue and humpback whales have been reduced by efficient modern whaling methods. The stocks of gray whales almost hit the extinction point because of heavy whaling in the past but have recovered dramatically following successful conservation measures.

Conservation of whales has been a leading issue of worldwide ecological concern. In 1931, the League of Nations established an international convention for the regulation of whaling. A major outcome of the convention was to halt the catching of female whales with calves. In the

Figure 13-14. Bequian whalemen flense blubber from a 10-meter humpback whale, which is easily identified by its long, white flipper. Butchering takes place in the water. (Photo courtesy of J. E. Adams, University of Minnesota, Duluth.)

TABLE 13-4 Whaling 1973–74

Country	*Number of Whales Caught*
Russia	15,083
Japan	10,095
South Africa	1,817
Norway	1,812
Australia	1,080
Somalia	451
Iceland	365
Spain	224
Brazil	32
U.S.A.	21
Denmark	13

SOURCE: United Nations Statistical Yearbook, 1975.

TABLE 13-5 Whaling by Species 1973–74

Species	*Number Caught*
Blue	7
Fin	2,075
Humpback	9
Sei/Bryde's	8,030
Other Baleen whales	21
Sperm	20,858
World	30,993

SOURCE: U. N. Statistical Yearbook, 1975.

TABLE 13-6 Basic Facts About Whale Population

Type	*Current Estimated Numbers (in thousands)*	*Current Remaining (in percent)*	*Quotas*[a] *(1975–1976)*	*Conservation Measures*
Blue	13	6	0	Now fully protected
Humpback	7	7	0	
Gray	11	73	0	
Bowhead	2		0	
Right	4		0	Not yet fully protected
Fin	100	22	585	
Sei	75	38	2,230	
Bryde's	40	n.d.	1,363	
Sperm (male)	230	43	11,070	
Sperm (female)	390	68	7,970	
Minne	360	83	9,360	

SOURCE: U.S. Marine Mammal Commission.
[a]Approved by the International Whaling Commission.

following years, the convention restricted hunting seasons for various species. The Antarctic season was limited to three months. Following World War II, the International Commission on Whaling (ICW) specified the various species of whales that needed to be protected. The ICW established a new method of determining yearly quotas for Antarctic pelagic whaling. According to the new method, a unit of catch called the *blue whale unit* (BWU), which equalled 1 blue whale, 2 fin whales, 2.5 humpback whales, or 6 sei whales, was used as an index of total annual whale catch. In the 1950s the commission set 10,000 BWU as the limit for the fin whale catch. Because of depletion of the fin whale, the commission subsequently revised the limit to 4,500 BWU and during 1967–1968 again to 3,200 BWU.

The United Nations Conference on Human Environment, held in Stockholm in 1972, unanimously recommended a ten-year halt to worldwide whaling. The United States ceased all whaling activities in 1971 and has since then supported various conservation measures. Recently Great Britain, Norway and Holland (who found it uneconomic to continue whaling) have supported the whaling moratorium by quitting major hunting grounds in the Antarctic and the Northern Pacific. At present, Russia and Japan are hearing considerable protest against their continued whaling activities.

13-6 VITAMINS AND DRUGS FROM OCEANS

One of the newer fields of oceanography, *marine pharmacology*, proposes to extract and identify substances from marine organisms. These would be used primarily for medical purposes, such as cures for diseases. Marine pharmacology is not a new field; marine organisms have always been thought to have curative properties. According to the ancient historian Pliny, water crabs, freshly softened and swallowed with water or in burned form, would be a good antidote for most poisons. And the Chinese are known to have used various seaweeds in the treatment of throat diseases.

Marine pharmacologists are currently investigating organisms that have unique physiological, chemical, and physical properties. Examples, in addition to those in the previous paragraph, include: (1) the sea cucumbers, which possess a nerve toxin capable of "freezing" nerves without damaging them; such toxins may be indispensable in post-operative treatment of amputations. (2) Barnacles possess a cement that can be employed as dental glue because it is waterproof and very strong. (3) The venom of a man-of-war fish is useful in preventing heart attacks. (4) Certain species of horseshoe crabs contain in their blood a unique substance, lysate, that has medical use. It reveals excess levels of bacterial *endotoxin** in vaccines and serums.

The National Institute of Health Research is at present investigating marine organisms for their value as sources for drugs and vitamins. Recently, scientists at the institute discovered that sharks can withstand more brain damage than most mammals and, for unknown reasons, seem to be immune to cancer. Such a discovery, of course, is of paramount importance to medical sciences.

Thus far only 3 percent of the plants of the earth have been chemically and pharmaceutically evaluated; yet 50 percent of all prescriptions filled a few years ago contained a drug of natural terrestrial origin. On the other hand, at present less than 1 percent of known marine organisms havé been evaluated for their use as drugs and vitamins.

SUMMARY

1. Marine biological resources include fish, shellfish, vitamins, drugs, and mammals.
2. Fishing is a very old profession. A few countries account for the main bulk of the world fish catch. In 1975, the world fish catch was 69 million metric tons.
3. Important fishing areas of the world include the waters of North America, northwestern Europe, South America, east India, and east Asia.
4. In 1969, the proportionate fish catch according to oceans was, in percent: Pacific, 53; Atlantic, 40; Indian, 5; and Mediterranean Sea, 2.
5. Mariculture is the technique of raising fish and shellfish in captivity until they reach commercial size. Mariculture is a slowly but steadily growing field of oceanography.
6. Ocean ranching and marine pharmacology are steadily growing research areas in oceanography.
7. Because several species of whales are on the verge of extinction, their protection has received considerable attention from conservationists.

*Some microorganisms such as bacteria produce poisonous substances that can be excreted into the surrounding medium (exotoxins) or retained within their own cells (endotoxins). The endotoxin is released when the cells disintegrate.

Suggestions for Further Reading

Banse, Karl. 1973. "Global Distribution of Organic Production in the Oceans." In *Ocean Resources and Public Policy* (T. S. English, editor). Seattle, Washington: University of Washington Press.

Bardach, J. E., H. E. Ryther, and W. O. McLarney. 1972. *Aquaculture.* New York: Wiley.

Chapman, V. J. 1970. *Seaweeds and Their Uses,* second edition. London: Methuen.

Chapman, W. M. 1973. "Food from the Sea and Public Policy." In *Ocean Resources and Public Policy* (T. S. English, editor). Seattle, Washington: University of Washington Press.

Cushing, D. H. 1975. *Fisheries Resources of the Sea and Their Management.* New York: Oxford University Press.

Genthe, H. C., Jr. 1975. "Sea Forest: Role of Kelp." *Oceans, 8,* 50–55.

Hill, P. J., and M. A. Hill. 1976. *The Edible Sea.* New York: Barnes and Noble.

Holt, S. J. 1969. "The Food Resources of the Ocean." *Scientific American, 221*(3), 178–94.

Laurie, A. H. 1972. *The Living Oceans.* New York: Doubleday.

Mackintosh, N. A. 1965. "The Stocks of Whales." London: *Fishing News* (Becks).

NOAA. 1974. *Fisheries of the United States.* Current Fishery Statistics 6700, National Oceanic and Atmospheric Administration, U.S. Department of Commerce. Washington, D.C.: U.S. Government Printing Office, 1975.

Ryther, John H. 1969. "Photosynthesis and Fish Production in the Sea." *Science, 166,* 72–76.

14 Physical Resources

14-1 INTRODUCTION

PLANETARY OCEANS CONTAIN MANY METALLIC AND NONMETALLIC MINERALS as well as energy in various forms: geothermal energy, tidal waves, and such currents as the warm Gulf Stream. In fact, water per se is the single largest *mobile ore,* with a volume of 350 million cubic miles. Each cubic mile of water contains 4.7 billion tons of water and carries 165 million tons of dissolved solids. Dissolved salts that are common seawater include common salt, magnesium, and bromine. Oceans contain 5×10^{16} tons of mineral matter. Seafloors are extensively dotted with manganese nodules. Recent trends in worldwide ocean mining and petroleum extraction testify to the rapidly growing exploitation of marine physical resources (Figures 14-1, 14-2; Table 14-1). This exploitation has also been boosted by the discovery of methane gas in the Adriatic Sea; of gold, silver, copper, and zinc in the hot brines of the Red Sea; and of chromium in the Indian Ocean. Although ocean mining is increasing relative to land mining, it is still very expensive. It costs, for example, about $5 per ton to mine on land. To mine an equal amount of marine ore may cost between $5 per ton (for salt) to as high as $100 per ton (for magnesium). As a result, present marine mining practices are restricted to minimum depths and few locations, such as the waters off Alaska and off South Africa, where gold and diamonds, respectively, are mined.

14-2 PETROLEUM

Major commercial marine production of oil began in 1930 at Lake Maracaibo, Venezuela. Since then, petroleum operations have grown at

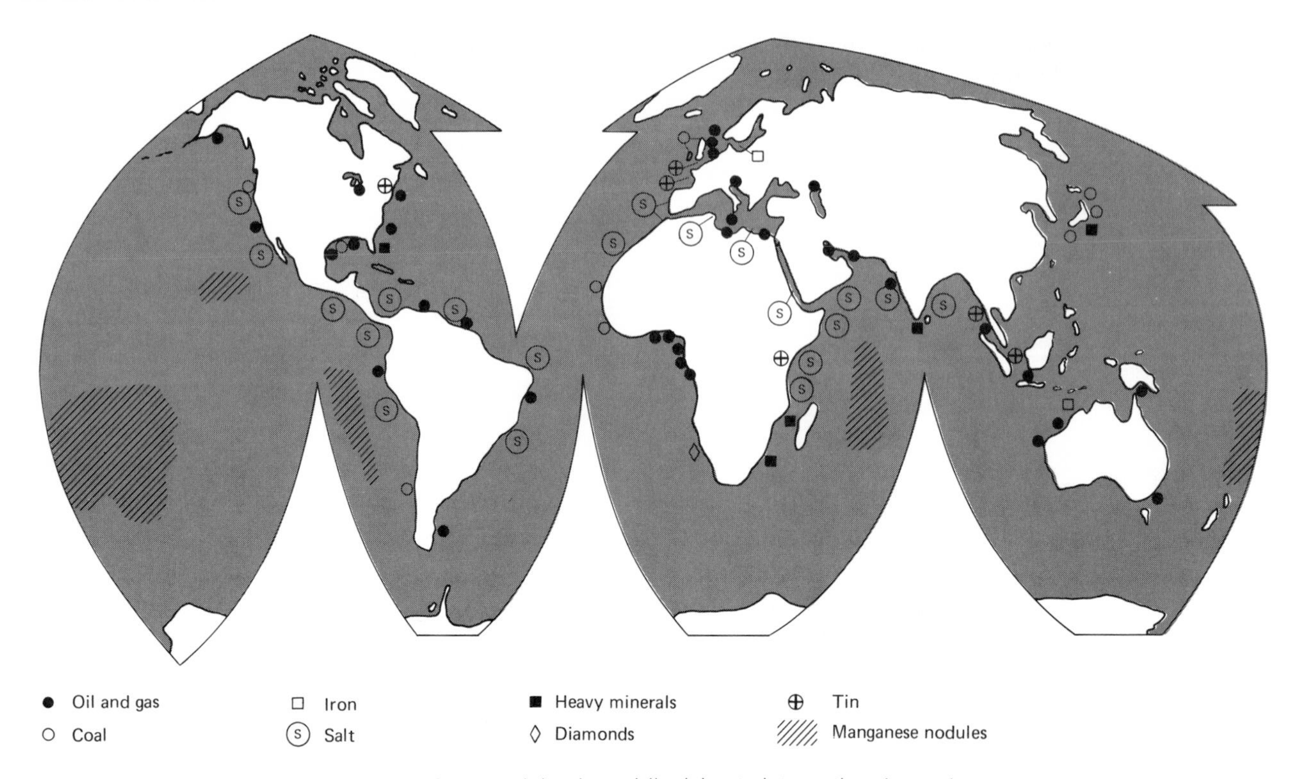

Figure 14-1. Ocean mining is rapidly rising to international prominence.

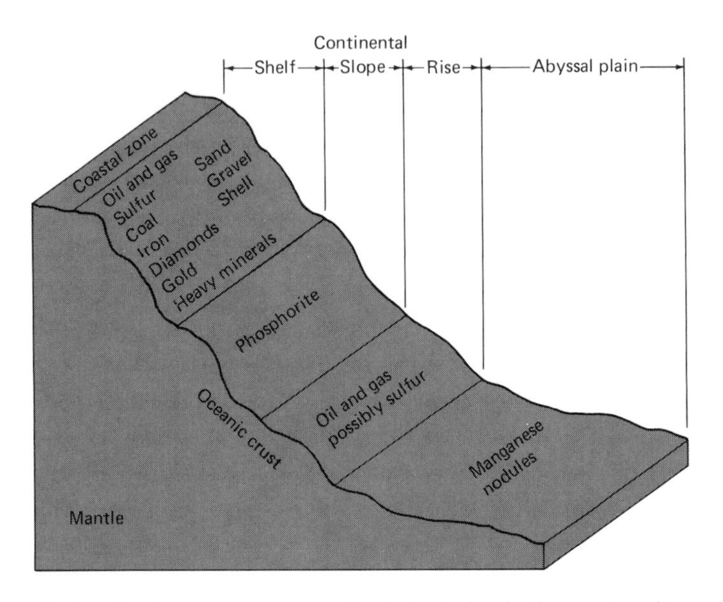

Figure 14-2. Geologic occurrence of principal physical resources in oceans.

TABLE 14-1 Selected Locations of Ocean Mining Activity in the United States and Abroad

Country	Water Depth (meters)	Minerals Dredged or Explored
United States		
off California	10±	Shells
off Alaska	60	Gold
Blake Plateau	100 to 500	Manganese nodules Phosphate
Japan	10	Iron sands
Indonesia	56	Tin
Thailand	50	Tin
South Africa	30	Diamonds
Great Britain	30	Tin and sand
Australia and New Zealand	100	Gold, phosphate and heavy metals
Indian and Pacific oceans	over 5,000	Manganese nodules

an unprecedented rate, especially in the Gulf of Mexico off the coasts of Texas and Louisiana. Up to 1970, U.S. oil companies invested over $13 billion for worldwide offshore drilling activity. In view of recent energy crises, this figure has changed considerably and now runs into several tens of billions of dollars. Offshore petroleum exploration and production have become a global phenomenon.

Some scientists estimate that more than 85 billion barrels of petroleum are deposited in offshore waters. Such reserves constitute about 20 percent of the world's total. The U.S. Department of Interior's estimates that, by 1980, 30 percent of the oil and 40 percent of the natural gas in the United States will be obtained from its offshore waters. Current *proven* reserves on the outer continental shelf are approximately 4.3 billion barrels of oil and 34.2 trillion cubic feet of natural gas. *Estimated* reserves, however, include 3 to 19 billion barrels of oil and 9.7 to 27 trillion cubic feet of natural gas.

Petroleum is found mainly in the continental shelves and slopes and in small ocean basins where thick Tertiary sediments are common. Abyssal plains contain thinner beds of sediment and therefore do not possess commercially feasible quantities of petroleum. At present, the offshore waters of approximately 75 countries are being explored for petroleum. Commercial quantities of petroleum have been definitely found in the North Sea, where Great Britain, Norway, and other European countries are already engaged in mining. Offshore petroleum exploration is also either planned or is already under way off the coasts of Greenland, Canada, the United States, Mexico, and China. In the United States, Georges Bank off the New England coast, and Baltimore

Canyon off New Jersey are two of the newest areas of preliminary offshore petroleum activity.

At present, offshore petroleum comes from areas within 150 kilometers of the coast and from depths of about 100 meters. However, by 1980 petroleum will be obtainable from depths of 2000 meters. The success of deep-water production is dependent not only on available technology but also on continuing high demand for petroleum on a worldwide scale. Current trends show that petroleum needs are presently being satisfied, so the global search for petroleum goes on.

The Geologic Aspect

Most petroleum begins as organic material in the form of microscopic plants, which are later deposited within the sediment on shallow to moderate seafloors. Requirements for the formation of petroleum include: (1) an abundant supply of marine life; (2) rapid deposition of organisms following death; (3) quick burial of the remains, which keeps them from decaying. Rapid burial aids in initiating the natural distillation of organic matter. The deposition of organically rich material should proceed in an enclosed basin where water circulation and oxidation are low. If these conditions are met, organic matter will be converted into liquid hydrocarbons (petroleum). The precise formation mechanisms of petroleum are still not known. Geologists believe that it takes at least 1 million years or so to form petroleum.

Once petroleum is formed, it must have free flow in order to accumulate in permeable strata such as sandstone or limestone. *Reservoir rock* is any rock that permits the free movement of petroleum (and thus its accumulation). Almost 60 percent of the world's petroleum reserves are confined in sandstone, whereas 40 percent is embedded in the open pores of limestone and dolomite. To prevent petroleum from escaping, permeable rock containing petroleum must be capped by nonpermeable rock. For example, a layer of slate will prevent any upward escape of petroleum.

Anticline *Up-folded structure.*

A suitable geologic structure called a *trap* slows the free migration of petroleum and stores it in limited space (Figure 14-3). Petroleum is generally found in a trap called an anticline. An *anticline* is a structure that has an upward fold with a crest, and flanks on either side. Petroleum migrates with water in anticlines, upward toward the crest, where it then concentrates and subsequently separates into gas–oil and water in accordance with density differences. When oil is struck, natural gas first escapes, along with an enormous amount of energy. Oil is next released, and then water is released. In some wells only gas is found; in others, oil or water alone may be found, depending on the geologic history of the area.

Sometimes petroleum migrates along a *fault,* which is a break in the earth's crust. The upward movement of buried *salt domes* deforms overlying rocks and provides traps for petroleum to migrate and accumulate. Sometimes petroleum concentrates where a porous sandstone layer is covered (above and under) by nonporous rocks such as shale. This type of geologic structure or trap is described as an unconformity.

Fault *A fracture in the earth's crust along which displacement has occurred.*

Composition

Chemically, petroleum is a complex mixture of compounds of carbon and hydrogen known as hydrocarbons. It can have either a paraffin or an asphalt base. A *paraffin base* means that the compound has a heavier hydrocarbon chain type; an *asphalt base* is characterized by ring structure. Crude oil may contain either of these hydrocarbons but more commonly contains their mixture. Crude oil is first transported over land or water by pipelines or supertankers. It is processed in refineries, where petrochemical products are differentiated into natural

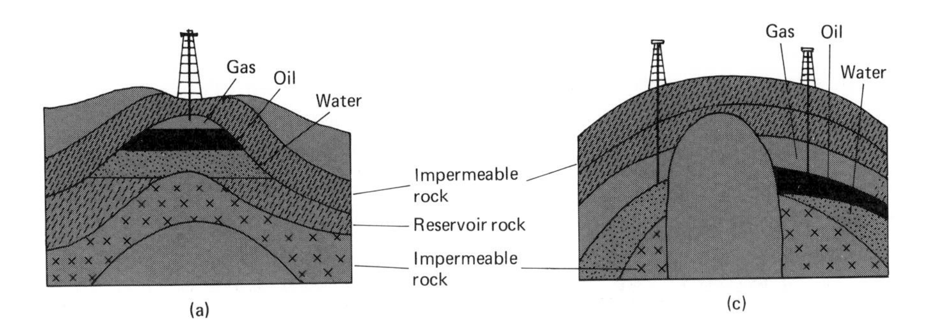

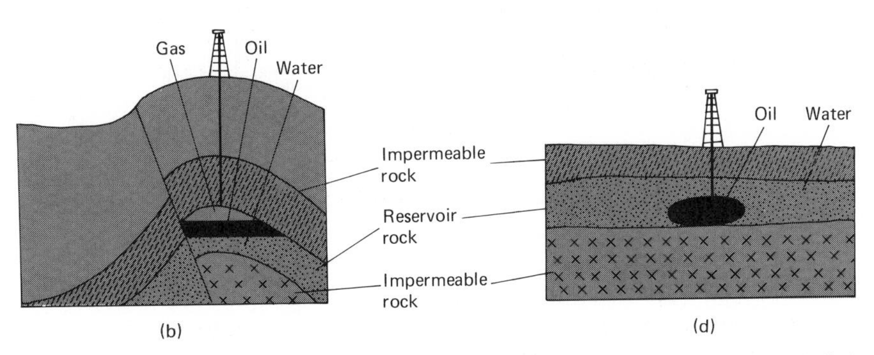

Figure 14-3. Major types of petroleum traps. (a) Anticline. (b) Fault. (c) Salt dome. (d) Unconformity.

gases such as butane, propane, and other valuable hydrocarbons, and into lower-grade products such as solid bitumens and tar.

14-3 MANGANESE NODULES

Manganese nodules *Deep ocean deposits consisting of oxides of iron, manganese, copper, and nickel.*

Manganese nodules are concretions of minerals that resemble potatoes; they are extensively distributed in major oceans at depths of 3500 to 4500 meters (Figure 14-4). Manganese nodules have also been discovered in shallow waters, such as those on the Blake Plateau off the eastern United States.

Manganese nodules vary in composition, size, shape, and occurrence. Nearshore nodules contain lower amounts of metals than those in deeper waters. The metal content also varies from sample to sample. For example, manganese content in some samples is as high as 50 percent. The average composition of nodules in terms of weight percent is: manganese, 30; iron, 24; nickel, 1.0; and cobalt, 0.5 (Table 14-2).

Manganese nodules grow in layers at an average rate of about 1 millimeter per 1000 years. They are believed to form at an annual rate of 10 million tons, and they presently exist in concentrations of up to 100,000 tons per square mile. Variations in grades of nickel, cobalt, and manganese occur over large lateral distances on the seafloor. According to J. L. Mero (1965), the aggregate deposit of nodules in the ocean is 1 to 7 trillion tons; it contains 400 billion tons of manganese, 16.4 billion tons of nickel, 8.8 billion tons of copper, and 9.8 billion tons of cobalt.

(a)

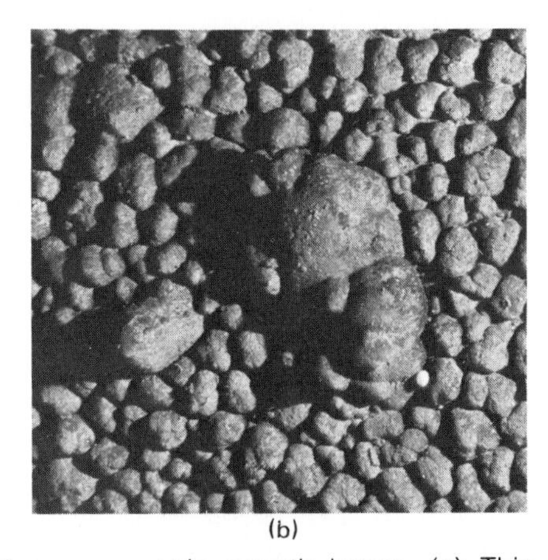

(b)

Figure 14-4. Manganese nodules show concentric growth layers. (a) This specimen was obtained from Georges Bank, 240 kilometers off New England, where offshore petroleum exploration is in progress. (Photo courtesy of S. McGinn; specimen courtesy of K. O. Emery, Woods Hole Oceanographic Institution.) (b) Manganese nodule deposit seen at great depth in the Pacific Ocean. (Photo courtesy of Lawrence Sullivan, Lamont-Doherty Geological Observatory of Columbia University.)

TABLE 14-2 Average Composition of Major Constituents of Ferromanganese Minerals

Constituents	*Weight Percentage*
MnO_2	31.7
Fe_2O_3	24.3
SiO_2	19.2
Al_2O_3	3.8
$CaCO_3$	4.1
Insoluble in HCl	26.8
H_2O	13.0

SOURCE: Adapted from J. L. Mero, 1965, *The Mineral Resources of the Sea.* New York, Elsevier.

A number of private and government-sponsored research projects are now in progress to establish commercial marine mining methods for manganese nodules. Large-scale mining of manganese nodules may become a commercial reality by the year 2000.

14-4 PHOSPHORITE

Phosphorite (phosphate rock) is a major industrial source of phosphate, a vital element in the life cycle of all living things. It is extensively used for the production of fertilizer. The use of phosphorus for fertilizer was recorded as early as 2000 B.C. In the past much of it was obtained from animal bones, fish, and guano. In the United States, 60 percent of this phosphate is sold as agricultural fertilizer; 25 percent goes to chemical industries; and the remaining 15 percent is exported.

At the 1970 rate of mining, more than 1000 years' supply of phosphorites in the land of the United States will be available. Most marine phosphorites are of a lower grade than are their terrestrial counterparts. The bulk of the marketable grade is 31 to 36 percent phosphate (Table 14-3).

Most phosphorite dredged from the sea bottom consists of nodules and crusts that contain a maximum of about 29 percent phosphate. This may be the maximum grade to be expected from the ocean. However, a higher phosphate content is found in many land deposits, mostly because of enrichment by weathering. Phosphorite deposits have been discovered off Peru, Chile, Mexico, the eastern and western United States, Argentina, South Africa, Japan, and several islands in the Indian Ocean. At least 50 percent of the continental shelf area of Australia is presently under concession for phosphate exploration.

K. O. Emery (1960) estimated that about 6,000 square miles of the southern California offshore area are covered by phosphorite nodules

TABLE 14-3 Chemical Composition of Marine and Terrestrial Phosphorites (In weight percentages)

Constituents	Marine: Forty Mile Bank off Southern California	Marine: Agulhas Bank off South Africa	Terrestrial: Florida	Terrestrial: Morocco (North Africa)
P_2O_5 (phosphate)	29.6	22.7	31.2	32.1
CaO (lime)	47.4	37.3	36.4	51.6
R_2O_3	0.43	9.4	12.7	—
CO_2 (carbon dioxide)	3.9	7.1	2.2	5.5
Fe	3.3	—	2.0	4.2
Organic	0.1	—	6.2	—
Total	84.7	76.5	90.8	93.4

SOURCE: J. L. Mero, 1965. *The Mineral Resources of the Sea*, New York, Elsevier.

amounting to an estimated reserve of 1 billion tons. Of this, according to John Mero (1965), if 10 percent of the phosphorite deposits are mined, it would provide a 200-year supply for California at a rate of a half million tons annually. Mero also estimated the *world* supply of phosphorite. The world holds about 50 billion tons of phosphate rock in reserve. Only eight countries hold more than 98 percent of it. Among the major producers of phosphate rock are the United States, the Soviet Union, Tunisia, and Morocco. Finally, Mero estimates that the continental shelves of the world occupy an area of about 10 million square miles. Assuming that 10 percent of this shelf area contains deposits of phosphorite similar to those off southern California, there should be approximately 3×10^{11} tons of phosphorite on continental shelves. If 10 percent of this material were economical to mine, the reserves of seafloor phosphorus material would be 3×10^{10} tons or about a 1000-year supply at the present rate of world consumption.

A new deposit of phosphorite has been discovered off the west coast of Mexico. This is a fine-grained, unconsolidated deposit of marine apatite, a phosphate mineral rich in calcium. The average grade of the deposit is about 15 percent and it reaches as high as 40 percent. It covers an area of at least 60 × 160 kilometers. The high-grade section of the deposit could recover at least 200 billion tons of phosphate minerals, which is about 4,000 years of reserves at the present rate of world consumption (Mero, 1965).

14-5 SULFUR

Sulfur is largely used in chemical industries. It occurs in the cap rock of salt domes buried within continental and seafloor sediments.

Sulfur is recovered from the oceans when heated water piped from the surface is injected into the deposit. The heated water melts the sulfur, which is forced upward with compressed air. About two-thirds of the world's production of sulfur is derived from bedded deposits or from those associated with salt domes.

Up to 1970, subsea production was confined to two salt dome deposits in offshore Louisiana, which accounted for about 20 percent of U.S. production. Sulfur is also obtained from two other sources as a by-product from sulfur-rich petroleum and coal and from terrestrial gypsum. A recent discovery of sulfur-rich salt domes in the deepest part of the Gulf of Mexico assures the future supply of sulfur in North America.

14-6 DISSOLVED SALTS

Of 92 naturally occurring elements, 60 are extracted from seawater. Most of these elements are present in infinitesimal amounts, especially certain critical elements, such as gold, silver, copper, zinc, and lead. These are mined expensively and on a limited scale (Table 14-4). Many other minor and trace elements, such as copper, molybdenum, and boron, are highly concentrated in the bodies of marine organisms (see Table 10-4).

Dissolved solids amount to 35 parts per thousand (35‰). Thus each cubic mile of sea water contains about 165 million tons of solids. Only common salt, bromine, and magnesium are presently extracted commercially. Sodium and chlorine are extracted by the conventional

TABLE 14-4 Selected Minerals in Seawater (per capita share for 6 billion people)

Mineral	*Amount*
Water	2×10^8 tons
Salt (Nacl)	6×10^6 tons
Magnesium	2×10^5 tons
Calcium	9×10^4 tons
Potassium	9×10^4 tons
Bromine	1×10^4 tons
Aluminum	200 tons
Manganese	2 tons
Copper	460 pounds
Silver	140 pounds
Gold	3 pounds

SOURCE: M. B. Schaefer, 1968, in *Ocean Engineering and Goals, Environment and Technology*, J. F. Brahtz (ed.), New York, Wiley, p. 17.

evaporation method and are used in the manufacture of common salt. Bromine accounts for two-thirds of the world's total production and is used largely by the auto industry to produce antiknock compounds for gasoline.

Magnesium is one of the most abundant dissolved salts in seawater. The bulk of U.S. production of magnesium since World War II has come from the sea. The Dow Chemical Company in Freeport, Texas, processes 1 million gallons per hour of seawater from the Gulf of Mexico to produce magnesium.

14-7 HEAVY MINERALS AND OTHER DEPOSITS

Heavy minerals include zircon, rutile, monazite, ilmenite, gold, platinum, and diamond. These minerals are obtained by beach mining throughout the world. Diamonds are mined in South Africa; ilmenite, rutile, zircon, and iron in India, and monazite, thorium, and gold and iron ores in Alaska (Figure 14-1).

Mud enriched with copper and zinc has recently been found in the Red Sea. Oceanographers have learned that red clays have an annual rate of accumulation of 5×10^8 tons at an average depth of 200 meters. Their aluminum content ranges from 5 to 15 percent; copper content approaches 0.20 percent. Recently, oceanographers have learned that submarine lava is enriched in iron and magnesium minerals (Figure 14-5).

14-8 WATER AS A MINERAL RESOURCE

Seawater becomes an even more vital mineral resource when it is used in the form of freshwater for industrial, agricultural, and domestic purposes.

Desalination is the process of converting seawater into freshwater. At present, desalination plants are in operation in over 200 locations throughout the world. At the largest desalination plant located in Saudi Arabia, 6 million gallons of seawater are processed daily. Over 50 different techniques of desalination are either proposed, planned, or already in operation in many parts of the world. Three commonly used techniques for desalting seawater are: (1) distillation, (2) electrodialysis, and (3) freezing.

Figure 14-5. Submarine pillow lava is relatively rich in iron and magnesium minerals. It acquires rounded shape on rapid cooling in water. Pillow lavas are commonly located near volcanic vents or fissures that are created as the sea floor spreads in the form of plates on either side of midoceanic ridges. This flow is in the North Atlantic. (Photo courtesy of J. G. Schilling, Graduate School of Oceanography, University of Rhode Island.)

Distillation

In *distillation,* seawater is evaporated, and freshwater vapors are condensed and collected on a slanted glass surface (Figure 14-6). The distillation method is economical, because it employs solar energy as a heat source. Most oceangoing ships, especially those sailing in the

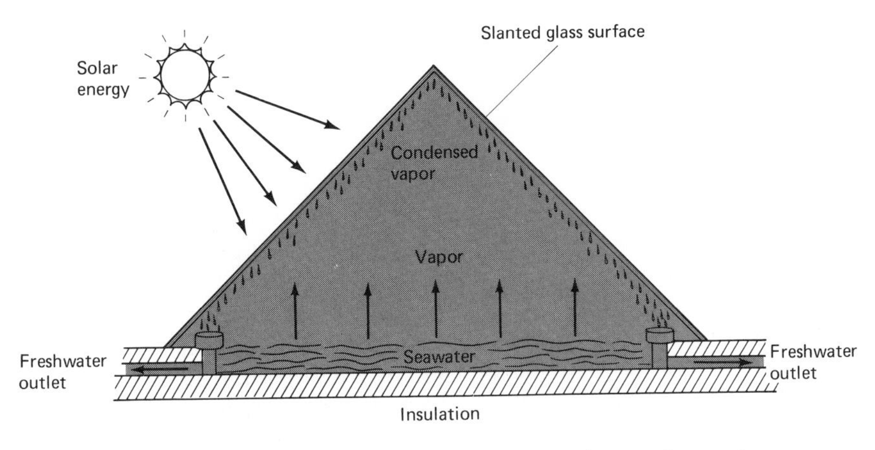

Figure 14-6. Desalination of seawater by distillation, using solar energy. (*Source:* J. J. Bhatt, 1975, *Environmentology: Earth's Environment and Energy Resources,* Cranston, R.I., Modern Press.)

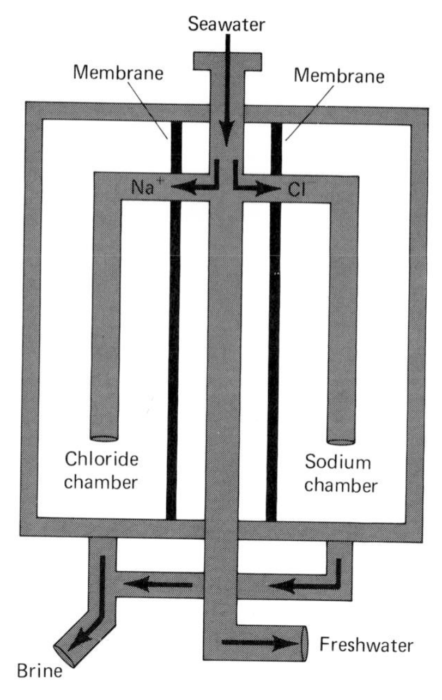

Figure 14-7. Desalination of seawater by electrodialysis. Sodium ions (Na^+) are directed toward the oppositely charged chloride (Cl^-) chamber. A reverse attraction also results. Because of polar ionization, freshwater is obtained. (*Source:* J. J. Bhatt, 1975, *Environmentology: Earth's Environment and Energy Resources*, Cranston, R.I., Modern Press.)

warmer regions of the oceans, obtain their freshwater supplies by this method. In the United States, distillation for desalination purposes is carried out at Key West, Florida, and Freeport, Texas. These process about 2 million gallons of freshwater daily.

Electrodialysis

In *electrodialysis*, freshwater is obtained by using electric currents to separate the ions of many salts, particularly sodium chloride. Seawater is passed through a tube dividing a container into two chambers, one for sodium ions and the other for chlorine ions (Figure 14-7). As the seawater flows through the tube, the sodium and chloride ions are systematically diverted to oppositely charged ions in each chamber; and salt is thus depleted. A desalination plant at Webster, South Dakota, processes 250,000 gallons of water daily, using electrodialysis.

Freezing

In the *freezing* process, freshwater is simply separated from saltwater at freezing temperatures (Figure 14-8). Ice is later thawed to obtain freshwater, and salt residues are disposed of as commercial salt. This

Figure 14-8. Icebergs are nature's way of desalinating seawater into freshwater, using the freezing method. These natural tanks contain millions of tons of freshwater. They may be transported to areas of water shortage for domestic, industrial, and agricultural purposes. (Photo courtesy of R. B. Theroux, National Marine Fisheries Service.)

Iceberg *A massive piece of glacier ice that has broken from the front of the glacier (calved) into a body of water. It floats with its tip at least 5 meters above the water's surface and at least 4/5 of its mass submerged.*

process is not very efficient, because the ice is slightly blackish, and the melted water may still have a salty taste. The freezing process is commonly used in cold countries areas such as Scandinavia and Russia. In the United States the Office of Saline Water (Department of the Interior) has set up a pilot plant at Wrightsville Beach, North Carolina. It processes about 200,000 gallons of freshwater daily. The general operation of a typical freezing process used for the desalination of seawater is shown in Figure 14-9.

14-9 POWER FROM OCEANS

Energy from Waves

When powerful sea waves pound the shore, they release large amounts of energy. These waves are capable of lifting ships weighing thousands of tons. Strong waves produce approximately 100,000 kilowatt-hours per year per meter of shoreline. If an appropriate engineering technique is used to translate the pistonlike movement of waves, it may be possible to tap energy from their dynamic motion. The up-and-down wave movements would provide levers with enough power to run a generator and subsequently to produce electricity. Ideally, the wave engine is supposed to drive a generator directly, but its actual operation is much more complex than the simple principle suggests. Because waves are irregular, the production of energy is not commercially rewarding.

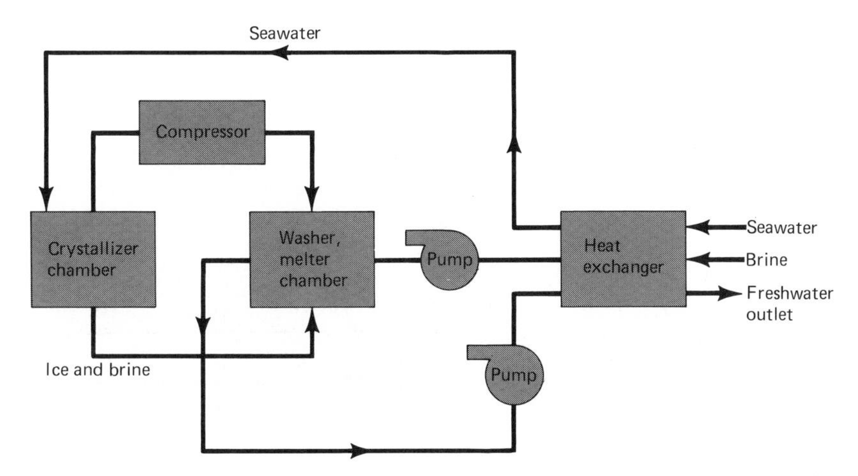

Figure 14-9. Desalination of seawater by freezing. (*Source:* J. J. Bhatt, 1975, *Environmentology: Earth's Environment and Energy Resources,* Cranston, R.I., Modern Press.)

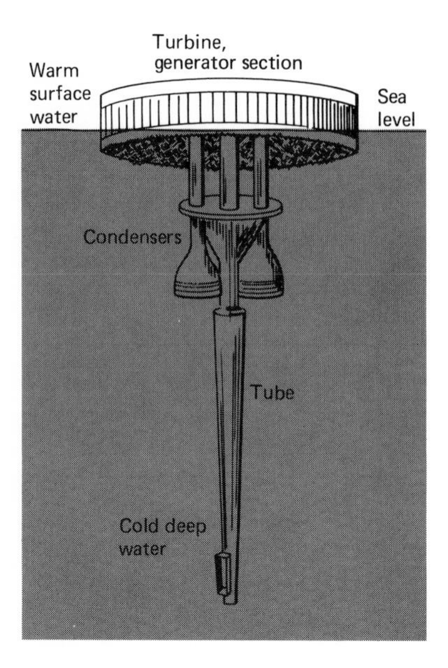

Figure 14-10. Artist's concept of harnessing energy from temperature differences of seawater.

Energy from Ocean Temperature Differences

A theoretical suggestion for harnessing the energy generated from the difference in hot and cold waters goes back to the nineteenth century. It was then realized that a giant natural heat engine exists with a boiler in the ocean at the equator and a condenser in the ice of Greenland, and that energy from the seawater can therefore be tapped. This idea was implemented in a pilot plant in Belgium. Later the scheme was put into practice when a larger unit was constructed in the Caribbean Sea. Thus the first *thalassothermic* (Greek: *thalasso,* sea; *therm,* heat) power plant began operating in Belgium; it produced 50 kilowatts. In subsequent years the plant in Cuba produced 22 kilowatts.

That there is a difference in the temperatures of surface and subsurface waters is the principle underlying the generation of power from seawater. The temperature difference is established with bottom cold waters and surface warm waters. The water at the surface of a tropical sea may be 25°C to 30°C; deep waters in the same locality are about 5°C. Hence, theoretically speaking, the difference of 25°C, a significant thermal gradient, should run a machine that would ultimately produce power.

In recent years, ocean engineers have proposed a floating generator system designed to produce power from the differences in ocean temperatures (Figure 14-10). The floating generator system would consist of three principal components: (1) a turbine generator turned by refrigerant gas, heated and vaporized by warm surface water; (2) condensers where deep, cold water would liquefy gas for return to the generator, and (3) a tube extending approximately 600 meters to pump up cold water for use in condensers.

Geothermal Energy

Geothermal energy in oceans is in the form of hot water and steam associated with underwater rift and fracture zones and with active volcanoes. It is especially important in Iceland. Geothermal energy can also be obtained from the trenches along the periphery of Pacific Ocean, where volcanic and earthquake zones coincide. Geothermal energy from coastal regions is primarily promising for the generation of electricity, the desalination of seawater, and the solution mining of sulfur and potash.

SUMMARY

1. Physical resources of the ocean are water itself, petroleum, other minerals, and alternate sources of energy.
2. At present, offshore areas of the world are under scrutiny for oil and gas deposits by private and governmentally supported oil companies.
3. By 1980 in the United States, 20 percent of all oil and 40 percent of all gas will come from offshore regions. Current proven U.S. reserves on the outer continental shelf are about 4.3 billion barrels of oil and 34.2 trillion cubic feet of gas.
4. Manganese nodules are generally found at depths of 3500 to 4500 meters in major oceans. The aggregate deposit of manganese nodules is between 1 and 7 trillion tons, which will last for several decades.
5. Phosphorite and sulfur are used in the manufacturing of fertilizers and chemicals.
6. Commercially important dissolved salts include common salt, bromine, and magnesium.
7. Desalination is a process of converting saltwater into freshwater by techniques of distillation, electrodialysis, or freezing.
8. Alternate sources of energy from the ocean include tidal power, power from waves, and power from temperature differences in seawater. Oceans contain potential geothermal areas, notably in Iceland and the Pacific, where volcanoes and deep trenches are densely distributed.

Suggestions for Further Reading

Emery, K. O. 1960. *The Sea Off Southern California.* New York: Wiley.

Emery K. O. 1975. "New Opportunities for Offshore Petroleum." *Technology Review, 77*(5), 31–33.

Hammond, A. L. 1974. "Manganese Nodules." *Science, 183,* 502–503, 644–646.

Hammond, A. L. 1975. "Minerals and Plate Tectonics." *Science, 189,* 779–781, 868–869.

Hedberg, H. D. 1976. "Ocean Boundaries and Petroleum Resources." *Science, 191,* 1009–1081.

King, H. M. 1971. "The Energy Resources of the Earth" In *Energy and Power.* San Francisco: Freeman.

Mero, J. L. 1965. *The Mineral Resources of the Sea.* New York: Elsevier.

National Academy of Sciences, Commission on Natural Resources. 1975. *Mineral Resources and Environment.* Washington, D.C.: Government Printing Office.

Park, C. F. Jr. 1975. *Earthbound–Minerals, Energy and Man's Future.* San Francisco: Freeman Cooper.

Skinner, B. J., and K. K. Turekian. 1973. *Man and the Ocean.* Englewood Cliffs, New Jersey: Prentice-Hall.

Weeks, L. A. 1973. "World Offshore Petroleum." In *Focus on Environmental Geology,* R. Tank (ed.). New York: Oxford, Chapter 25, 270–282.

Wenk, Edward. 1969. "The Physical Resources of the Ocean." *Scientific American, 221,* 166–76.

15
Environmental Oceanography

15-1 INTRODUCTION

Environmental oceanography IS A NEW BRANCH OF OCEANOGRAPHY that has come under considerable public attention in recent years. It deals mainly with the distribution of the causes and consequences arising from pollutants such as oil, radioactive wastes, DDT, and various metals such as lead and mercury that are poisonous to marine life (Figure 15-1). Environmental oceanography seeks, among other things, legal, scientific, and technological measures to remedy rising marine pollution. *Marine pollution* refers to the continual deterioration of the ocean environment, induced largely by man-made processes and occasionally by sporadic, natural processes such as volcanoes and earthquakes.

Figure 15-1. Man invades the ocean. (Photo courtesy of R. B. Theroux, National Marine Fisheries Service.)

15-2 OIL POLLUTION

Pollution of the oceans by oil became sensational global news in 1967 when the Liberian ship *Torrey Canyon* went aground off England's Cornwall coast, spilling 34 million gallons of oil. The tragedy of the *Torrey Canyon* became worse when massive destruction of marine organisms was brought about not by spilled oil but by detergents used to disperse it. Public concern over oil pollution in oceans again was raised by the Santa Barbara oil spill in 1969 when 83,000 gallons of crude oil were spilled. In subsequent years, concern over the oil spills subsided somewhat, partly perhaps because concern over energy crisis supplanted it. But the drama of oil pollution from tankers was again epitomized in late 1976 when a dozen or so oil tankers simultaneously spilled oil either within or nearby U.S. waters. The Liberian ship *Argo*

Merchant was wrecked near Nantucket Island, Massachusetts, and spilled 7.5 million gallons of oil (Figure 15-2). The Panamanian ship *Grand Zenith* mysteriously sank, carrying with it 8.2 million gallons of oil and a crew of 38. The Panamanian tanker *Olympic Games* grounded in the Delaware River near southeast Philadelphia, spilling about 134,000

Figure 15-2. The *Argo Merchant*, a drama of obsolescence, negligence, and inevitable circumstances. (a) U.S. Coast Guard rescues the crew. (b) Earliest phase of grounding (white represents spilling oil). (c) Intermediate phase climaxes *Argo Merchant's* battle for survival. (d) The ship is completely incapacitated. (Photos courtesy of U.S. Coast Guard, First Coast Guard District, Boston.)

TABLE 15-1 Summary of Major Oil Spills Caused by Oil Tankers

Date	Location	Cause	Spill	Methods Used to Collect Oil	Damage
1967	English Channel	*Torrey Canyon* ran aground	34 million gallons	Chalk treated with stearic acid	Detergent killed marine life
1968	San Juan Bay	*Ocean Eagle* broke in half as it ran aground	3.5 million gallons	Hydrophobic powder	No information
1969	Santa Barbara	Well blowout due to faulty blowout preventer	83,000 barrels per hour	Well plugged with 1150 bags of cement slurry	Birds, sea lions, and life damaged
1971	San Francisco Bay	Careless navigation	Unknown	No information	No information
1976	Off Nantucket	*Argo Merchant* ran aground	7.5 million gallons	Combustion method using silica; unsuccessful	Not yet determined

gallons of oil. At about the same time in Los Angeles the wreck of a blast-shattered tanker *Sansinena* further compounded the spillage (Table 15-1).

Besides these spectacular oil spills, small-scale oil pollution goes on as usual. Over 3500 tankers of 1000 tons or more are in operation, and they dump over 5 million tons of oil into the ocean annually. Spills from tankers account for one-third of this amount. The problem of oil spills is more acute because of the fact that only 20 percent of the operating tankers discharge over 90 percent of this oil. The remaining 80 percent of ocean tankers take necessary measures to assure cleaner discharges. In 1971, there were 2500 oil spills in U.S. waters (Table 15-2).

As demand for fossil fuels, particularly petroleum, continues to rise, a higher frequency of oil spills would appear to be inevitable. But

TABLE 15-2 Oil Spills in U.S. Waters[a]

Source	1968	1969
Vessels	347	532
Shore facilities	295	331
Unidentified	72	144
Total	714	1,007

[a]Covers reported spills of over 100 barrels.

SOURCE: U.S. Federal Water Quality Administration, 1970, *Clean Water for the 1970s*.

oil spills are also caused by natural seepages, by occasional well blowouts, and by oil slush discharges. Every time a tanker prepares to reload, it has to flush the residue of its previous load. This flushing, of course, guarantees continual oil pollution albeit in small amounts.

Causes

Oil spills from ships are caused by a combination of several factors: (1) obsolescence (although a generation of new tankers is replacing old ships, those still in existence account for most of the spills mentioned); (2) malfunctioning of navigational instruments (the captain of the *Argo Merchant* admitted at a hearing in New York that his ship carried no LORAN—long-range navigation—equipment and that the ship's gyrocompass was not functioning prior to the wreck); and (3) careless navigation (the captain of the *Argo* also admitted that he was using water current charts for November during the month of December).

Consequences

Immediate and long-term impacts of oil pollution on the marine environments are: (1) ecological hazards; (2) damage to fisheries; (3) damage to recreational resources; and (4) overall environmental degradation of the coastal zones (Figure 15-3). The case of the *Argo Merchant* vividly exemplifies these possible consequences. The spill drove hundreds of sea birds ashore. It aroused immediate fear that oil would threaten humpback whales, which migrate in the spilled area, and would quite possibly damage the rejuvenated population of gray seals. These fears were well-founded because of the fact that the *Argo's* oil (number 6 weight) dissolves rapidly, and because, since it has a higher density than seawater, it will sink faster (Figure 15-4). Moreover, number 6 oil is highly toxic. Fortunately stormy weather conditions carried the oil away from the New England shore and into the Gulf Stream Current. But the greatest concern over the oil spill from the *Argo Merchant* came from New England's fishing industry. The fishery industry is one of the major economic resources of New England, employing over 30,000 men in Massachusetts alone. As a result, the fishermen filed a $60-million suit against the *Argo Merchant's* owner.

Suggested Measures

The immediate concern is how to overcome these offshore spills, which seem to be rising. A few possible measures include: (1) Improved coastal navigation systems and stricter regulation of marine traffic. In the Canso Strait, Nova Scotia, a model system has been set up to regulate ship traffic by a sophisticated radar system. Canada requires all foreign vessels entering or departing the strait to take on pilots.

(2) The Coast Guard should be permitted to enforce the *Waterways Safety Act of 1972* very strictly in order to upgrade the standards of foreign ships entering U.S. waters, but the Coast Guard should establish a navigational traffic system beyond the Canso Strait, which would allow for more efficient enforcement of the law. The *Federal Water Pollution Act* permits enforcement of U.S. laws within 12 miles of this country's shore. The act imposes strict liability on any ship that releases oil; it requires offenders to pay for cleanup. The captain of *Olympic*

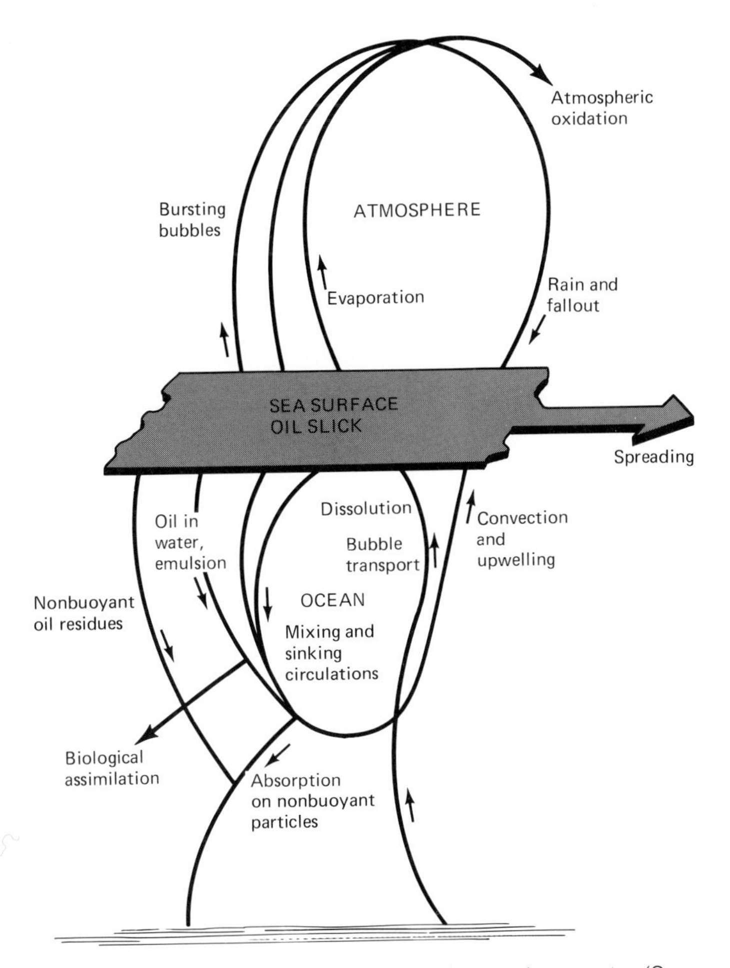

Figure 15-3. Dynamics of oil pollution in marine environments. (*Source:* J. G. Widmark et al., 1970, "Organic Chemicals," in E. D. Goldberg, ed., *A Guide to Marine Pollution,* New York, Gordon and Breach Scientific Publications.)

Games was arrested and held on $50,000 bail under provisions of this law.

(3) The International Maritime Consultative Organization (IMCO) should seek mandatory compliance with improved standards, such as the requirement of double bottoms for newly constructed tankers. Another example includes the building of new tankers with separate tanks for ballast water.

The problem of oil pollution by ships will be partially resolved when obsolete tankers are permanently taken out of the oceans and when offshore oil terminals are built.

15-3 RADIOACTIVITY

Radioactivity *The spontaneous disintegration of the nucleus of an atom with the emission of radiation.*

The world became aware of radioactive hazards for the first time when Hiroshima and Nagasaki were bombed in the 1940s. Soon after

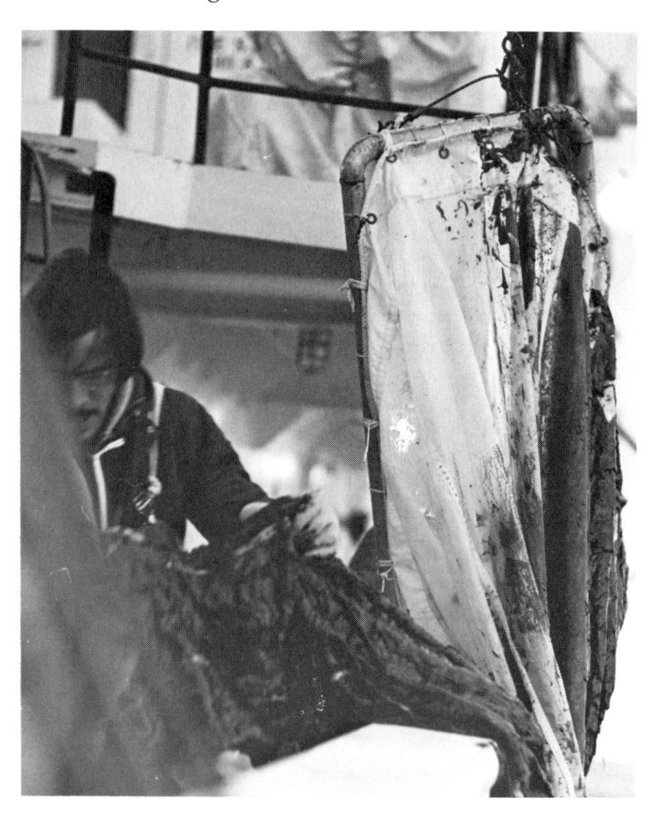

Figure 15-4. An oceanographer assesses the amount of freshly spilled oil (black). (Photo courtesy of National Marine Fisheries Service.)

this, many nuclear explosions have been conducted experimentally in the atmosphere, underground and undersea, particularly during the 1950s and 1960s (Figure 15-5). Today the nuclear industry is rapidly expanding on a worldwide level. Over a dozen nations have nuclear reactors. However, most use their facilities solely for the purpose of generating electricity.

In 1973, 28 reactors having a total capacity of 31,000 megawatts were sold to U.S. utility companies. By the end of the century, about 1000 nuclear power plants will be operating in this country. As of June 1973, there were 174 reactors in operation, under construction, or on order in the United States. The number of nuclear-powered ships in the world is expected to reach 500. The demand for nuclear energy in the future will increase rapidly since fossil fuels, particularly petroleum, will dwindle by the year 2000. By the twenty-first century, it is estimated that nuclear power from fission (the splitting of heavy radionuclides into lighter ones in order to generate energy) will account for 20 percent of the total energy output.

Despite the controversy over the hazards of nuclear energy, certain of its important advantages must not be ignored. For example, the dividends of nuclear energy include utilization of nuclear batteries installed

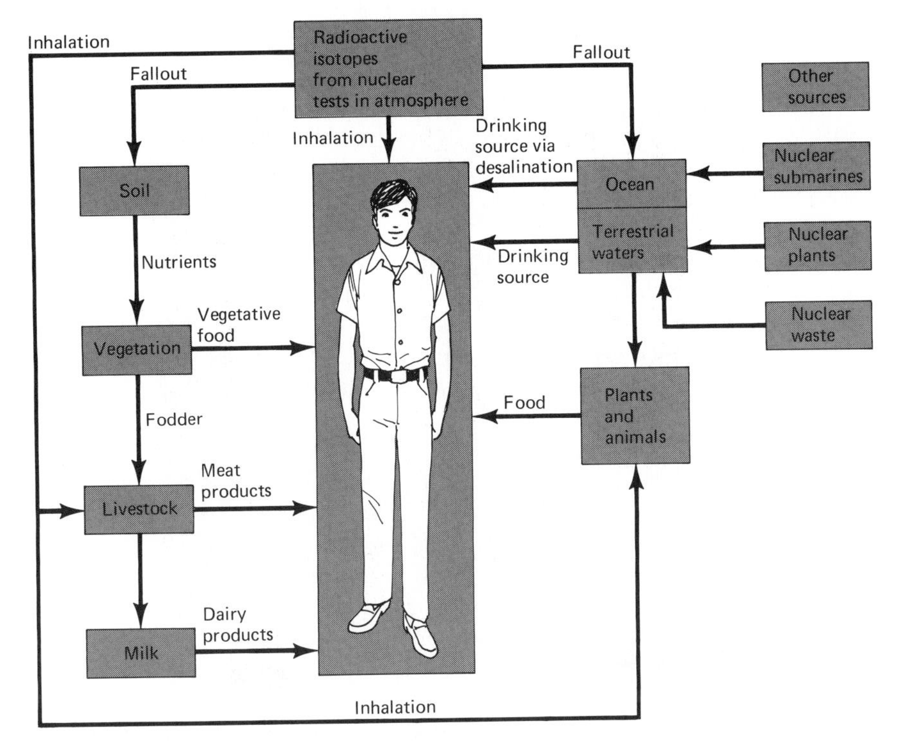

Figure 15-5. Dynamics of radioactivity in the environment, including in man.

in artificial hearts or for treatment of weak hearts, of radioactive cobalt for curing certain malignant tumors, and of deuterium for fusion power.

Causes

The fear of radioactive contamination, particularly in oceans, stems largely from the fact that their enormous volume, size, and depth make them a vulnerable repository for radioactive wastes, excellent sites for nuclear tests, and a suitable medium for nuclear ships.

Radioactive contamination of the oceans and the atmosphere may be caused by the spread of radionuclides. Natural radionuclides are the decay products of naturally occurring radioactive elements such as uranium and are also formed by the interaction of sunlight with the atmosphere; radiocarbon (^{14}C) is a notable radionuclide. Artificial radionuclides are produced by breaking heavy nuclides such as uranium or plutonium (fission), by bombarding stable nuclei with neutrons (neutron induction), and by combining the nuclei of light isotopes such as deuterium and tritium.

Sources

The release of artificial radionuclides into air and water is causing grave concern in the world today. Three major sources of ocean contamination by artificial radioactive material include: (1) nuclear explosions; (2) nuclear power plants; and (3) nuclear-driven ships. Radionuclides from nuclear explosions, particularly from testing of nuclear weapons, enter oceans in many ways. Most radioactive or debris is dissolved in seawater by undersea explosions. Atmospheric explosions deposit radionuclides in the air; subsequently, radioactive fallout may accumulate in oceans via rain water, runoff, and underground aquifers. Nuclear tests on small islands can cause significant contamination of the oceans.

In the mid-1950s, a series of nuclear tests at Bikini and Eniwetok atolls caused considerable radioactive contamination of oceans, particularly in the North Pacific (Figure 15-6). Scientists detected high radioactivity in seawater samples collected near the test sites.

Fallout *The shower of radioactive debris, usually from the earth's atmosphere.*

Nuclear explosions inject large amounts of radioactive debris directly into the atmosphere. This debris later falls to the earth's surface. Because oceans occupy two-thirds of the earth's surface, they accommodate greater quantities of such debris. The return of radioactive particles from the atmosphere to the surface is *fallout*. The intensity of radioactive fallout depends on several factors; for example, on the original amount of nuclear material used; on the latitude and altitude of the test site, on the season, and on meteorological conditions existing at the time

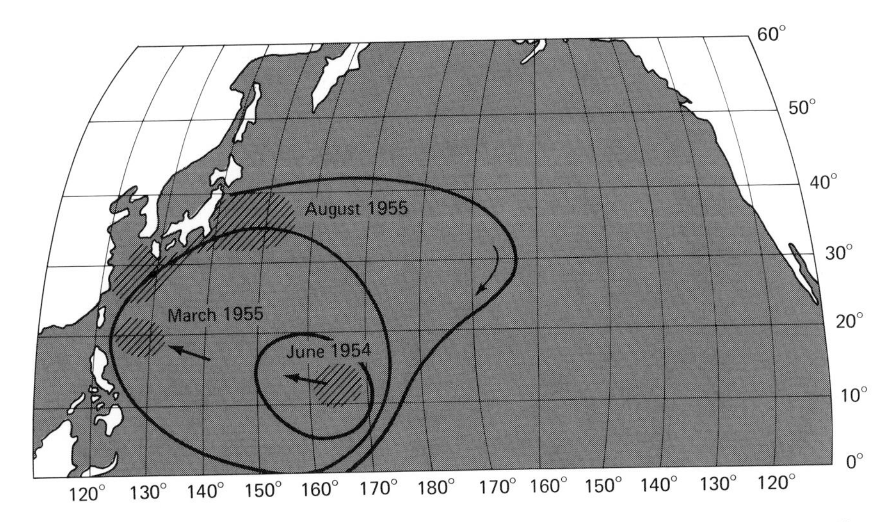

Figure 15-6. Radioactive contamination of waters in the northern Pacific. (*Source:* Y. Miyak, 1971, D. W. Wood, ed., *Impingement of Man on the Oceans*, New York, Wiley-Interscience.)

of the blast. Common radionuclides include radio-iron, strontium, and cesium. Of lesser importance are manganese, zinc, zirconium, nibium, and rubidium. Worldwide distribution of radium (^{226}Ra) in seawater is summarized in Table 15-3.

Three categories of fallout material are recognized: (1) Local fallout, (2) tropospheric fallout, and (3) stratospheric fallout. *Local fallout* marks the immediate geographic region where the test was conducted. Radioactive fallout is heaviest in particle size and greatest in quantity; it settles immediately after the explosion.

Table 15–3 Worldwide Distribution of Radium (^{226}Ra), in Seawater[a]

	Depth (in meters)		
	0–500	*500–2000*	*2000 to sea floor*
North Pacific	2–10	4–13	6–16
South Pacific	2–10	5–15	8–17
North Atlantic	2–8	4–9	5–12
Indian Ocean	3–6	5–9	6–12
Caribbean Sea	2–6	3–7	4–7
Mediterranean Sea	about 6	5–12	about 10
Providence channel	3–6	3–6	
Florida Straits	3–5	about 5	
Black Sea	7–12	about 10	
Red Sea	4–6	about 6	
Baltic Sea	3–15		

[a] $\times 10^{14}$ grams per liter.

SOURCE: B. J. Szabo, 1967, "Radium Content in Plankton and Sea Water in the Bahamas," *Geochim Cosmochim Acta, 31*, 1321–1331.

Lighter radioactive debris travels upward by the thrust of the blast into the atmosphere, where it is distributed into the troposphere and the stratosphere. *Tropospheric fallout* is whirled around the globe in an easterly direction; it will settle down in a few months or so at about the same latitude where the nuclear test took place. *Stratospheric fallout,* characterized by circulation patterns, remains there for a relatively longer time before returning to the surface. Stratospheric fallout flows laterally from the equatorial to the polar regions and falls through the stratosphere into the troposphere above the upper latitudes of temperate zones. For this reason, radioactive fallout concentration is heaviest between 50°N and 50°S.

Nuclear Power Plants

Radioactive substances from nuclear plants enter the oceans chiefly in packaged disposal dumped to the deep seafloor and from reprocessing plants such as those used for resin. This source supplies from 100 to over 1000 curies annually through outlets such as pipelines that are hooked up to the oceans. The liquid waste from reprocessing plants includes radioactive strontium, cesium, rubidium, rhodium, and radon.

Nuclear-Driven Ships

Radioactive material may be directly released to oceans through leakage in nuclear-powered ships. In some cases, leakage ensues from nuclear warheads lost from ships or aircraft. The loss of and fortunate recovery of a hydrogen bomb by the United States in waters off the coast of Spain is a notable example. Occasionally, the disappearance of nuclear-driven submarines will contaminate the oceans significantly. In the 1960s, the *Thresher* and the *Scorpion* were mysteriously lost. The amount of contamination that these submarines generated is not known.

Consequences

The greatest impact of radiation in oceans is the inevitability of genetic changes in the organisms that are irradiated. Experiments with mice have shown that mutational changes take place even under chronic low-level radiation exposure. A total of about 100 rems would be needed to double natural mutational changes. At an acute dose of high-level exposure only 30 rems are needed to obtain the same result. A generation of 30 to 100 rems can bring about drastic consequences.

15-4 CHLORINATED HYDROCARBONS

Chlorinated hydrocarbons are synthetic organic chemical compounds familiar as pesticides. *Pesticides* is a general term applied to *insecticides, fungicides, herbicides, fumigants,* and *rodenticides. DDT* is a short name for a complex pesticide: dichloro-diphenyl-trichloroethane.

DDT *Dichloro diphenyl Trichloroethane* $(C_6H_4Cl)_2 \cdot CH \cdot CCl_3 \cdot$ *Manufactured as a contact insecticide.*

After World War II the annual production of chlorinated hydrocarbons, particularly of DDT, steadily increased because of their application in agriculture for the control of disease-bearing pests and for malaria eradication. Use of DDT has recently fallen off though.

The malaria eradication campaign was sponsored by the World Health Organization of the United Nations to improve the living conditions of people in malarial countries. The results of DDT are both positive and negative. On the one hand, malaria has been brought under considerable control; on the other, DDT residues have continually increased in the environment (Figure 15-7).

Among the several types of synthetic organic chemicals, only a few have proved to be dangerous to both aquatic and terrestrial organisms. DDT is the most typical example. DDT is more persistent than other pesticides. Residues of DDT in the sea environment are selectively absorbed by plankton, which in turn are swallowed by small fish, which in turn are ingested by larger fish, thus eventually transferring DDT to higher organisms including man (Table 15-4).

15-5 METALLIC POLLUTION

Heavy metals such as lead, copper, zinc, cadmium, nickel, chromium, lead, and mercury are responsible not only for degrading the quality of seawater but for killing a number of marine organisms (Figure 15-8). Most heavy metals are practically indestructible in an aqueous environment. Many organisms can concentrate large amounts of heavy metal in their bodies and later pass these on through the food chain from fish to man (Table 15-5). Heavy metals are often delivered into the sea as domestic and industrial effluents.

Metals are often associated with chlorinated hydrocarbons because hydrocarbons constitute a part of the metal, as napthenate in copper; as sulfate or oxide in zinc; as methyl, ethyl, alkoxyethyl, or mercuric chloride in mercury. Finally, heavy metals are also supplied to the ocean from mine residues, which are carried by streams and rivers.

Some heavy metals may come from the crude or heavy fuel oil leaks (or spills) from ships, pipelines, offshore wells, and refineries. Vana-

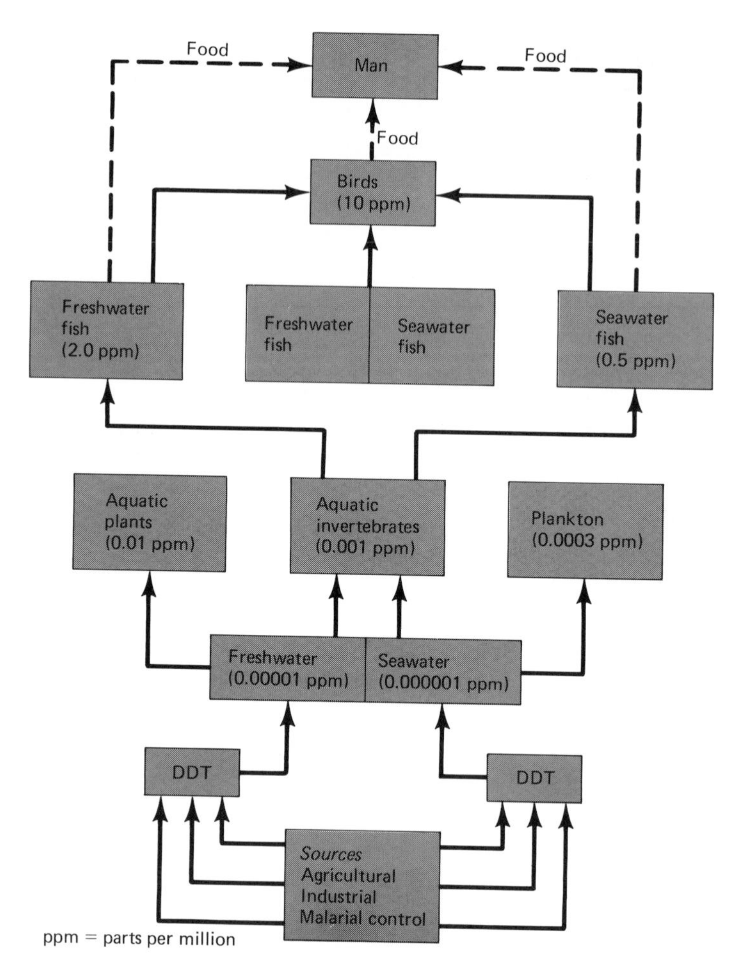

Figure 15-7. Dynamics of DDT in the aquatic environment and in birds and man. (Modified after C. A. Edwards, 1971, *Persistence of Pesticides in the Environment*, Cleveland, CRC Press.)

TABLE 15-4 Effects of DDT on Fish

Organism Tested	*Lethal Concentration*	*Exposure Time in hours*
Goldfish	0.027 mg/l	96
Rainbow trout	0.5–0.32 mg/l	24–35
Salmon	0.08 mg/l	36
Brook trout	0.032 mg/l	36
Guppy	0.75 ppb	29
Stoneflies species	0.32–1.8 mg/l	96

SOURCE: Simplified after P. H. McGauhey, 1968, *Engineering Management of Water Quality*, New York, McGraw-Hill.

Figure 15-8. Massive fish kill caused by pollution of normal water. (Photo courtesy of Erie County Health Department, New York.)

dium, manganese, cobalt, and nickel are found in crude oils as nonvolatile porphyrins.

15-6 POLLUTION IN COASTAL ZONES

A coastal zone is a transitional zone where land and man meet the sea. Oceanic pollution is more concentrated along this zone primarily

TABLE 15-5 Consequences of Metallic Pollution on Fish

Compound	*Organism Tested*	*Concentration (in milligrams per liter)*	*Exposure time (in hours)*
Cadmium chloride	Goldfish	0.3 Cd	190
Copper sulfate	Stickleback	0.03 Cu	160
Copper nitrate	Stickleback	0.02 Cu	192
Lead nitrate	Minnows, stickleback, brown trout	0.33 Pb	No data
Mercuric chloride	Stickleback	0.01 Hg	204
Nickel nitrate	Stickleback	1.0 Ni	156
Zinc sulfate	Stickleback	0.32 Zn	120
	Rainbow trout	0.52 Zn	64

SOURCE: Simplified after P. H. McGauhey, 1968, *Engineering Management of Water Quality*, New York, McGraw-Hill.

because of high population concentration and heavy density of industrial complexes. Along the coastal zone, the construction of factories and the existence of power stations, refineries, harbor installations, and municipal waste dumps constantly contribute to the degeneration of the oceanic environment. The major sources of this degeneration are the municipal wastes and industrial effluents (Table 15-6).

The future trend of coastal pollution in the United States is not promising. It is estimated that by the year 2000, the U.S. population will reach 312 million, and most Americans will be living in the three megalopolis areas located along water bodies: San Francisco to San Diego, New York to Miami, and Chicago to Syracuse. This trend of population dovetails with the increase of offshore petroleum, mining, and nuclear-energy-development activities. In short, coastal pollution will inevitably increase in the future.

TABLE 15-6 Examples of Industrial Pollutants Discharged into the Ocean

Type	*Examples*
Inorganic compounds	Agricultural runoff Acid manufacturing Mining and quarrying Iron and steel and related industries
Organic and inorganic compounds	Textiles Paper and pulp industries Paint industry
Organic compounds	Petrochemicals and fuel Dyes Breweries Fish processing factories

SUMMARY

1. Environmental oceanography is the study of causes and consequences of pollution of the oceanic environment.
2. Increasing global population and industrialization have accelerated usage of oceans.
3. Oil pollution has significantly increased in recent years primarily because of the expanding offshore petroleum exploration and tanker spills.
4. Because of the growth of nuclear energy, oceans have become contaminated. Despite several preventive measures, the amount of radioactive nuclides in oceans is increasing.

5. Chlorinated hydrocarbons, notably DDT, are persistent and dangerously polluting pesticides. DDT, selectively absorbed by plankton, enters the bodies of many sea organisms in the food chain.
6. Heavy-metal contamination by lead, copper, zinc, cadmium, nickel, chromium, lead, and mercury is present in the coastal waters where industrial effluents are most plentiful.
7. Coastal pollution will increase with rising population and industrial ocean-related activities.

Suggestions for Further Reading

Bartlett, Jonathan (editor). 1977. *The Ocean Environment.* New York: H. W. Wilson.

Bhatt, J. J. 1975. *Environmentology: Earth's Environment and Energy Resources.* Cranston, Rhode Island: Modern Press.

Brown, R. A., and H. L. Huffman. 1976. "Hydrocarbons in Open Ocean Waters." *Science, 191,* 847–849.

Fisher, N. S. 1975. "Chlorinated Hydrocarbon Pollutants of Marine Phytoplankton: A Reassessment." *Science, 189,* 463–464.

Friedman, W. A. 1972. *The Future of the Oceans.* New York: Dobson.

Heyerdahl, Thor. 1975. "How to Kill an Ocean." *Saturday Review,* November 29, 12–18.

Lowenstein, J. M. 1976. "Nuclear Pollution of the Seas." *Oceans, 9,* 60–65.

Mathews, W. I. T., F. E. Smith, and E. D. Goldberg (editors). 1972. *Man's Impact on Terrestrial and Oceanic Ecosystems.* Cambridge, Massachusetts: M.I.T. Press.

National Research Council. 1975. Study Panel on Assessing Potential Ocean Pollutants. National Academy of Sciences. Washington, D.C.: Government Printing Office.

Nelson-Smith, A. 1973. *Oil Pollution and Marine Ecology.* New York: Plenum.

16

Ocean Management and Conservation

16-1 INTRODUCTION

THE MANAGEMENT OF THE WORLD'S OCEANS, PARTICULARY FOR conservation of their resources, including minerals, food, energy, and environment, is one of the greatest challenges of this century. The task is even more difficult because: (1) The size of the oceans is a formidable management problem; it blankets two-thirds of the earth's surface and touches the national boundaries of over 150 independent nations. (2) Many member nations of the United Nations claim portions of ocean as part of their territorial jurisdiction. Nations claim different ranges of territorial rights; for example, Peru and Chile each claim 200 miles. Iceland and Britain are engaged in a cod war for conflicting claims on their respective territorial waters. (3) Landlocked nations such as Switzerland should be allocated a share of the ocean. (4) The bulk of oceanic territory (85 percent) falls outside the jurisdiction of all nations; who should own these waters is therefore a difficult question. (5) The rate of pollution of the marine environment, particularly in coastal zones, varies considerably. An industrially advanced country discharges more pollutants than does a less industrialized neighbor. France and Germany dump more industrial pollutants into the Atlantic Ocean than do Spain and Portugal. A similar situation exists between the United States and Mexico, and Japan and Malaysia. (6) Updated data on marine resources are still lacking on a global scale. It is difficult, therefore, to make important decisions concerning ocean mining and fisheries.

16-2 OCEAN MANAGEMENT: LAWS OF THE SEA

Nations have conflicted over territorial claims of oceans for centuries. The issue of maintaining the freedom of the high seas has been consistently fought throughout history. In 1608, Hugo Grotius in his book *Mare Liberum* upheld the Dutch protest against Spain, Portugal, and Britain for their claims over the high seas and asserted that all nations have equal access to the high seas. The modern legal framework for the management of world's oceans is based on the fact that the high seas belong to all nations.

Ocean-related issues of environment, territorial claims, and ownership of resources are currently being discussed in the International Laws of the Seas (Table 16-1). The greatest legal debate among nations of the world hinges on the question of who owns how much of the sea, both horizontally and vertically. In recent years several conventions have been debated and formulated at various United Nations conferences. In 1958, over 81 countries participated in the Geneva Conference, and four important conventions, which have since been ratified by several other nations, were adopted: (1) the Convention on the Territorial Sea and Contiguous Zone, (2) the Convention on the Continental Shelf, (3) the Convention on the High Seas, and (4) the Convention on Fishing and Conservation of the Living Resources of the High Seas (Figure 16-1).

TABLE 16-1 Laws of the Sea

Place and Year	*Remarks*
Holland, 1609	In his book *Mare Liberum* Grotius emphasized the principle of the freedom of the seas.
Washington, D.C. 1945	The *Truman Proclaimation-2667* gave U.S. the sovereignty over the oceanic resources found in the continental shelf.
South American countries 1952	Chile, Ecuador, and Peru proclaimed their territorial jurisdiction and sovereignty over the sea, seabed, and subsoil up to 200 miles.
Geneva 1958 and 1960	88 countries participated at the conference. Agreed upon four conventions (see Section 16-2).
New York 1967	At the United Nations, *The Malta Resolution* was proposed to set up a global ocean policy that would proclaim oceans as common heritage of mankind (see Section 16-3).
Caracas, Venezuela 1974 Geneva 1975 New York 1976	Series of debates on such issues as freedom of scientific exploration of oceans; antipollution measures; management of the high seas; management of living resources and other related matter. In progress since 1974.
Washington, D.C. 1977	*The Fisheries Management and Conservation Act* became law on March 1, 1977.

Territorial Sea and Contiguous Zone

The territorial seas of a nation exist beyond the internal waters of bays, sounds, and estuaries. According to the Convention on the Territorial Sea and the Contiguous Zone, each nation will have sovereign power to a band of sea adjacent to its coast (the *territorial sea*) and to the air space over the territorial sea as well as to its bed and subsoil. There is disagreement among nations concerning the exact width of the territorial sea, but it seems to vary up to 200 nautical miles.

The contiguous zone exists within the 12-mile limit, that is, 12 miles from the baseline that marks the point from which the extent of the territorial sea is measured. The Convention on the Territorial Sea and Contiguous Zone also established that each coastal nation would have a right within the contiguous zone to carry out immigration, customs, and sanitary policies, and would have the privilege to impose penalties on those who violate these and other regulations.

Recently the United States extended the limit of its territorial seawaters from the 3 to 12 miles, in order to protect the fishery industries. In 1976, the United States extended its economic zone to 200 miles. This decision was made in response to heavy fishing by Japan, Russia, and some European countries, which was hurting U.S. fishermen.

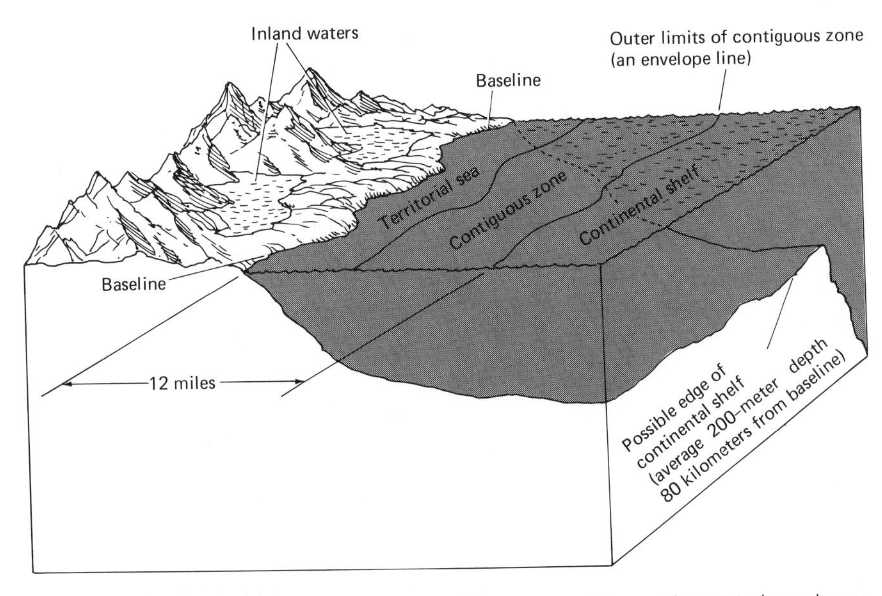

Figure 16-1. Major geologic elements of the ocean. These elements have been debated at Law of the Sea conferences since 1958.

The Continental Shelf

According to the Convention on the Continental Shelf, that shelf is defined as "the seabed and subsoil of the submarine areas adjacent to the coast but outside the area of the territorial sea to a depth of 200 meters [656 feet] or beyond that limit, to where the depth of the superadjacent waters admits of the exploitation of the natural resources of the said areas," as well as "the seabed and subsoil of similar submarine areas adjacent to the coasts of islands."

The High Seas

According to the Convention on the High Seas, the "high seas" encompasses all portions of the sea outside territorial seas or the internal waters of a nation. The high seas belong to all nations and are therefore under no jurisdiction of an individual nation. Access to the high seas is open to coastal and noncoastal nations. The convention guarantees freedom of navigation, fishing, construction of submarine cables and pipelines, and flight.

Fishing and Conservation

The Convention on Fishing and Conservation of the Living Resources of the High Seas also adopted this provision:

> All states have the duty to adopt or to cooperate with other states in adopting such measures for their respective nationals as may be necessary for the conservation of the living resources of the high seas.

At present, the territorial sea and the continental shelf are the major issues of the International Laws of the Sea. According to the recent report on the territorial sea by the Food and Agricultural Organization of the United Nations, out of 106 countries, 40 nations claimed 12 miles as their territorial jurisdiction, 29 claimed 3 miles, 14 claimed 6 miles, and 4 claimed 10 miles. The remaining nations, including most South American nations, claimed 200 miles as the limit of their territorial sea.

Because of these varying claims for territorial sea limits, international disputes are frequent. Russian fishing vessels operate within the United States' limits of its newly declared 200-mile economic zone. Consequently, Russian fishing vessels off New England have recently been captured by the U.S. Coast Guard, and it is likely that the Russian fishermen will have to pay the necessary fines to recover its vessels. Similarly, a number of Korean fishing vessels have been detained by the

U.S. Coast Guard for infringing on territorial waters in the Gulf of Alaska.

Fish wars exist between Great Britain and Iceland. Iceland lays territorial claim to 200 miles of sea, whereas Britain recognizes only 12 of these miles and therefore continues to send her fishing vessels within the territorial sea limits of Iceland. In 1973, the Icelandic Coast Guard cut the British trawling lines and ships were rammed. Britain sent three frigates to protect the trawlers. Iceland lodged a complaint against Great Britain with the United Nations and with NATO. The issue was subsequently settled when Britain agreed to curtail fishing in Icelandic waters.

The International Convention on the Continental Shelf provides to each coastal nation *sovereign rights* over the continental shelf for purposes of exploration and exploitation of its natural resources. But the definition of continental shelf varies. Each coastal nation is accorded the right of permanent and exclusive access to the natural resources of the continental shelf up to the 200-meter depth limit (Figure 16-2). In general, international consensus favors the 200-meter isobath.

The continental shelf convention has been implemented with some success, however, in the North Sea. The North Sea continental shelf has been divided between bordering nations according to an *equidistance principle.* According to this principle, the continental shelf of each nation is determined to be equidistant from the nearest points of the baselines from which the breadth of the territorial sea of each is measured. However, there are some disagreements about how to determine equidistance, particularly between Germany and Denmark.

The general principles of the International Laws of the Sea also accord to each coastal nation the right to permanent and exclusive access to mineral resources found in its territorial waters or in their beds or subsoil. Again, disagreement exists concerning the extent of the territorial sea.

16-3 THE MALTA RESOLUTION

During the twenty-second session of the United Nations Assembly in 1967, Malta proposed to the General Assembly *"an Examination of the question of the reservation exclusively for peaceful purposes of the seabed and ocean floor and subsoil thereof underlying the high seas beyond the limits of present national jurisdiction and the uses of their resources in the interest of mankind."* The following year the General Assembly established a permanent standing committee on the peaceful uses of the seabed, beyond national jurisdiction. Forty-two nations are

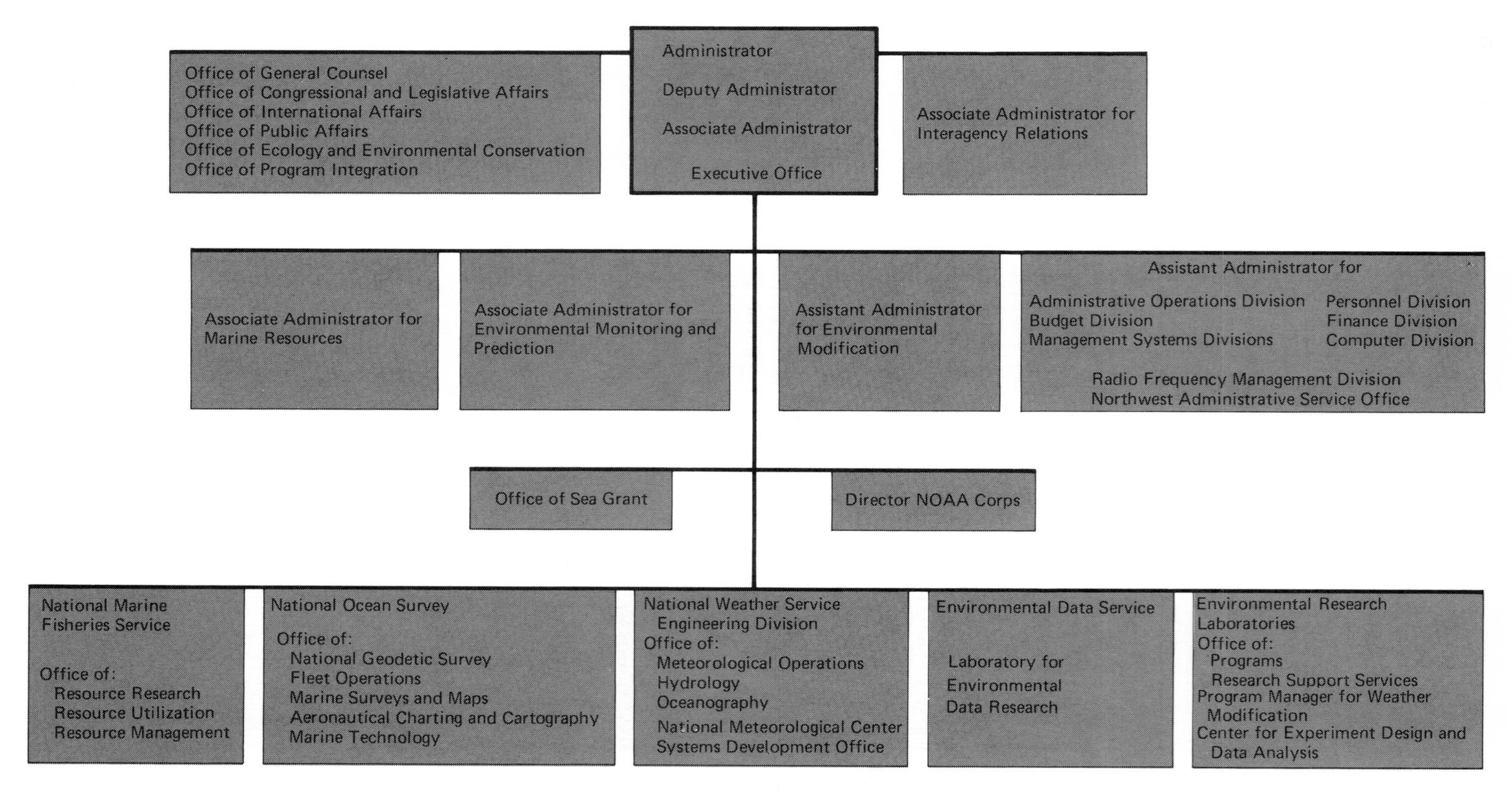

Figure 16-2. Organizational chart of the National Oceanographic and Atmospheric Administration as of March 1972. (*Source:* U.S. Department of Commerce.)

represented on this committee, and their representatives have a six-year term of office.

In essence the Malta Resolution was geared to give the United Nations a greater role in managing the world's oceans, particularly the resources of seabed and subsoil which fall beyond national jurisdiction. Since then there have been many proposals for international marine jurisdiction, including a recommendation in 1969 by the U.S. Commission on Marine Science and Engineering Resources that all claims to explore or exploit deep-sea mineral resources must be registered. The International Registry Authority would ask that claims be registered for specific mineral resouces, such as oil and natural gas for all minerals exclusive of oil and gas, or for all mineral resources. The nation registering the claim would have to meet all requirements set by the International Registry Authority and would have to be financially and technically able and willing to conduct such work. If a nation seeking mineral resources were to discover adequate mineral veins during its exploration stage, the International Registry Authority would then permit it to register a claim for exploitation. This type of license would be valid for a specific period of time; at the end of that period, it could be renewed or cancelled.

In 1969, the COMSER also recommended that an international fund be set up. Every nation registering a claim to exploit would be asked to allocate a certain amount of its profit to the international fund. An international fund agency would in turn use this revenue to support oceanic research in resource development involving obtaining more food from oceans and providing necessary monetary compensation to developing nations for depletion of their marine resources. The COMSER also recommended that coastal nations have exclusive jurisdiction of coastal waters running to a depth of either 200 meters or 50 miles, whichever is further; it established an intermediate zone beyond the 50-mile mark to the 2500-meter isobath or 100-mile point, whichever is further. Within the coastal zone, nations would be responsible for the administration of resources; beyond that zone, in intermediate waters, they would be required to contribute to an international fund.

16-4 THE INTERNATIONAL LAW COMMISSION

The International Law Commission is an important organ of the United Nations. Its responsibilities include maintaining international criminal jurisdiction and administering legal codes for offenses against peace and security. This commission also looks into the legal aspects of disputes among nations concerning their claims on territorial waters and resources.

16-5 U.S. OCEAN MANAGEMENT

The United States' water-related management activity began in 1790 when the Coast Guard was established. The progress of the water management was rejuvenated when the *Refuse Act* was passed in 1899. This act gave Army Corps of Engineers authority to enforce appropriate regulations for navigational and pollution controls of rivers and estuaries. Subsequent progress of ocean management was slow until 1937, when the National Academy of Sciences undertook a systematic reconnaissance study of the oceans. The Academy recommended creation of oceanographic study centers. As a result, the nation's three leading oceanographic institutions—Scripps Institution of Oceanography at La Jolla, California (now a part of the University of California at San Diego); Woods Hole Oceanographic Institution at Cape Cod, Massachusetts; and Lamont-Doherty Geological Observatory at Palisades, New York—were established. In 1959, a new era in ocean management began when the U.S. Navy outlined its program entitled *Ten Years in Oceanography* (TENOC). But the most significant development came during the 1960s when a series of ocean-related legislative acts were introduced into Congress. In particular, the *Marine Resources and Engineering Act of 1966* set the tone for U.S. oceanographic public policy. This act is defined as "An Act to provide a comprehensive long range co-ordinated National program in Marine Science to establish a National Council on Marine Resources and Engineering Development, and a Commission on Marine Science, Engineering and Resources and for other purposes . . . ." the COMSER report provided the necessary framework to develop a strong ocean-related program and public policy. After about three years of intensive investigation, the commission made 121 recommendations that included provisions for management of coastal zones, pollution, the global environment, and marine resources. In addition, the commission recommended the creation of the National Oceanic and Atmospheric Administration (NOAA). The NOAA was originally favored as an independent agency reporting directly to the President. In 1970, it was absorbed into the Department of Commerce. In 1969, the NOAA was authorized $773 million to begin its operation. The commission projected that the NOAA will have an operating budget of $2 billion dollars annually by 1980. At present, the NOAA is one of the primary governmental agencies entrusted with the overall management of ocean-related matters. The NOAA also works with other federal agencies, including the Department of the Interior and the Environmental Protection Agency.

The 200-Mile Economic Zone

U.S. ocean management assumed a far greater responsibility with the passage of the *Fishery Conservation and Management Act of 1976*. As a result, Congress declared a 200-mile fishery conservation zone. The essence of the 200-mile economic zone was to give the United States exclusive fishery management authority. Among other things, the act made necessary provisions to establish national standards for fishery conservation and management; regional fishery management councils; civil penalties and criminal offenses; permits for foreign fishing and enforcement of the act by the Coast Guard. The act became law on 1 March 1977.

No sooner had the act become effective than the U.S. Coast Guard seized two Russian fishing vessels about 130 miles southeast of Nantucket, Massachusetts, on the grounds of violation of the Fishery Conservation and Management Act. At present the case is pending in the U.S. Attorney's Court in Boston. If found guilty, the captain could be imprisoned for up to one year and possibly fined $100,000, and his vessels could be confiscated. The outcome of this incident is important, because it will set the precedent for possible violations in the future.

Preliminary conservation measures for living and nonliving ocean resources have begun in recent years on a small and isolated scale. Some of the measures include attempting to gain a better knowledge of the life of sea animals and their migratory patterns (Figure 16-3), identifying and classifying mineral resources (Figure 16-4), and taking a systematic inventory of fish catches (Figure 16-5). The effectiveness of these various conservation measures, however, depends upon management skills, the well-planned use of ocean resources, and global cooperation.

Figure 16-3. The systematic study of sea mammals using tagging techniques is an integral aspect of the conservation of ocean resources. (Photo courtesy of National Marine Fisheries Service.)

Figure 16-4. Careful identification, classification, and inventory of economic minerals is an integral aspect of the conservation of ocean resources. (Photo courtesy of Woods Hole Oceanographic Institution.)

SUMMARY

1. The management and conservation of oceans and their resources are challenging problems. Ocean management is complicated by lack of data, territorial disputes, and the size of the ocean.
2. The Laws of the Sea, formulated under the auspices of the United Nations, embody the principle of freedom of the high seas.
3. Conferences on the Laws of the Sea have adopted four conventions: (1) the Convention on the Territorial Sea and Contiguous Zone, (2) the Convention on the Continental Shelf; (3) the Convention on the High Seas; and (4) the Convention on Fishing and Conservation of Living Resources of the High Seas.

Figure 16-5. Taking correct inventories of fish catches on local to global levels is an integral aspect of the conservation of ocean resources. (Photo courtesy of National Marine Fisheries Service.)

4. The Malta Resolution was proposed to give the United Nations a greater participation in the management of ocean resources.
5. The International Law Commission of the United Nations is entrusted with the legal management of the oceans.
6. In the United States, the National Oceanic and Atmospheric Administration (NOAA), an agency of the Department of Commerce, oversees most ocean-related matters, including the new 200-mile fishing limit.

Suggestions for Further Reading

Alexander, L. M. (editor). 1970. *The Law of the Sea: National Policy Recommendation*. Proceedings of the Fourth Conference on the Law of the Sea Institute, University of Rhode Island.

Anderson, E. V. 1974. "The Law of the Sea Battle—An Overview." *Marine Technology Society Journal, 8,* 4–14.

Borgese, Elisabeth. 1973. *Pacem in Maribus*. New York: Dodd–Mead.

Cox, Vic. 1976. "Oceanic Gamble: UN Law of the Sea Conference." *Newsweek, 87,* 13 (March 22).

Fishery Conservation and Management Act of 1976, Public Law 94–265, 94th Congress, H.R. 200, April 13, 1976.

Gardner, R. N. 1976. "Offshore Oil and the Law of the Seas." *New York Times,* page 14, March 14.

Hollings, E. F. 1975. "National Ocean Policy Study: A New Senate Initiative." *Oceans, 8,* 6–7 (January).

Laurd, D. E. 1974. *The Control of the Sea-Bed: A New International Issue.* New York: Taplinger.

McCloskey, William, Jr. 1976. "The 200-Mile Fishing Limit: United States Girds for Jurisdiction," *Oceans, 9,* 60–63.

Shapley, Deborah. 1976. "Sea Law Treaty: Amid U. S. Gains, the Prospects for Science Are Sinking." *Science, 192,* 980–981.

Wenk, Edward, Jr. 1972. *The Politics of the Ocean.* Seattle: University of Washington Press.

Appendix

TABLE A-1 Common Conversion Factors

1 fathom (fm) = 6 feet (ft) = 1.829 meters (m)
1 meter (m) = 100 centimeters (cm) = 39.37 inches = 3.281 feet
1 kilometer (km) = 0.6214 miles (mi)
1 kilogram (kg) = 1000 grams (g) = 2.205 pounds (lb)
1 liter (l) = 1000 cubic centimeters (cm^3) = 1.057 quarts
42 U.S. gallons = 1 barrel

TABLE A-2 Measures of Length and Weight

10 millimeters (mm)	= 1 centimeter (cm)
10 centimeters	= 1 decimeter (dm)
10 decimeters	= 1 meter (m)
1000 meters	= 1 kilometer (km)
10 milligrams (mg)	= 1 centigram (cg)
10 centigrams	= 1 decigram (dg)
10 decigrams	= 1 gram (g)
10 grams	= 1 decagram (dg)
10 decagrams	= 1 hectogram (hg)
10 hectograms	= 1 kilogram (kg)
1000 kilograms	= 1 (metric) ton (T)

TABLE A-3 Common Metric Prefixes and Their Equivalent Numerical Values

Prefix	*Numerical Value*	*Exponential Value*
mega	1,000,000	10^6
kilo	1,000	10^3
deci	$\frac{1}{10}$	10^{-1}
centi	$\frac{1}{100}$	10^{-2}
milli	$\frac{1}{1000}$	10^{-3}
micro	$\frac{1}{1000000}$	10^{-6}

TABLE A-4 Facts about the Earth

Size	
Equatorial diameter	7,927 miles
Equatorial circumference	24,902 miles
Polar diameter	7,900 miles
Polar circumference	24,860 miles
Mass	6,000 billion billion tons
Surface	
Area	197,000,000 square miles
Oceans	139,400,000 square miles
Land	57,500,000 square miles
Highest point	Mt. Everest, 29,028 ft
Lowest point	Mariannas Trench, 36,198 ft

Glossary

Abyssal Referring to the ocean zone of depths 2000 to 6000 meters.

Abyssal hills Relatively small topographic features of the deep ocean floor ranging from 600 to 1000 meters high and a few kilometers wide.

Abyssal plains Flat, nearly level areas occupying the deepest portions of many ocean basins.

Acidic solution A liquid with greater H^+ than its OH^- concentration; ph less 7.0

Acropora palmata Technical name for staghorn corals.

Aerosol An environmental term referring to a gaseous suspension of solid or liquid particles in the atmosphere.

Agar-agar A gelatin-like, colorless, tasteless, and odorless substance extracted from certain seaweeds.

Algae Primitive plants including seaweed and diatoms. Algae contain chlorophyll and are capable of photosynthesis.

Alginic acid A complex organic compound related to the carbohydrates, present in certain seaweeds.

Alkaline solution A liquid with greater OH^- than its H^+ concentration; pH over 7.0.

Amino acid Nitrogenous organic units of protein and vital constituents of all life.

Anaerobic Oxygenless conditon.

Anion A negatively charged ion. Examples include Cl^-, $SO_4^=$, and $0^=$.

Annelida A phylum of animals that includes highly developed segmented worms. Examples: earthworms, seaworms, etc.

Antibiotics Chemical substances synthesized by microorganisms (bacteria) having an ability to prevent growth of harmful bacteria by destroying them. Penicillin and streptomycin are two well-known examples.

Anticline Up-folded structure.

Aphotic zone Deeper parts of the ocean where there is not enough light to permit photosynthesis by plants.

Aquaculture A technique of raising organisms such as fish, shellfish, and algae under controlled conditions until they reach marketable size.

Arthropoda A phylum of segmented animals with jointed appendages on each segment. Examples are insects, crabs, and lobsters.

Asthenosphere A plastic mobile zone within the mantle about 100 to 400 kilometers below the earth's surface.

Astrolabe An instrument once used to measure the altitude of heavenly bodies, now replaced by the sextant.

Atoll A ring-like "coral" island or group of islands of reef origin encircling a lagoon in which there are no islands of noncoral origin.

Atom The smallest unit of an element. An atom consists of protons, electrons, and neutrons.

Atomic mass Mass number or the mass of an individual atom.

Atomic number The number of positive charges on the nucleus of an atom. The number of protons in the nucleus.

Atomic weight Average relative weight of the atoms of an element referred to an arbitary standard of 16.0000 for the atomic weight of oxygen.

ATP Adenosine triphosphate, an important energy yielding organic compound essential for all life.

Autotrophic Referring to organisms that use only inorganic materials as a source of nutrients.

Aves The birds. A class of warm-blooded vertebrates characterized by feathers and wings developed from forelimbs.

Baleen (whalebone) The filtering organ, consisting of numerous plates with fringed ends, of plankton-feeding whales.

Barrier reef A coral reef that runs parallel to the land but is separated from it by a lagoon.

Basalt A dark-colored igneous rock commonly found on the ocean floor, mineral enriched or relatively high in iron and magnesium content.

Bathyal An oceanic zone between depths of 200 to 2,000 meters.

Bathymetry Relating to the measurement of ocean depth.

Bathythermograph An instrument that measures temperature at various depths in the ocean.

Beach The deposit of sand and gravel along a shore that is usually the result of active transport by waves and currents.

Beaufort Scale A numerical scale for the estimation of wind force, based on its effect on common objects.

Benthic Referring to that part of the ocean bottom populated by organisms.

Benthos Organisms that live on the ocean bottom.

Biogenous sediment Sediments composed of organic debris from dead animals and plants.

Bioluminescence Light production by certain marine organisms due to biochemical reaction within the cells or organs, or in some form of secretion.

Biomass An index of the amount of living organisms in grams per unit area or unit volume.

Biosphere The total sphere of life in, on, and above the earth.

Boiling point The temperature at which a liquid begins to boil or to be converted into a gaseous or vapor state. The boiling point for pure water is 100°C or 212°F at room pressure.

Bore See *Tidal bore*.

Brachiopoda The phylum comprising bivalved marine shellfish such as lampshells.

Breaker zone The point at which waves break at the seaward margin of the surf zone.

British thermal unit (Btu) A unit usually defined as the amount of heat required to raise the temperature of one pound of water one degree Fahrenheit.

Bryozoa A phylum comprising minute colonial animals with calcareous skeletons.

Calcium carbonate A white insoluble solid occurring naturally as chalk, limestone, marble, and calcite. Used in the manufacture of lime and cement.

Calorie A unit defined as the amount of heat required to raise the temperature of 1 gram of water 1°C.

Carbon-14 A radioactive isotope of carbon with atomic weight 14, produced by collision between neutrons and atmospheric nitrogen. Useful in determining the age of objects younger than 30,000 years. Half-life is 5,565 years.

Catastrophic waves Sudden violent and temporary waves caused by earthquakes, volcanic activity, etc.

Cation Positive chemical ions such as sodium and calcium.

Celsius temperature scale A thermometric scale proposed by Anders Celsius in 1742. The present scale is caliberated at 0°C for the freezing point and 100°C for the boiling point of water (0° = 273.16°K).

Cephalopoda A class of the molluscs that includes the squid, nautilus, and octopus.

Cetacea An order of sea mammals that includes whales and dolphins.

Chaetognath Elongate transparent worm-like pelagic animals such as arrowworms.

Chlorophyll A green pigment found in plants that absorbs energy from sunlight, making it possible to synthesize organic matter from atmospheric carbon dioxide and water through photosynthesis.

Chlorophyta Green algae possessing chlorophyll and other pigments.

Chordata Animals possessing a notochord. Includes all vertebrates such as fish, amphibians, reptiles, birds, and mammals.

Cilia Hair-like structures used by certain invertebrate animals for locomotion and gathering of food.

Clastic Unconsolidated fragments of pre-existing rocks.

Clay Very fine geologic material with particle size between silt and colloid.

Coccoliths Microscopic calcareous plates, usually oval and perforated, borne on the surfaces of some marine flagellate organisms.

Coccolithophores Microscopic forms of algae that produce coccoliths.

Coelenterata Animals characterized by radial symmetry and a single body cavity performing all the vital functions. Examples include all corals, jellyfish, and sea anemones.

Colonial animals Animals living together in groups either attached or as separate individuals.

Commensalism An association between two organisms in which one is benefited by the other without any hurt or benefit to the other organism.

Compensation depth The depth at which the amount of a given matter (e.g. O_2 or $CaCO_3$) formed by the organisms in the overlying water column is equal to amount of the same matter dissolved by the water column.

COMSER An acronym for Commission on Marine Sciences and Engineering Resources.

Continental drift The concept that, due to the weakness of the suboceanic crust, continents can drift on earth's surface much as ice floats through water.

Continental margin Seaward extended portion of the landmass.

Continental rise A gently sloping surface at the base of the continental slope.

Continental shelf A gently seaward sloping surface extending between the shoreline and the continental slope.

Continental slope A relatively steeply sloping surface running seaward of the continental shelf.

Copepod Any one of a group of tiny marine crustaceans.

Coral A group of benthic animals belonging to the phylum Coelenterata. These animals live as individuals or in colonies and secrete calcium carbonate skeletons. In collaboration with certain algae, and under favorable conditions, corals can build reefs.

Core Innermost portion of the earth. Also a cylindrical sample of sediments.

Coriolis force A force produced as a consequence of earth's rotation. It causes objects in motion to be deflected to the right in the Northern Hemisphere and to the left in the Southern Hemisphere.

Covalent bond The linkage between two atoms in a molecule resulting from the sharing of electrons.

Crust Uppermost and thinnest segment of the earth. Thickness varies from 5 to 50 kilometers. Enriched in basaltic content in the oceanic crust and granitic in the continental crust.

Crustacea A class of the arthropods that includes barnacles, lobsters, crabs, and shrimps.

Curie A measuring unit of radioactivity quantitively defined as 3.7×10^{10} disintegrations per second. Named after Marie Curie (1867–1934).

DDT Dichloro diphenyl Trichloroethane $(C_6H_4Cl)_2 \cdot CH \cdot CCL_3$. Manufactured as a contact insecticide.

Decomposers Ability on the part of organisms such as bacteria to break down nonliving organic matter.

Deep-scattering layer Sound (or sonic) reflecting layer above the sea floor formed by a dense concentration of zooplankton and fish.

Delta A triangular-shaped deposit at the mouth of a river.

Density Mass per unit volume of a substance. In the Metric system expressed as grams per cubic centimeter (g/cc).

Demersal fish Those species living near the sea bottom.

Desalination Process of converting seawater into freshwater.

Detritus Unconsolidated mixture of inorganic and organic matter produced by weathering of rocks.

Diatom Group of algae. Microscopic, one-celled plants usually covered by siliceous matter.

Dinoflagellates Group of algae. Microscopic, one-celled organisms with pairs of hair-like flagella used for locomotion.

Diurnal tides Tides with one high and one low water mark during a tidal day.

Earthquake A vibrating or shaking state of the earth caused by the sudden release of stored strain.

Ebb current The movement of a current seaward.

Echinodermata A phylum of spiny skinned animals including star fish, sea urchins, and sea lilies.

Echo sounder Electronic device employed to determine and record water depth in the ocean.

Echo sounding A technique for determining the depth of water by measuring the time required for a sound signal to travel to the bottom and back to the ship that emitted the signal.

Ecology The study of the relationship between organisms and their environment.

Ekman spiral A theoretical model to explain the influence of a steady wind dragging over an ocean of unlimited depth and breadth and of uniform viscosity. As a consequence, a surface flow at 45° to the right of the wind is formed in the Northern Hemisphere. Water at increasing depth will drift in directions increasingly to the right until, at about 100 meters depth, the water is moving in a direction opposite to that of the wind. The net water transport is 90° to the wind, and velocity decreases with depth.

Environment The total physical, chemical, and biological system in which life perpetuates.

Enzymes A large group of proteins produced by living cells that help carry out various chemical reactions upon which life depends.

EPA An acronym for Environmental Protection Agency.

Equinox Refers to the times when the sun is directly over the equator. Day and night are of equal duration throughout the earth. The vernal equinox occurs about March 21 as the sun is moving into the Northern Hemisphere. The autumnal equinox occurs about September 21 as the sun in moving into the Southern Hemisphere.

Erosion The transportation of weathered (broken) material by a moving agent such as wind, water, or ice.

ERTS An acronym for Earth Resources Technology Satellite. A sophisticated earth-orbiting satellite that assesses various planetary environmental, agricultural, and mineralogical resources.

Estuary The mouth of a river valley where marine influence is manifested as tidal effects and increased salinity of the river water.

Euphausids Planktonic crustaceans (5–25 centimeters long). Some are the principal food of baleen whales.

Euphotic zone The surface layer of the ocean where enough light is received to support photosynthesis. This zone is usually 80 to 100 meters below sea level.

Evaporation The physical process whereby a liquid is converted to a gas at a temperature below the boiling point of the liquid.

Excess volatiles Elements like water, carbon dioxide, nitrogen, and sulfur that are more abundant in the ocean, atmosphere, and sediments than can be accounted for by breakdown of rocks.

Fahrenheit temperature scale (°F) A scale used for measurement of temperature. Freezing point of water is 32°; boiling point of water is 212°.

Fallout The shower of radioactive debris, usually from the earth's atmosphere.

Fan A gently sloping, fan-shaped feature usually located near the lower end of a canyon.

FAO Acronym for Food and Agricultural Organization (an agency within the United Nations).

Fathom A unit of depth in the ocean equal to 1.83 meters or 6 feet.

Fault A fracture in the earth's crust along which displacement has occurred.

Fauna The animal population.

Fjord A long, narrow, deep, U-shaped inlet that usually represents the seaward end of a glacial valley that has become partially submerged after the melting of the glacier.

Flagellum A whiplike living process used by some cells for locomotion.

FLIP Acronym for Floating Instrument Platform.

Flora The plant population.

Foraminifera Mostly microscopic, single-celled animals that possess shells composed of calcium carbonate.

Foreshore Lower shore zone between normal high and low water marks.

Fringing reef A reef attached directly to the shore.

Gastropoda A phylum comprising soft-bodied animals that are covered by a hard calcareous shell.

Geostrophic current A current that emerges from the earth's rotation and is the product of a balance between Coriolis effect and gravitational force.

Gondwana Theoretical ancient southern continent encompassing India, Australia, Antarctica, Africa, and South America.

Graded bedding A special type of stratification in which a gradation in grain size is displayed, with a concentration of coarser grains at the bottom and finer grains at the top.

Granite An igneous rock consisting of feldspar, quartz, and minor amounts of ferromagnesian minerals. Most common rock of the continental crust.

Guyot A flat-topped sea mountain.

Gyre A circular spiral motion of water.

Habitat A geographic area occupied by a particular plant or animal.

Hadal Refers to the deepest marine environment, over 6000 meters deep.

Half-life The amount of time required for half the atoms of a radioactive isotope to decay to an atom of another element.

Halocline A layer of water in which salinity changes drastically.

Heat capacity The amount of heat required to raise the temperature of a substance.

Herbivore An animal that feeds on plants.

Heterotrophs Animals and bacteria that depend on the organic compounds produced by other animals and plants for food. Such organisms are not capable of producing their own food by photosynthesis.

High water (HW) The maximum level reached by the rising tide before it begins to recede.

Higher high water (HHW) The higher of two high waters occurring during a tidal day where tides are mixed.

Higher low water (HLW) The higher of two low waters occurring a tidal day where tides are mixed.

Holoplankton Organisms that permanently follow a planktonic life.

Horse latitudes The latitude belts between 30° and 35° north and south where winds are light and variable. The principal movement of air masses at these latitudes is one of vertical descent. The climate is hot and dry.

Hurricane A tropical cyclone in which winds reach velocities in excess of 120 kilometers per hour (73 miles per hour). Generally applied to such storms in the North Atlantic Ocean, eastern North Pacific Ocean, Caribbean Sea, and Gulf of Mexico. In the western Pacific Ocean such storms are called typhoons.

Hydrologic cycle The circulation of water from the ocean to the atmosphere and land and back through evaporation, precipitation runoff, and subsurface percolation.

Hydrozoa A class of coelenterates that characteristically exhibits alternation of generations, with a sessile polypoid colony giving rise to a pelagic medusoid form by a sexual budding.

Hypertonic The property of an aqueous solution having a higher osmotic pressure (salinity) than another aqueous solution from which it is separated by a semipermeable membrane that will allow osmosis to occur. The hypertonic fluid will gain water molecules from the other fluid through the membrane.

Hypotonic The property of an aqueous solution having a lower osmotic pressure (salinity) than another aqueous solution from which it is separated by a semipermeable membrane that will allow osmosis to occur. The hypotonic fluid will lose water molecules to the other fluid through the membrane.

Iceberg A massive piece of glacier ice that has broken from the front of the glacier (calved) into a body of water. It floats with its tip at least 5 meters above the water's surface and at least 4/5 of its mass submerged.

Igneous rock Rock formed from the cooling of molten or partly molten material (magma).

IGY Acronym for the International Geophysical Year.

In situ In place or in position. From the Latin.

Interface A boundary between two substances having different properties such as density, salinity, or temperature. In oceanography it usually refers to a separation of two layers of water of different densities caused by significant differences in temperature and/or salinity.

Internal wave A wave that forms at the boundary of two water layers having different densities.

Intertidal wave Littoral zone, the foreshore. The ocean floor covered by the highest normal tides and exposed by the lowest normal tides and the water environment of the tide pools within this region.

Ion An atom that becomes electrically charged by gaining or losing one of more electrons. The loss of electrons produces a positively charged cation, and the gain of electrons produces a negatively charged anion.

Ionic bond A bond resulting from the electrical attraction that exists between cations and anions.

IPOD An acronym for International Phase of Ocean Drilling.

Isotherm Imaginary line joining points of equal temperature.

Isotonic The property of two fluids having equal osmotic pressure. If two such fluids are separated by a semipermeable membrane that will allow osmosis to occur, there will be no net transfer of water molecules across the membrane.

Isotope Atoms possessing nuclei with the same number of protons but a different number of neutrons.

Jellyfish Free-swimming coelenterates having a jellylike bell-shaped body.

JOIDES An acronym for Joint Oceanographic Institutions Deep Earth Sampling.

Juvenile water Water that is derived from the earth's interior. It is water brought to the surface for the first time, not recycled water.

Knot A unit of velocity equal to 1 nautical mile per hour (1.15 statute miles per hour).

Lagoon Body of shallow water characterized by a restricted connection with the sea.

Laurasia Theoretical ancient northern continent encompassing North America and Eurasia.

Lava Hot and molten fluid rock issuing from a volcano or a fissure in the earth's surface, or the same material after it solidifies.

Lithosphere The solid outer shell of the earth from 20 to 50 kilometers thick that includes the crust and upper part of the mantle. This layer is characterized by the interlocking of plates.

Littoral The nearshore environment.

Longshore current A nearshore current that flows parallel to the shore.

Lunar tide The tide produced by the force of the moon.

Magma Hot mobile rock material generated within the earth from which igneous rock results from cooling and crystallization.

Magnetic anomaly Distortion of the regular pattern of the earth's magnetic field resulting from the various magnetic properties of local concentrations of ferromagnetic minerals in the earth's crust.

Magnetometer An instrument that determines the magnetic intensity of rocks rich in iron content.

Manganese nodules Deep ocean deposits consisting of oxides of iron, manganese, copper, and nickel.

Mantle The intermediate layer between the earth's core and crust.

Marginal sea Seas that are adjacent to and widely open to the ocean.

Mariculture See *Aquaculture*.

Marsh A shallow wetland commonly found adjacent to the sea.

Mass spectrometer An instrument for classification and measurement of isotopic species by their mass.

Mass transport The net transfer of water by wave action in the direction of wave travel.

Mean high water The average height of all high waters over a 19-year period.

Mean low water The average height of all low waters over a 19-year period.

Mean sea level The average height of the sea for all stages of the tide.

Meridian An imaginary north–south line from which longitudes are known.

Meroplankton Organisms that have floating stages in their early development before becoming benthic.

Mesosphere A hypothetical thick and solid layer within the earth's mantle that separates the asthenosphere from the core. Also, the outer part of atmosphere between 400–1000 kilometers elevation.

Metamorphic rocks Altered or modified rocks. Modification in the mineral composition of these rocks is caused by the action of heat, pressure, and chemical fluids.

Metazoan Any animal with many cells that constitute the tissues or organs.

Meteoric water Water derived from the atmosphere.

Midocean ridge Great linear arch that extends through all the major oceans. The total length of the ridge is about 47,000 miles.

Mineral A naturally occurring substance with a definite chemical composition and characteristic physical properties by which it may be defined.

Minor elements Elements found in concentration of 10 to 100 parts per million.

Mohorovicic discontinuity (Moho) Seismic discontinuity indicating compositional change between the crust and mantle of the earth. It is found beneath the crust at depths ranging from 5 to 50 kilometers.

Molecule Smallest unit of a substance formed by two or more atoms. One molecule of water is composed of two atoms of hydrogen and one atom of oxygen.

Mollusca The phylum of invertebrates characterized by highly developed soft-bodied animals covered by hard calcareous shells and possessing a muscular foot for locomotion. Examples include the clam, oyster, snail, slug, squid, and nautilus.

Monsoon Arabic for season; referred to the winds of the Arabian Sea in the Indian Ocean that flow from the ocean to the land (i.e., from southwest) during summer and from land to the ocean (i.e., northeast) in winter time.

Mud A general term for the mixture of water with silt and clay-sized particles (less than 0.06 millimeter) together with materials of other dimensions.

Mussel A pelecypod, generally one that is not attached like an oyster.

Mutation An inherited alteration resulting from the modification of the hereditary material in the reproductive cells. The alteration may be insignificant or drastic.

Nansen bottle A specially designed bottle to obtain samples of ocean water from beneath the surface.

Nautical mile A length of mile used in ocean navigation is internationally defined as being the equivalent of 1852 meters.

Nekton Pelagic animals that are able to swim independently.

Nucleic acid A complex organic substance that constitutes the genetic material in all life. DNA (deoxyribonucleic acid) and RNA (ribonucleic acid) are examples.

Nutrients Inorganic and organic compounds of ions utilized by plants in food manufacturing.

Ocean basin The part of the floor of the ocean that is deeper than 2000 meters below sea level.

Oceanic crust The ocean part of the earth's crust enriched in basaltic rocks and about 5 kilometers thick.

Oceanic rise A large area of the deep ocean that is not a part of a mid-ocean ridge.

Oceanography The scientific study of the planetary oceans and seas.

Ooze Fine-grained pelagic deposits containing a minimum of 30 percent organic material. Oozes may be classifed according to their chemical makeup—as siliceous (silica-containing) or calcareous (limy)—or often according to the organism whose remains dominate the deposit—such as radiolarian ooze and foraminiferal ooze.

Osmosis The passage of a solvent through a membrane from a dilute solution into a more concentrated one, the membrane being permeable to molecules of solvent but not to molecules of solute.

Ostracoda A subdivision of crustaceans characterized by small bivalved animals living in salt and fresh water.

Paleomagnetism The residual magnetism in fossilized rocks.

Pangaea The hypothesized ancient single supercontinent that comprised all continents.

Parasitism An involuntary association between two or more organisms in which the host is harmed and the parasite is benefited.

Pelagic environment The open ocean environment, comprised of the *neritic* (depth 0 to 2000 meters) and the *oceanic* provinces (depth over 200 meters).

Pelagic sediment Refers to sediments of the deep ocean in contrast to those derived from the land.

Pelecypoda A class of the phylum Mollusca. Examples include clams, scallops, and so on.

Photosynthesis The process in which the energies of light and chlorophyll are used by plants to manufacture food from carbon dioxide and water.

Phytoplankton Collective term for planktonic plants.

Plankton bloom An enormous growth of plankton in a given specific area due to the abundant supply of nutrients by upwelling currents.

Plankton net A plankton trapping device that is towed through the water.

Plate tectonics The concept that the crust and upper mantle of the earth are divided into segments or plates that are always in friction with each other, consequently generating earthquakes, mountain ranges, and so on.

Pleistocene Subdivision of the Quarternary Period of geologic time. It began one million years ago and terminated 11,000 years ago. During its span, widespread continental glaciation occurred periodically.

Polar wandering The apparent shifting of the magnetic poles throughout geologic time.

Primary productivity The amount of organic matter manufactured by organisms from inorganic substances for a given volume of water or habitat in a given time.

Project Mohole The United States Marine project to drill a hole into and sample the earth's mantle. It was abandoned in the mid-60s.

Protozoa A phylum of primitive animals having one cell only. Examples include foraminifera, radiolarians, and so on.

Pycnocline That portion of the ocean zone where water density increases rapidly in response to changes in temperature and salinity.

Radioactive isotope See *Isotope*.

Radioactivity The spontaneous disintegration of the nucleus of an atom with the emission of radiation.

Radiolarians Group of protozoan plankton possessing delicate siliceous skeletons.

Rare gases All the gases (such as argon, helium, krypton, radon, and xenon) that are found in insignificant quantity in the earth's atmosphere.

Red tide The red coloration of seawater caused by an overgrowth of certain organisms (e.g., dinoflagellates). It is poisonous to many organisms, particularly those that feed upon dinoflagellates.

Reef A predominantly organic deposit made by living or dead organisms that forms a mound or ridge-like elevation.

Residence time The ratio of the total amount of an element in the ocean at a given time to its rate of replacement.

Respiration Oxidation- and energy- providing process of living organisms in which oxygen is consumed and carbon dioxide is removed from their body system.

Reversing thermometer Permanently recording thermometer attached on a Nansen bottle.

Salinity Index of the amount of total dissolved solids in seawater. It is expressed in parts per thousand by weight in 1 kilogram of seawater.

Salinometer A type of *hydrometer* deployed to measure salinity.

Sargassum Drifting seaweed found in the Sargasso Sea.

SCUBA Acronym for Self-Contained Underwater Breathing Apparatus.

Sea Generally chaotic waves produced by wind.

Seamount A submarine mountain rising more than 1,000 meters above the ocean floor.

Sediment General term applied to loose solid material made available from the breakdown of pre-existing rocks and organic matter.

Sedimentary rock Rocks formed of sediment.

Seiche A stationary wave.

Seismic Relating to earthquakes.

Seismic wave See *Tsunamis*.

Seismogram The record made by a seismograph.

Seismograph An instrument that records seismic waves.

Sessile Refers to an organism permanently stationary or fixed.

Sextant A navigational instrument that measures angles.

Shoreline The boundary between sea and land.

Sial A layer of the crust beneath all continents rich in silica and aluminum.

Sill A ridge.

Silt All sediments having a particle size between 1/16 to 1/256 of a millimeter.

Sima A layer of the crust beneath continents and oceans rich in silica and magnesium.

Sonar An acronym for Sound Navigation and Ranging.

Sounding A measured depth of water.

Specific Gravity The ratio of the mass of a body to the mass of an equal volume of water at a given temperature of 1 gram of a given substance 1°C.

Spring tides High tides that occur twice a week when the moon is full or new.

Storm surge Usually high water waves resulting from strong wind action.

Submarine canyon A steep, narrow canyon cut into the continental shelf or slope.

Substrate A base upon which organisms thrive.

Surf The breaking waves in a coastal region.

Swell Long-period waves (as opposed to short-period waves that are characteristic of a storm).

Symbiosis A mutually beneficial relationship between two or more organisms.

Taxonomy The science of classification of organisms based as far as possible on natural relationships.

Tectonics The study of the origin and history of earth's structure and deformation of its crust.

Terrigenous sediments Land-derived material deposited in the ocean.

Tethys Sea An ancient sea that divided Laurasia from Gondwana along the present Alps—Himalayan mountain belt.

Thermocline A layer of water in which rapid changes in temperature occur in the vertical column.

Thermohaline circulation Vertical circulation of ocean water brought about by density differences resulting from temperature and salinity variations.

Tidal bore A steep-nosed tide crest rushing (along with a high tide) upstream.

Tide The periodic rise and fall of the planetary ocean level in response to the gravitational interaction of earth, moon, and the sun.

Tombolo A sand or gravel deposit that links an island with the mainland or joins two islands.

Trade winds The nearly constant current of air that blows toward the equator in the tropics. The winds are northwesterly in the Northern Hemisphere and southeasterly in the Southern Hemisphere.

Trench A long, narrow and steep-sided depression on the ocean bottom.

Tsunami A long-period sea wave produced by an earthquake or volcanic eruption.

Turbidite A sedimentary rock formed by turbidity currents and characterized by graded bedding.

Turbidity current A downslope movement of dense, sediment-bearing water produced when sand and mud on the continental shelf are first dislodged and then let loose in suspension.

Viscosity A measure of a fluid's resistance to flow.

Water The normal *oxide* of hydrogen. Natural water is never pure but contains dissolved substances. Pure water is colorless, odorless, and tasteless.

Wave An oscillatory movement in a body of water manifested by an alternate rise and fall of the surface.

Wave amplitude One-half the wave height.

Wave base The depth at which wave action ceases to stir the sediments.

Wave-built terrace An embankment extending seaward from the shore line produced by wave deposition.

Wave period The time for a wave crest to traverse a distance equal to one wavelength.

Wave refraction The process by which the direction of a train of waves moving in shallow water at an angle to the contours is changed. The part of the wave train advancing in shallower water moves more slowly than that part still advancing in deeper water, causing the wave crests to bend toward alignment with the water contours.

Weathering The distintegrating or decomposing of earth's material through physical, chemical, and biological processes.

Westerly winds Characteristic winds of atmospheric zones north and south of the trade wind zones where the direction of air flow is reversed.

Zooplankton Animal plankton.

Index